Advanced Distillation Technologies

Advanced Distillation Technologies

Design, Control and Applications

Anton Alexandru Kiss

WILEY

A John Wiley & Sons, Ltd., Publication

This edition first published 2013

Registered office
John Wiley & Sons Ltd, The Atrium, Southern Gate, Chichester, West Sussex, PO19 8SQ, United Kingdom

For details of our global editorial offices, for customer services and for information about how to apply for permission to reuse the copyright material in this book please see our website at www.wiley.com.

Library of Congress Cataloging-in-Publication Data
Kiss, Anton Alexandru.
Advanced distillation technologies : design, control, and applications / Anton Alexandru Kiss.
pages cm
Includes index.
ISBN 978-1-119-99361-2 (hardback)
1. Distillation. I. Title.
TP156.D5K54 2013
542'.4–dc23

2012047063

Set in 10.5/13pt, Sabon by Thomson Digital, Noida, India.

Reprinted with corrections December 2013.

Dedicated to the loving memory of my grandparents, and to all who contributed so much to my work over the years.

Contents

Preface xiii

Acknowledgements xv

1 Basic Concepts in Distillation **1**

1.1 Introduction 1
1.2 Physical Property Methods 2
1.3 Vapor Pressure 6
1.4 Vapor–Liquid Equilibrium and VLE Non-ideality 8
1.4.1 Vapor–Liquid Equilibrium 8
1.4.2 VLE Non-ideality 11
1.5 Relative Volatility 13
1.6 Bubble Point Calculations 14
1.7 Ternary Diagrams and Residue Curve Maps 16
1.7.1 Ternary Diagrams 16
1.7.2 Residue Curve Maps 18
1.8 Analysis of Distillation Columns 24
1.8.1 Degrees of Freedom Analysis 26
1.8.2 McCabe–Thiele Method 27
1.8.3 Approximate Multicomponent Methods 33
1.9 Concluding Remarks 34
References 35

2 Design, Control and Economics of Distillation **37**

2.1 Introduction 37
2.2 Design Principles 38
2.2.1 Operating Pressure 39
2.2.2 Heuristic Optimization 40

2.2.3 Rigorous Optimization 41
2.2.4 Feed Preheating 42
2.2.5 Intermediate Reboilers and Condensers 42
2.2.6 Heat Integration 43
2.3 Basics of Distillation Control 44
2.3.1 Single-End Control 46
2.3.2 Dual-End Control 49
2.3.3 Alternative Control Structures 52
2.3.4 Constraint Control 53
2.3.5 Multivariable Control 54
2.4 Economic Evaluation 55
2.4.1 Equipment Sizing 56
2.4.2 Equipment Cost 59
2.4.3 Utilities and Energy Cost 62
2.4.4 Cost of Chemicals 63
2.5 Concluding Remarks 63
References 64

3 Dividing-Wall Column 67

3.1 Introduction 67
3.2 DWC Configurations 70
3.3 Design of DWCs 75
3.3.1 Heuristic Rules for DWC Design 77
3.3.2 Approximate Design Methods 78
3.3.3 V_{min} Diagram Method 79
3.3.4 Optimal Design of a DWC 82
3.4 Modeling of a DWC 83
3.4.1 Pump-Around Model 84
3.4.2 Two Columns Sequence Model 84
3.4.3 Four Columns Sequence Model 85
3.4.4 Simultaneous Models 86
3.4.5 Simulation of a Four-Product DWC 86
3.4.6 Optimization Methods 86
3.5 DWC Equipment 87
3.5.1 Liquid/Reflux Splitter 89
3.5.2 Column Internals 91
3.5.3 Equipment Sizing 91
3.5.4 Constructional Aspects 94
3.6 Case Study: Separation of Aromatics 97
3.7 Concluding Remarks 103
References 107

4 **Optimal Operation and Control of DWC** **111**
4.1 Introduction 111
4.2 Degrees of Freedom Analysis 112
4.3 Optimal Operation and V_{min} Diagram 114
4.4 Overview of DWC Control Structures 117
4.4.1 Three-Point Control Structure 118
4.4.2 Three-Point Control Structure with Alternative Pairing 120
4.4.3 Four-Point Control Structure 121
4.4.4 Three-Point Control Structure with Nested Loops 121
4.4.5 Performance Control of Prefractionator Sub-system using the Liquid Split 122
4.4.6 Control Structures Based on Inferential Temperature Measurements 123
4.4.7 Feedforward Control to Reject Frequent Measurable Disturbances 126
4.4.8 Advanced Control Techniques 127
4.5 Control Guidelines and Rules 128
4.6 Case Study: Pentane–Hexane–Heptane Separation 129
4.7 Case Study: Energy Efficient Control of a BTX DWC 132
4.7.1 Energy Efficient Control Strategies 135
4.7.2 Dynamic Simulations 139
4.8 Concluding Remarks 148
References 149

5 **Advanced Control Strategies for DWC** **153**
5.1 Introduction 153
5.2 Overview of Previous Work 154
5.3 Dynamic Model of a DWC 156
5.4 Conventional versus Advanced Control Strategies 163
5.4.1 PID Loops within a Multi-loop Framework 163
5.4.2 Linear Quadratic Gaussian Control 165
5.4.3 Generic Model Control 167
5.4.4 Multivariable Controller Synthesis 167
5.5 Energy Efficient Control Strategies 171
5.5.1 Background of Model Predictive Control 173
5.5.2 Controller Tuning Parameters 175
5.5.3 Dynamic Simulations 176
5.6 Concluding Remarks 180

Notation 181
References 183

6 Applications of Dividing-Wall Columns 187

6.1 Introduction 187
6.2 Separation of Ternary and Multicomponent Mixtures 188
6.3 Reactive Dividing-Wall Column 195
6.4 Azeotropic Dividing-Wall Column 198
6.5 Extractive Dividing-Wall Column 199
6.6 Revamping of Conventional Columns to DWC 203
6.7 Case Study: Dimethyl Ether Synthesis by R-DWC 205
6.8 Case Study: Bioethanol Dehydration by A-DWC and E-DWC 212
6.9 Concluding Remarks 223
References 223

7 Heat Pump Assisted Distillation 229

7.1 Introduction 229
7.2 Working Principle 231
7.3 Vapor (Re)compression 232
7.3.1 Vapor Compression 233
7.3.2 Mechanical Vapor Recompression 233
7.3.3 Thermal Vapor Recompression 234
7.4 Absorption–Resorption Heat Pumps 234
7.4.1 Absorption Heat Pump 234
7.4.2 Compression–Resorption Heat Pump 235
7.5 Thermo-acoustic Heat Pump 236
7.6 Other Heat Pumps 240
7.6.1 Stirling Cycle 240
7.6.2 Vuilleumier Cycle 241
7.6.3 Brayton Cycle 241
7.6.4 Malone Cycle 242
7.6.5 Solid–Sorption Cycle 242
7.7 Heat-Integrated Distillation Column 244
7.8 Technology Selection Scheme 245
7.8.1 Energy Efficient Distillation Technologies 246
7.8.2 Multicomponent Separations 249
7.8.3 Binary Distillation 254
7.8.4 Selected Scheme Applications 263

7.9 Concluding Remarks 265
References 265

8 Heat-Integrated Distillation Column 271
8.1 Introduction 271
8.2 Working Principle 273
8.3 Thermodynamic Analysis 277
8.4 Potential Energy Savings 280
8.4.1 Partial Heat Integrated Distillation Column (p-HIDiC) 280
8.4.2 Ideal Heat Integrated Distillation Column (i-HIDiC) 281
8.5 Design and Construction Options 282
8.5.1 Inter-coupled Distillation Columns 284
8.5.2 Distillation Column with Partition Wall 285
8.5.3 Concentric Distillation Column 287
8.5.4 Concentric Column with Heat Panels 288
8.5.5 Shell & Tube Heat-Exchanger Column 289
8.5.6 Plate-Fin Heat-Exchanger Column 290
8.5.7 Heat Transfer Means 292
8.6 Modeling and Simulation 295
8.7 Process Dynamics, Control, and Operation 297
8.8 Applications of HIDiC 300
8.9 Concluding Remarks 304
References 305

9 Cyclic Distillation 311
9.1 Introduction 311
9.2 Overview of Cyclic Distillation Processes 313
9.3 Process Description 316
9.4 Mathematical and Hydrodynamic Model 319
9.4.1 Mathematical Model 319
9.4.2 Hydrodynamic Model 321
9.4.3 Sensitivity Analysis 323
9.5 Modeling and Design of Cyclic Distillation 327
9.5.1 Modeling Approach 329
9.5.2 Comparison with Classic Distillation 331
9.5.3 Design Methodology 331
9.5.4 Demonstration of the Design Procedure 333
9.6 Control of Cyclic Distillation 335

9.7 Cyclic Distillation Case Studies 338
9.7.1 Ethanol–Water Stripping and Concentration 338
9.7.2 Methanol–Water Separation 341
9.8 Concluding Remarks 347
References 349

10 Reactive Distillation 353
10.1 Introduction 353
10.2 Principles of Reactive Distillation 354
10.3 Design, Control and Applications 357
10.4 Modeling Reactive Distillation 362
10.5 Feasibility and Technical Evaluation 364
10.5.1 Feasibility Evaluation 364
10.5.2 Technical Evaluation 367
10.6 Case Study: Advanced Control of a Reactive Distillation Column 371
10.6.1 Mathematical Model 371
10.6.2 Open-Loop Dynamic Analysis 374
10.6.3 Closed-Loop Performance 374
10.7 Case Study: Biodiesel Production by Heat-Integrated RD 378
10.8 Case Study: Fatty Esters Synthesis by Dual RD 383
10.9 Concluding Remarks 387
References 388

Index 393

Preface

Our modern society is currently facing an energy revolution, and it needs to identify properly the potential threats and use all the opportunities to meet the needs of the growing population. Accordingly, chemical engineers have embarked on a quest for shaping a much needed sustainable future—especially considering that chemical industry is among the most energy demanding sectors. Distillation is a thermal separation method widely applied in the chemical process industry as the separation technology of choice, despite its very low thermodynamic efficiency. Remarkably, almost every single product on the market includes components that went through a distillation column. Even now, when changing from fossil fuels to a bio-based economy, it is clear that in the next two decades distillation will retain its significance as the main method for separating mixtures—although this old workhorse of the chemical industry is facing some new big and bold challenges.

Owing to the limitation of fossil fuels, the need for energy independence, and the environmental problem of the greenhouse gas effect, there is a considerable increasing interest in the research and development of integrated chemical processes that require less capital investment, reduced operating costs, and have high eco-efficiency. Energy efficient distillation is a hot topic in separation technology due to the key advantages of the integrated technologies, such as reduced investment costs and low energy requirements, as well as an increasing number of industrial applications. Although the research and development carried out at universities and industrial companies in this exciting field is expanding quickly, there is still no book currently available focusing on this important area in distillation technology—the largest consumer of energy in the chemical process industry.

Therefore, we feel that there is a significant gap that can be addressed with this book and it will be of immense interest to a readership across the world. The book provides engineers with a wide and relatively deep insight into integrated distillations using non-conventional arrangements. Readers can learn from this material the background, recent developments, fundamental principles, design and simulation methods, detailed case studies of distillation processes, as well as expected future trends. We believe that the abundant valuable resources included here—relevant equations, diagrams, figures, and references that reflect the current and upcoming integrated distillation technologies—will be of great help to all readers from the (petro-)chemical industry, bio-refineries, and other related areas.

This book is the first comprehensive work about advanced distillation technologies, covering many important topics such as key concepts in distillation technology, principles of design, control, equipment sizing and economics of distillation, DWC design and configurations, optimal operation, controllability and advanced control strategies, industrial and pilot-scale DWC applications (in ternary separations, azeotropic distillation, extractive distillation, and reactive distillation), HIDiC design and configurations, heat pump assisted applications, cyclic distillation, and reactive distillation. Each chapter is independently written and consists typically of an introduction, working principle, process design, modeling and simulation, process control and operation, specific equipment, industrial and applied research examples, concluding remarks, as well as a comprehensive list of useful references for further reading.

Note that the author is aware about the unavoidable presence of some minor mistakes. That is why I would like to express my gratitude for every observation and suggestion towards further improving this material.

Acknowledgements

This book is the result of several years of dedicated work on various distillation topics, and I truly hope that the reader will find it useful and readable. Clearly, I am very grateful to everyone who has contributed in one way or another to make it a possible success. Therefore, I wish to express my deep gratitude and thanks to many of my collaborators and co-authors of scientific articles covering almost all topics in this book: Sorin Bildea, Radu Ignat and Ionela Lita (*University "Politehnica" of Bucharest, RO*), Eugeny Kenig and Ömer Yildirim (*University of Paderborn, DE*), David Suszwalak (*Ecole Nationale Supérieure de Chimie de Mulhouse, FR*), Carlos Infante Ferreira and Ruben van Diggelen (*Delft University of Technology, NL*), Rohit Rewagad (*University of Twente, NL*), Alexandre Dimian and Gadi Rothenberg (*University of Amsterdam, NL*), André de Haan, Servando Flores-Landaeta, Mayank Shah, and Edwin Zondervan (Eindhoven University of Technology, NL), Florin Omota (*Fluor, NL*), Zoltan Nagy (*Loughborough University, UK*), Juan Gabriel Segovia-Hernández (*Universidad de Guanajuato, MX*), Vladimir Maleta (*MaletaCD, UA*), as well as Hans Pragt and Cornald van Strien (*AkzoNobel, NL*).

Furthermore, I had the privilege during this time to benefit from smart and appropriate observations, during personal discussions or indirect contacts with some remarkable persons from academia and industry, to whom I am also truly indebted: Žarco Olujić, Andrzej Stankiewicz, and Johan Grievink (*Delft University of Technology, NL*), Igor Dejanović (*University of Zagreb, HR*), Sigurd Skogestad (*NTNU Trondheim, NO*), Ivar Halvorsen (*SINTEF ICT, NO*), Norbert Asprion (*BASF, DE*), Levente Simon (*BASF, CH*), Thomas Gruetzner (*Lonza, CH*), Björn Kaibel (*Julius Montz GmbH, DE*), Jeffrey Felix (*Sulzer Chemtech, CH*), Henry Kister (*Fluor/FRI, US*), William Luyben (*Lehigh University, US*),

Larry Biegler (*Carnegie Mellon University, US*), Dolf Bruinsma and Simon Spoelstra (*ECN, NL*), Yuji Tanaka (*Cosmo Research Institute, JP*), Aris de Rijke (*DSM Research, NL*), Jan Harmsen (*Shell Global Solutions, NL*), Andrzej Górak (*Technical University of Dortmund, DE*), Panos Seferlis (*Aristotle University of Thessaloniki, GR*), Donato Aranda (*Federal University of Rio de Janeiro, BR*) Sascha Kersten (*University of Twente, NL*), as well as a number of other colleagues.

In addition, the excellent support and valuable help from the editors Rebecca Stubbs, Sarah Tilley, Emma Strickland, Jasmine Kao and cover designer Dan Jubb (*John Wiley & Sons, UK*) is greatly acknowledged. My thankfulness is extended also to my friends around the world for their moral support and for keeping in touch with me during these busy and turbulent times. And last but not least, my special thanks go to my loving family—especially my better half—for their tender affection, understanding, relentless support, and continuous encouragement.

1

Basic Concepts in Distillation

1.1 INTRODUCTION

Distillation is a thermal separation method for separating mixtures of two or more substances into its component fractions of desired purity, based on differences in volatilities of components—which are in fact related to the boiling points of these components—by the application and removal of heat. Note that the term distillation refers to a physical separation process or a unit operation. Remarkably, distillation can be combined with another distillation operation, leading to a dividing-wall column (Harmsen, 2010), or with a chemical reaction, leading to reactive distillation (Sundmacher and Kienle, 2003; Sundmacher, Kienle, and Seidel-Morgenstern, 2005; Luyben and Yu, 2008; Sharma and Singh, 2010), and/or other chemical process operations (Schmidt-Traub and Gorak, 2006).

At the commercial scale, distillation has many applications, such as the separation of crude oil into fractions (e.g., gasoline, diesel, kerosene, etc.), water purification and desalination, the splitting of air into its components (e.g., oxygen, nitrogen, and argon), and the distillation of fermented solutions or the production of distilled beverages with high alcohol content (Forbes, 1970). Distillation underwent enormous development due to the petrochemical industry, and as such it is one of the most important technologies in the global energy supply system (Harmsen, 2010). Essentially, all transportation fuel goes through at least one distillation column on its way from crude oil to readily usable fuel, with tens of thousands of distillation columns in operation worldwide. In view of the foreseen depletion of fossil fuels and the switch to renewable sources of energy such as biomass, the most likely transportation fuel will be ethanol,

Advanced Distillation Technologies: Design, Control and Applications, First Edition.
Anton Alexandru Kiss.

methanol, or derivatives. The synthesis of alternative fuels leads typically to aqueous mixtures that require distillation to separate ethanol or methanol from water. Consequently, distillation remains the separation method of choice in the chemical process industry. The importance of distillation is unquestionable in providing most of the products required by our modern society (e.g., transportation fuel, heat, food, shelter, clothing, etc.).

The analysis, design, operation, control, and optimization of distillation columns were studied extensively in the last century but, until the introduction of computers, only hand calculations and graphical methods were developed and applied in distillation studies. As distillation analysis involves many iterative vapor–liquid phase equilibrium calculations, and tray-to-tray component balances that are ideal for digital computation, the use of computers has had a beneficial effect in recent decades (Luyben, 2011). Many companies still have their own in-house process simulators, although commercial steady-state and dynamic process simulators (e.g., Aspen Plus®, Aspen Dynamics®, ChemCAD, Aspen HYSYS®, PRO/II, etc.) are now available and dominate the field—with distillation playing a key role in these simulators.

The topic of distillation is very broad and it would require many volumes to cover it in a comprehensive manner. Consequently, for more details the reader is directed to several good books, which cover this subject in great detail: Kister (1992a), Kister (1992b), Taylor and Krishna (1993), Stichlmair and Fair (1998), Seader and Henley (1998), Doherty and Malone (2001), Mujtaba (2004), Petlyuk (2004), Lei, Chen, and Ding (2005), and more recently Luyben (2006, 2011).

It is important to note that distillation can separate chemical components only if the compositions of the vapor and liquid phases that are in equilibrium with each other are different. Therefore, a practical understanding of vapor–liquid equilibrium (VLE) is essential for the analysis, design, and control of distillation columns. This introductory chapter presents in a structured and convenient way the basic concepts of distillation: property methods, vapor pressure, bubble point, relative volatility, VLE, vapor–liquid–liquid equilibrium (VLLE), ternary diagrams, residue curve maps (RCM), and theoretical stage and short-cut design methods for distillation.

1.2 PHYSICAL PROPERTY METHODS

An extremely important issue in distillation calculations is the selection of an appropriate physical property method that will accurately describe the phase equilibrium of the chemical system. Missing or inadequate physical properties can undermine the accuracy of a model or even prevent one from performing the simulation. For this reason, finding good values for inadequate or missing physical property parameters is crucial to a successful simulation. Nevertheless, this depends strongly upon choosing the right estimation methods—an issue already recognized in the world of chemical processes modeling by the axiom "garbage in, garbage out" which means that the simulation results have the same quality as the input data/parameters (Carlson, 1996). In most design situations there is some type of data—for example, VLE reported in the literature, experimental measurements, and data books (Gmehling *et al.*, 1993; Perry and Green, 1997)—that can be used to select the most appropriate physical property method. The process simulators used nowadays (e.g., Aspen Plus, ChemCAD, HYSYS, PRO/II) have libraries with numerous alternative methods—the most commonly used being NRTL, UNIQUAC, UNIFAC, Chao–Seader, van Laar, Wilson, Grayson, Peng–Robinson, Soave–Redlich–Kwong (SRK), and derivatives of them.

Figure 1.1 provides a very convenient scheme that can be used for the quick and easy selection of an appropriate property model for virtually any chemical system (Aspen Technology, 2010a, 2010b). The property model names used here are given as in the Aspen Plus process simulator. Table 1.1 summarizes the commonly used property methods available in Aspen Plus (Aspen Technology, 2010b).

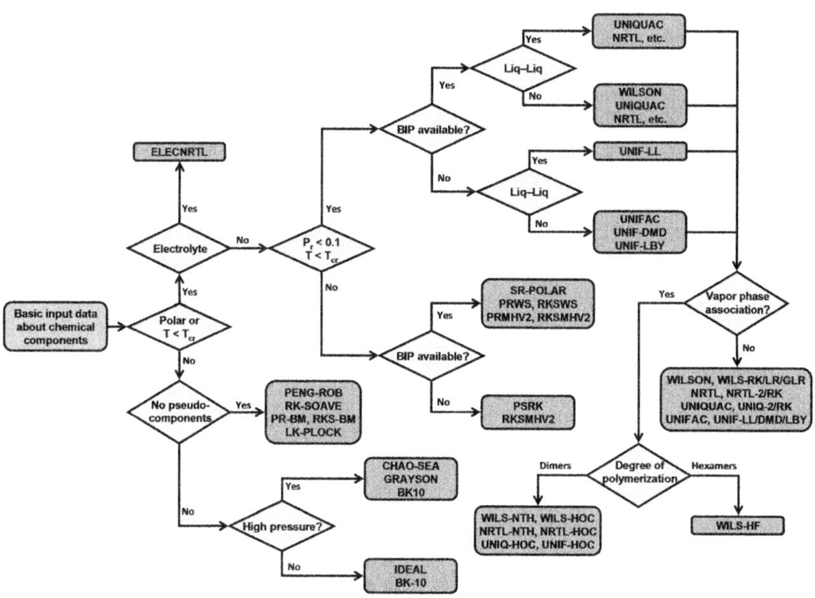

Figure 1.1 Property methods selection scheme

Table 1.1 Summary of commonly used property methods available in Aspen Plus

Abbreviation	*K*-value method	
IDEAL	Ideal gas/Raoult's law/Henry's law	
SOLIDS	Ideal gas/Raoult's law/Henry's law/solid activity coefficients	
BK-10	Braun K-10	
CHAO-SEA	Chao–Seader corresponding states model	
GRAYSON	Grayson–Streed corresponding states model	
STEAM-TA	ASME steam table correlations	
STEAMNBS	NBS/NRC steam table equation of state	
AMINES	Kent–Eisenberg amines model	
APISOUR	API sour water model	
	Equation of state	
BWR-LS	BWR Lee–Starling	
LK-PLOCK	Lee–Kesler–Plöcker	
PENG-ROB	Peng–Robinson	
PR-BM	Peng–Robinson with Boston–Mathias alpha function	
PRWS	Peng–Robinson with Wong–Sandler mixing rules	
PRMHV2	Peng–Robinson with modified Huron–Vidal mixing rules	
PSRK	Predictive Redlich–Kwong–Soave	
RKSWS	Redlich–Kwong–Soave with Wong–Sandler mixing rules	
RKSMHV2	Redlich–Kwong–Soave with modified Huron–Vidal mixing rules	
RK-ASPEN	Redlich–Kwong–ASPEN	
RK-SOAVE	Redlich–Kwong–Soave	
RKS-BM	Redlich–Kwong–Soave with Boston–Mathias alpha function	
SR-POLAR	Schwartzentruber–Renon	
Abbreviation	Liquid activity coefficient	Vapor fugacity coefficient
PITZER	Pitzer	Redlich–Kwong–Soave
PITZ-HG	Pitzer	Redlich–Kwong–Soave

B-PITZER	Bromley–Pitzer	Redlich–Kwong–Soave
ELECNRTL	Electrolyte NRTL	Redlich–Kwong
ENRTL-HF	Electrolyte NRTL	HF hexamerization model
ENRTL-HG	Electrolyte NRTL	Redlich–Kwong
NRTL	NRTL	Ideal gas
NRTL-HOC	NRTL	Hayden–O'Connell
NRTL-NTH	NRTL	Nothnagel
NRTL-RK	NRTL	Redlich–Kwong
NRTL-2	NRTL (using dataset 2)	Ideal gas
UNIFAC	UNIFAC	Redlich–Kwong
UNIF-DMD	Dortmund-modified UNIFAC	Redlich–Kwong–Soave
UNIF-HOC	UNIFAC	Hayden–O'Connell
UNIF-LBY	Lyngby-modified UNIFAC	Ideal gas
UNIF-LL	UNIFAC for liquid–liquid systems	Redlich–Kwong
UNIQUAC	UNIQUAC	Ideal gas
UNIQ-HOC	UNIQUAC	Hayden–O'Connell
UNIQ-NTH	UNIQUAC	Nothnagel
UNIQ-RK	UNIQUAC	Redlich–Kwong
UNIQ-2	UNIQUAC (using dataset 2)	Ideal gas
VANLAAR	Van Laar	Ideal gas
VANL-HOC	Van Laar	Hayden–O'Connell
VANL-NTH	Van Laar	Nothnagel
VANL-RK	Van Laar	Redlich–Kwong
VANL-2	Van Laar (using dataset 2)	Ideal gas
WILSON	Wilson	Ideal gas
WILS-HOC	Wilson	Hayden–O'Connell
WILS-NTH	Wilson	Nothnagel
WILS-RK	Wilson	Redlich–Kwong
WILS-2	Wilson (using dataset 2)	Ideal gas
WILS-HF	Wilson	HF hexamerization model
WILS-LR	Wilson (liquid enthalpy reference state)	Ideal gas
WILS-GLR	Wilson (ideal gas, liquid enthalpy reference state)	Ideal gas
WILS-VOL	Wilson with volume term	Redlich–Kwong

1.3 VAPOR PRESSURE

Distillation is based on the fact that the vapor of a boiling mixture will be richer in the components with lower boiling points. Consequently, when this vapor is sufficiently cooled the condensate will contain more volatile (e.g., light, low-boiling) components, while at the same time the original mixture will contain more of the less volatile (e.g., heavy, high-boiling) components.

Vapor pressure—a physical property of a pure chemical component—is the pressure that a pure component exerts at a given temperature when both liquid and vapor phases are present. In other words, the vapor pressure of a liquid at a particular temperature is the equilibrium pressure exerted by the molecules leaving and entering the liquid surface. Here are some key issues to consider:

- Vapor pressure is related to boiling, and it increases with the energy input.
- A liquid boils when its vapor pressure equals the ambient pressure.
- The ease with which a liquid boils depends on its volatility. Distillation occurs because of the differences in volatility of the components in the liquid mixture.
- Liquids with a high vapor pressure (volatile liquids) boil at low temperatures, and vice versa.
- The vapor pressure (and also the boiling point) of a liquid mixture depends on the relative amounts of the components in the mixture.

Table 1.2 provides the vapor pressure of some common substances at ambient temperature. Note that chemicals with a non-zero vapor

Table 1.2 Vapor pressure of some common substances at 20 °C

Chemical component	Vapor pressure (bar)	Vapor pressure (mmHg)
Ethylene glycol	0.005	3.75
Water	0.023	17.5
Propanol	0.024	18.0
Ethanol	0.058	43.7
Methyl isobutyl ketone (MIBK)	0.265	198.6
Freon 113 (1,1,2-trichlorotrifluoroethane)	0.379	284
Acetaldehyde	0.987	740
Butane	2.2	1650
Formaldehyde	4.357	3268
Carbon dioxide	57	42753

pressure lower than atmospheric are liquids, while those with a vapor pressure higher than atmospheric are gases, under normal conditions.

Raoult's law states that the vapor pressure of an ideal solution is dependent on the vapor pressure of each chemical component and on the mole fraction of the component present in the solution. Once the components in the solution have reached equilibrium, the total vapor pressure (p) of the solution is:

$$p = \sum_{j=1}^{N_C} p_j = \sum_{j=1}^{N_C} p_j^* x_j \tag{1.1}$$

with the individual vapor pressure for each component defined as: $p_j = p_j^* x_j$, where p_j is the partial pressure of component j in the mixture (in the solution), p_j^* is the vapor pressure of the pure component j, and x_j is the mole fraction of component j in the mixture.

The vapor pressure depends only on temperature and not on composition, since it is a pure component property. The dependence on temperature is usually a strong one, with an exponential increase of the vapor pressure at higher temperatures. Figure 1.2a shows some typical vapor pressure curves, for benzene, toluene, and xylene—with the exponential increase clearly observable at high temperatures. Figure 1.2b plots the natural log of the vapor pressure versus the reciprocal of the absolute temperature. It can be seen that as temperature increases (to the left of the plot) the vapor pressure is higher. In both plots of Figure 1.2, a vertical (constant temperature) line shows that, at a given temperature, benzene has a higher vapor pressure than toluene and xylene. Therefore, benzene is the *lightest* component, while xylene is the *heaviest* component—from the volatility (not density) standpoint. Correspondingly, a horizontal (constant-pressure)

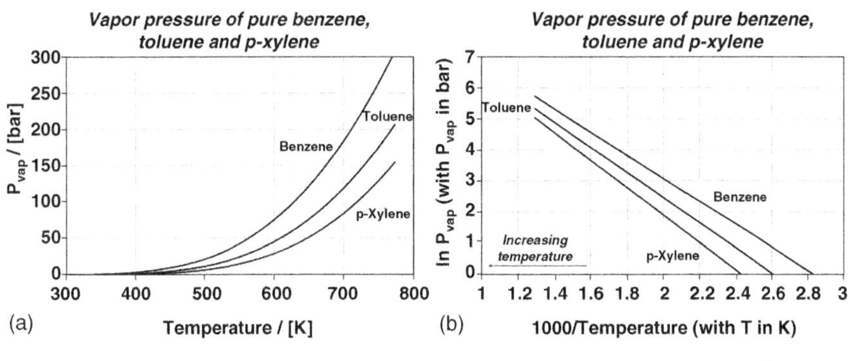

Figure 1.2 Vapor pressure of pure components: benzene, toluene, and p-xylene

line shows that benzene boils at a lower temperature than does toluene or xylene. Therefore, benzene is the *lowest boiling* component, while xylene is the *highest boiling* component. Note also that in Figure 1.2a the vapor pressure lines for benzene, toluene, and xylene are fairly parallel, meaning that the ratio of the vapor pressures does not change much with the temperature or pressure. Consequently, the ease or difficulty of benzene/toluene/xylene separation—directly translated into the energy requirements for the specified separation—does not change much with the operating pressure. However, other chemical components can have temperature dependences of the vapor pressure that are quite different to this example (Luyben, 2006).

In the case of distilling the binary mixture benzene–toluene, the concentration of the lighter (low-boiling) benzene in the vapor phase will be higher than that in the liquid phase—while the reverse is true for the heavier (high-boiling) toluene. As a result, benzene and toluene can be separated in a distillation column into a top distillate stream that is almost pure benzene and a bottoms stream that is fairly pure toluene. Using experimental vapor pressure data for each component, equations can be fitted by means of two, three, or more parameters. The Antoine equation—derived from the Clausius–Clapeyron relation—relates the vapor pressure and temperature for pure components:

$$\log p = A - B/(C + T) \tag{1.2}$$

where p is the vapor pressure, T is temperature, and A, B, and C are constants specific for each pure chemical component—their numerical values depend on the units used for vapor pressure (e.g., bar, mmHg, kPa) and on the units used for temperature (°C or K). The simplified form with the constant C set to zero ($\log p = A - B/T$) is known as the August equation, and describes a linear relation between the logarithm of the pressure and the reciprocal temperature—assuming that the heat of vaporization is independent of temperature.

1.4 VAPOR–LIQUID EQUILIBRIUM AND VLE NON-IDEALITY

1.4.1 Vapor–Liquid Equilibrium

Vapor–liquid equilibrium data for two-component (binary) systems is commonly represented by means of T–xy and xy diagrams—where T is the temperature, and x, y are the liquid and vapor composition, respectively, expressed in mole fraction. Basically, the *T–xy* diagram plots the temperature versus the liquid and vapor composition, while the *xy* diagram plots only *y* versus *x*. Although these types of diagrams are generated at a constant pressure, the *T–xy* an *xy* diagrams are extremely convenient for the analysis of binary distillation systems—especially since the operating pressure is relatively constant in most distillation columns.

Figure 1.3 shows the *T–xy* diagram (also known as the boiling point diagram) for the benzene–toluene system at atmospheric pressure—that is, how the equilibrium compositions of the components in a liquid mixture vary with temperature at a fixed pressure. The boiling point of benzene is that at which the mole fraction of benzene is 1, while the boiling point of toluene is that at which the mole fraction of benzene is 0. As illustrated by the *T–xy* diagram, benzene is the more volatile component and therefore has a lower boiling point than toluene. The lower curve in the *T–xy* diagram is called the *bubble-point curve* (saturated liquid curve), while the upper one is known as the *dew-point curve* (saturated vapor curve). The saturated liquid/lower curve gives the mole fraction of benzene in the liquid phase (x) while the saturated vapor/upper curve gives the mole fraction of benzene in the vapor phase (y). Drawing a horizontal line at a certain temperature and reading off the intersection of this line with the two curves give the compositions of the two liquid and vapor phases. Note that the *bubble-point* is defined as the temperature at which the liquid starts to boil, while the *dew-point* is the temperature at which the saturated vapor starts to condense. The

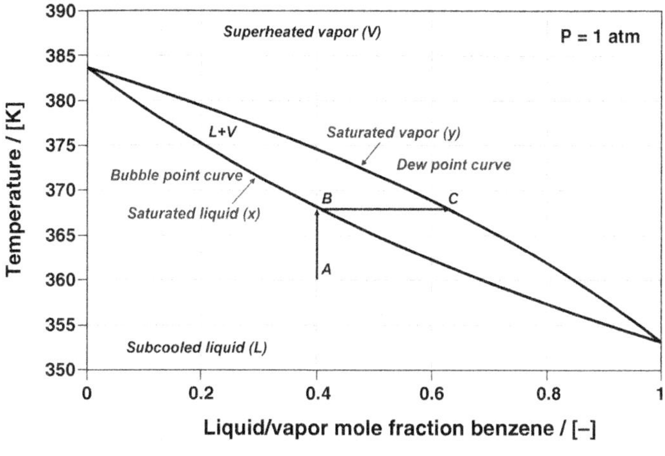

Figure 1.3 *T–xy* diagram for the mixture benzene–toluene at atmospheric pressure

region below the bubble-point curve shows the equilibrium composition of the subcooled liquid, while the region above the dew-point curve shows the equilibrium composition of the superheated vapor. Note that in the region between the lower and upper curves, there are two phases present—both liquid and vapor. For example, when a subcooled liquid is heated (point A, at 0.4 mole fraction of benzene) its concentration remains constant until it reaches the bubble-point (point B) when it starts to boil. The vapors produced during the boiling have the equilibrium composition of point C (ca. 0.65 mole fraction of benzene), and are thus over 60% richer in benzene than the original liquid mixture. This difference between the liquid and vapor compositions is in fact the basis for distillation operations.

The *T–xy* diagram can be easily generated in process simulators such as Aspen Plus, and the results at several pressures can be plotted (Figure 1.4). It is important to note that the higher the pressure, the higher the temperatures.

The *xy* diagram is also an effective tool in the analysis of distillation systems. Figure 1.5 illustrates the *xy* diagrams for the binary mixture benzene–toluene (Figure 1.5a) and propane–propylene (Figure 1.5b). As benzene and toluene have a relatively large difference in boiling points, the curve is noticeably shifted from the diagonal ($x = y$). However, propylene and propane have quite close boiling points, which leads to a very difficult separation—as illustrated in the *xy* diagram by the fact that the curve is very close to the diagonal ($x = y$). Remarkably, both *T–xy* and xy diagrams

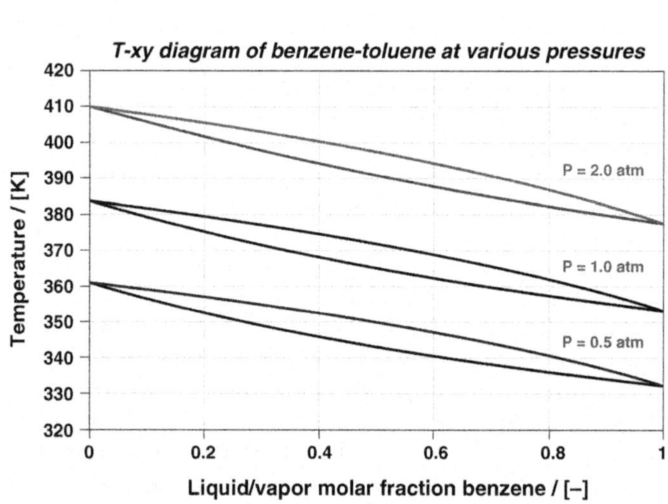

Figure 1.4 *T–xy* diagram for the mixture benzene–toluene at various pressures

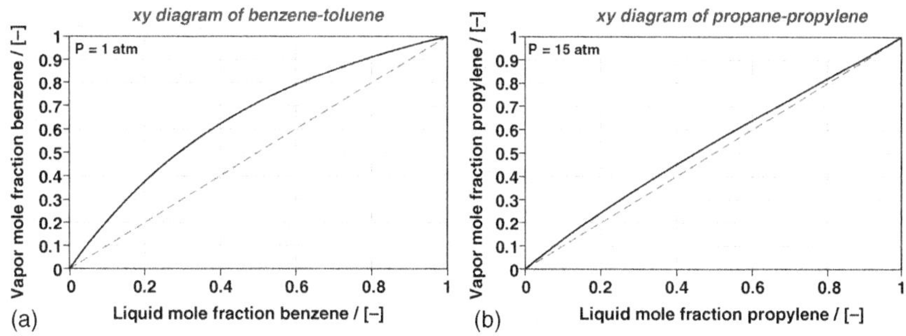

Figure 1.5 The *xy* diagram for the mixture benzene–toluene (a) and propane–propylene (b)

provide valuable insight into the phase equilibrium of binary systems, as they can be used for quantitative analysis of distillation (Luyben, 2006).

1.4.2 VLE Non-ideality

Liquid-phase ideality—equivalent to activity coefficients $\gamma_j = 1$ (unity)—occurs only when the components are very similar. The benzene–toluene system described earlier is a common example, where the activity coefficients of both components are very close to unity. However, if the components are quite different then non-ideal behavior occurs. For example, let us consider a methanol–water mixture; here water is very polar but methanol is polar only at the -OH end of the molecule while the -CH_3 end is non-polar. This difference results in some non-ideality (Figure 1.6). Figure 1.6a gives the *T–xy* curve at atmospheric pressure,

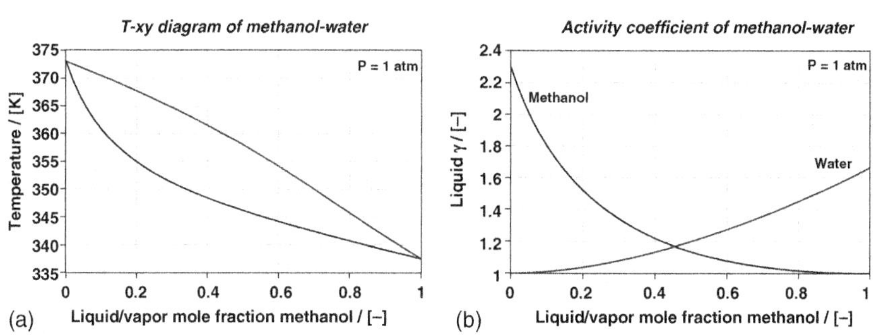

Figure 1.6 *T–xy* diagram (a) and activity coefficient plot (b) for methanol–water

while Figure 1.6b shows the variation of activity coefficients for both water and methanol over the composition space. The NRTL physical property method was used in this example to generate these plots. The activity coefficient values range up to $\gamma = 2.3$ for methanol at the $x = 0$ limit and $\gamma = 1.66$ for water at $x = 1$ (Luyben, 2006).

Let us consider now an ethanol–water mixture, in which case the $-CH_2CH_3$ (ethyl) end of the ethanol molecule is more non-polar than the $-CH_3$ end of methanol. As expected, the non-ideality is more pronounced—as clearly illustrated by the *T–xy* and *xy* diagrams shown in Figure 1.7. Note that the *xy* curve shown in Figure 1.7b crosses the diagonal (45° line, where $x = y$) at about 90 mol.% ethanol—this clearly indicates the presence of an azeotrope. Note also that the temperature at the azeotropic composition (351.0 K) is slightly lower than the boiling point of ethanol (351.5 K).

In fact, the most intriguing VLE curves are generated by azeotropic systems that give rise to VLE plots where the equilibrium curves crosses the diagonal (on the *xy* diagram). Note that an azeotrope is defined as the composition at which the liquid and vapor compositions are equal. When this occurs in a distillation column, there is no further change in the liquid and vapor compositions from tray to tray—hence the azeotrope represents a *distillation boundary*. Azeotropes can be classified according to the phase as *homogeneous* (single liquid phase) or *heterogeneous* (two liquid phases), according to the boiling temperature as *minimum-boiling* or *maximum-boiling*, and they can occur in binary, ternary, and multicomponent systems. The ethanol–water mixture described in the previous example has a minimum-boiling homogeneous azeotrope (single liquid phase boiling at 78 °C, with the composition of 89.3 mol.% ethanol). The VLE non-ideality and the types of azeotropic systems are tackled in more detail by Stichlmair and Fair (1998).

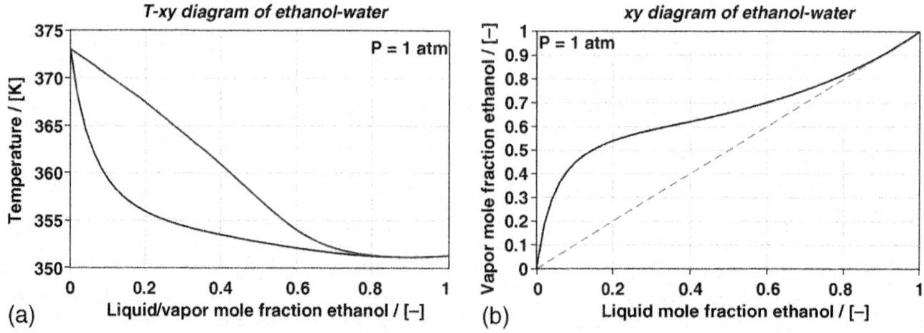

Figure 1.7 *T–xy* diagram, activity coefficient plot (a) and *xy* diagram (b) for ethanol–water

1.5 RELATIVE VOLATILITY

Relative volatility is a measure of the differences in volatility (or boiling points) between two components, indicating how easy or difficult a particular separation will be. The golden rule for distillation is that the larger the relative volatility, the easier the separation. The relative volatility of component L (light) with respect to component H (heavy) is defined as the ratio of the *y*/*x* values (vapor mole fraction divided by the liquid mole fraction):

$$\alpha_{\mathrm{LH}} = \frac{y_{\mathrm{L}}/x_{\mathrm{L}}}{y_{\mathrm{H}}/x_{\mathrm{H}}} \tag{1.3}$$

Note that relative volatilities can be applied to both binary and multicomponent systems. In binary systems, the relative volatility α between the light and heavy components can be used to give a simple relationship between the composition of the liquid phase and the vapor phase, with x and y being the mole fraction of the light component in the liquid and vapor phase, respectively:

$$y = \frac{\alpha x}{1 + (\alpha - 1)x} \tag{1.4}$$

Figure 1.8 gives the *xy* curves for several values of α—under the assumption that α is constant over the entire composition space. For multicomponent systems, a similar relationship can be derived. Consider N_{C}, the number components, with component 1 being the lightest,

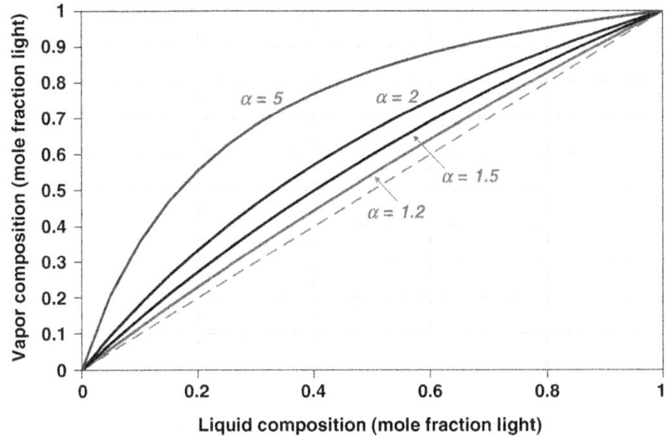

Figure 1.8 The *xy* curves for various relative volatilities: $\alpha = 1.2$, 1.5, 2, and 5

component 2 is the next lightest, and so forth down to component N_C, the heaviest (H) of all. The relative volatility of component j with respect to component H is defined as (Luyben, 2006):

$$\alpha_j = \frac{y_j/x_j}{y_H/x_H} \tag{1.5}$$

Solving for y_j and summing all the y values, which must add to unity, leads to:

$$y_j = \alpha_j x_j \frac{y_H}{x_H} \tag{1.6}$$

$$\sum_{j=1}^{N_C} y_j = \sum_{j=1}^{N_C} \alpha_j x_j \frac{y_H}{x_H} = \frac{y_H}{x_H} \sum_{j=1}^{N_C} \alpha_j x_j = 1 \tag{1.7}$$

Then, solving for y_H/x_H and substituting this ratio into the first equation above give:

$$\frac{y_H}{x_H} = 1 / \sum_{j=1}^{N_C} \alpha_j x_j \tag{1.8}$$

$$y_j = \alpha_j x_j / \sum_{j=1}^{N_C} \alpha_j x_j \tag{1.9}$$

The last equation relates the vapor composition to the liquid composition for a constant relative volatility multicomponent system. If the relative volatilities are not constant, this equation cannot be used and bubble point calculations are required instead (Luyben, 2006).

1.6 BUBBLE POINT CALCULATIONS

When heating a liquid consisting of two or more components, the bubble point is the temperature at which the first bubble of vapor is formed. Since the vapor will probably have a different composition than that of the liquid, the bubble point—as well as the dew point—at different compositions is very valuable data when designing distillation systems. Note that for single-component mixtures the bubble point and the dew point are the same, and are commonly referred to as the *boiling point*.

The most common VLE problem is to calculate the temperature and vapor composition (y_j) that is in equilibrium with a liquid of known composition (x_j), at a known total pressure (P) of the system. At equilibrium, the *chemical potential* of each component in the liquid and vapor phases are equal:

$$\mu_j^{\mathrm{L}} = \mu_j^{\mathrm{V}} \tag{1.10}$$

The liquid-phase chemical potential of component j can be expressed in terms of liquid mole fraction (x_j), vapor pressure (P_j^{S}), and activity coefficient (γ_j):

$$\mu_j^{\mathrm{L}} = x_j P_j^{\mathrm{S}} \gamma_j \tag{1.11}$$

Similarly, the vapor-phase chemical potential of component j can be expressed in terms of vapor mole fraction (y_j), total system pressure (P), and fugacity coefficient (σ_j):

$$\mu_j^{\mathrm{V}} = y_j P \sigma_j \tag{1.12}$$

Thus, at equilibrium the general relationship between vapor and liquid phases is (Luyben, 2006):

$$y_j P \sigma_j = x_j P_j^{\mathrm{S}} \gamma_j \tag{1.13}$$

The fugacity coefficient is unity ($\sigma_j = 1$) if the total pressure of the system is not too high. Moreover, if the liquid phase is ideal—meaning that there is no interaction between the molecules—then the activity coefficient is unity ($\gamma_j = 1$). However, the occurrence of an ideal liquid phase is much less common than the ideal gas phase, because the components interact in liquid mixtures—they can either attract or repulse (Luyben, 2006). Assuming that both liquid and vapor phases are ideal (i.e., $\sigma_j = 1$ and $\gamma_j = 1$), the bubble point calculation involves an iterative calculation to find the temperature T that satisfies the equation:

$$P = \sum_{j=1}^{N_{\mathrm{C}}} x_j P_{j(T)}^{\mathrm{S}} \tag{1.14}$$

Note that the total pressure P and all x_j values are known, while equations for the vapor pressures of all components as functions of temperature T are also known. Since an analytical derivative of the temperature-dependent vapor pressure functions can be used, the Newton–Raphson convergence method is very convenient and efficient in this iterative calculation (Luyben, 2006).

1.7 TERNARY DIAGRAMS AND RESIDUE CURVE MAPS

Ternary diagrams and reside curve maps are extremely valuable tools for the design of distillation systems, especially when VLE non-ideality is involved (e.g., phase splitting and/or azeotropes).

1.7.1 Ternary Diagrams

Using ternary diagrams, a three-component system can be represented in only two dimensions. Although there are three components, the sum of the mole fractions must add to unity and, as such, specifying two mole fractions is sufficient to define the composition (Luyben, 2006). Figure 1.9a gives a typical rectangular ternary diagram. The mole fraction of component 1 is shown on the abscissa, while the mole fraction of component 2 is shown on the ordinate. Since these axes represent mole fractions, both of these dimensions lie in the interval from zero to one. The three corners of the triangle represent the three pure components, while the edges represent binary mixtures. Any point located within the triangle represents a ternary mixture (Doherty and Malone, 2001; Dimian, 2003; Luyben, 2006).

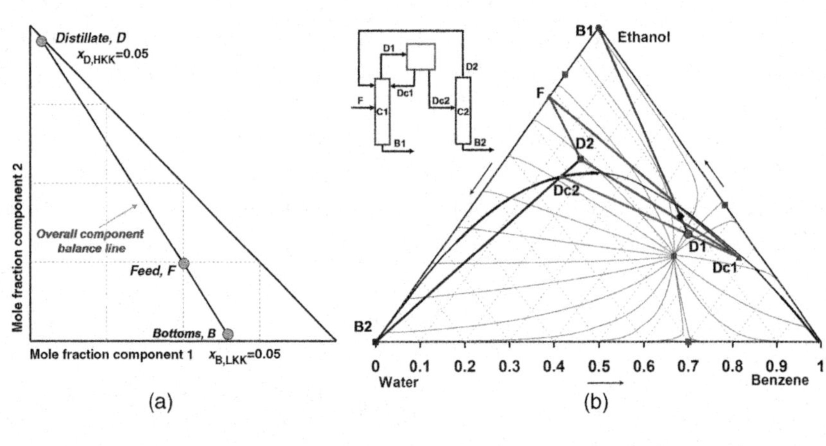

Figure 1.9 Ternary diagram and mixing rules (a); conceptual design of an azeotropic distillation system: ethanol–benzene–water (b)

Since only two compositions are sufficient to define the composition of a stream, the stream can be easily located on this diagram by entering the appropriate coordinates. For example, Figure 1.9a shows the location of stream *F*, which is a ternary mixture of 50 mol.% component 1, 25 mol.% component 2, and the remaining 25 mol.% is component 3. A useful aspect of ternary diagrams is the *ternary mixing rule*, which states the following. If two ternary streams are mixed together—one being stream D (x_{D1}, x_{D2}) and the other stream B (x_{B1}, x_{B2})—the mixture has a composition (z_1, z_2) that lies on a straight line in an x_1–x_2 ternary diagram that connects the x_D and x_B points (Luyben, 2006). This property is very useful in separations.

Figure 1.9a also illustrates the application of this mixing rule to a distillation column. Note that instead of mixing, a distillation column separates a feed (F) into two product streams (D and B), but the geometry is exactly the same. The two products D and B have compositions located at point (x_{D1}, x_{D2}) and (x_{B1}, x_{B2}), respectively. The feed F has a composition located at point (z_1, z_2) that lies on a straight line joining D and B. This geometric relationship is derived from the overall molar balance and the two overall component balances around the column:

$$F = D + B \tag{1.15}$$

$$Fz_1 = Dx_{D1} + Bx_{B1} \tag{1.16}$$

$$Fz_2 = Dx_{D2} + Bx_{B2} \tag{1.17}$$

Substituting the first equation into the second and third gives:

$$(D + B)z_1 = Dx_{D1} + Bx_{B1} \tag{1.18}$$

$$(D + B)z_2 = Dx_{D2} + Bx_{B2} \tag{1.19}$$

Rearranging these two equations to solve for the ratio of B/D gives:

$$\frac{D}{B} = \frac{z_1 - x_{D1}}{x_{B1} - z_1} \tag{1.20}$$

$$\frac{D}{B} = \frac{z_2 - x_{D2}}{x_{B2} - z_2} \tag{1.21}$$

Equating these two equations and rearranging, results in:

$$\frac{z_1 - x_{D1}}{x_{B1} - z_1} = \frac{z_2 - x_{D2}}{x_{B2} - z_2} \tag{1.22}$$

$$\frac{x_{D1} - z_1}{z_2 - x_{D2}} = \frac{z_1 - x_{B1}}{x_{B2} - z_2} \tag{1.23}$$

This straight-line relationship is very useful in representing what is going on in a ternary distillation system, giving a visual representation of the mass balance of the distillation column. Figure 1.9b demonstrates the use of these rules in a ternary diagram (showing the liquid split envelope) combined with a residue curve map (RCM) for the conceptual design of an azeotropic distillation system: ethanol–benzene–water. The near azeotropic stream F is fed to the first distillation column, together with D2 and Dc1—their mix resides in the triangle F–D2–Dc1. The products of the first column are the bottoms B1 (ethanol) and the distillate D1. The D1 stream is fed to a decanter that separates (according the liquid split tie lines) an organic phase Dc1 (recycled to the first column C1) and an aqueous phase Dc2. The Dc2 stream is fed to a second distillation column C2 that produces the bottoms B2 (water) and a distillate D2 that is recycled to the first column.

1.7.2 Residue Curve Maps

Residue curve maps (RCMs) are a powerful tool for better understanding of the design and operation of distillation columns, especially when the separation involves azeotropic mixtures. Residue curves can be used to determine which separations are possible by distillation. Moreover, compared to stage-to-stage column-composition-profile calculations, residue curves are mathematically much easier to work with and they can provide a great deal of insight into the separation of a mixture. However, despite the current increased computing power, checking the separation feasibility for applications involving mixtures that form two liquid phases is still a relatively complex task. In such heterogeneous systems, RCMs can be used to exploit the two-liquid-phase behavior to facilitate a desired separation. In addition, a RCM can be also used to check the feasibility of separation of homogeneous mixtures and for developing the conceptual design of distillation tower sequences. Separation synthesis based on a RCM enables engineers to develop the best distillation designs, even for complex, highly non-ideal systems that are found in chemical and specialty chemical plants. The ability to visualize the physical constraints of the separation process helps engineers to generate better design alternatives more quickly, resulting in significant time and cost savings.

The simplest form of distillation is a process in which a multicomponent liquid mixture is boiled in an open vessel and the vapor is continuously removed as formed. At any instant in time the vapor is in equilibrium with the remaining liquid. Because the vapor is always richer in the more volatile components than the liquid, the composition of the liquid changes continuously over time, becoming more and more concentrated in the least volatile species. A simple distillation residue curve is a graph showing how the composition of the liquid residue changes over time. A RCM is a collection of liquid residue curves originating from different initial compositions. It contains the same information as phase diagrams, but represented in a more useful way for understanding how to synthesize a distillation sequence. Similarly, for a distillation column equipped with trays, a distillation curve is defined as the locus of the tray compositions at total reflux. A *distillation curve map* (DCM) can be generated easily by choosing a tray liquid composition and stepping up and down by a series of bubble and dew points. Numerical investigation shows that distillation and residue curves are, in general, close to each other. In fact, both are related to the variation of concentration in a distillation column operated at infinite reflux—RCM for a packed column and DCM for a tray column (Doherty and Malone, 2001; Dimian, 2003; Luyben, 2006).

Figure 1.10 shows a typical RCM and the types of characteristic points that can be encountered. All of the residue curves originate at the light (lowest boiling) pure component in a region, move towards the intermediate boiling component, and end at the heavy (highest boiling) pure component in the same region. The lowest temperature nodes are denoted as *unstable nodes* (UNs), as all trajectories leave from them, while the highest temperature points in the region are termed *stable nodes* (SNs), as all trajectories ultimately reach them. The points at which the trajectories approach from one direction and end in a different direction (as always is the point of an intermediate boiling component) are termed *saddle points* (S). Residue curves that divide the composition space into different distillation regions are called distillation boundaries. The concept of characteristic points is important in classifying azeotropic mixtures.

A better understanding of the residue curve map is illustrated in Figure 1.11a. In this example of a zeotropic system, benzene is an unstable

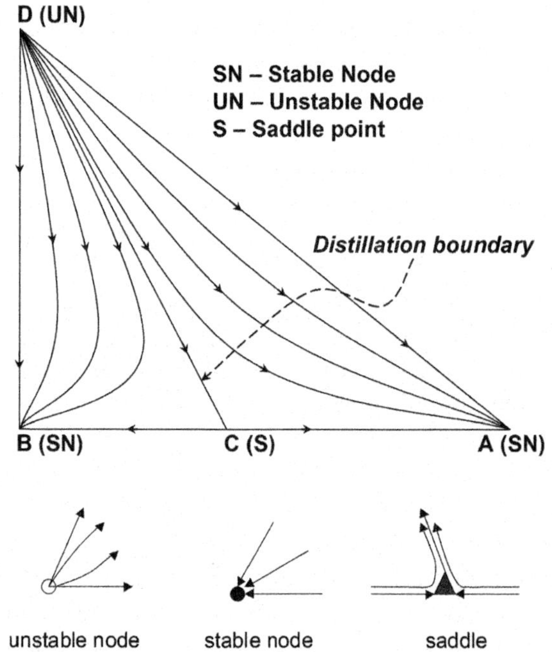

Figure 1.10 Residue curve map (RCM) and types of characteristic points

node, while ethyl-benzene is a stable node and toluene is a saddle point. Note that trajectories move from the lowest temperature towards the highest.

Using various references, the simple distillation process can be described by the set of equations:

$$\frac{dx_i}{d\tau} = x_i - y_i; \quad \text{for } i = 1 \ldots n \text{ (number of components)} \tag{1.24}$$

where x_i and y_i are the mole fractions of component i in liquid and vapor phase, respectively, and τ is the nonlinear time scale. Research studies (Stichlmair and Fair, 1998; Doherty and Malone, 2001) have also determined the relationship between the number of nodes (stable and unstable) and saddle points that can exist in a validly drawn RCM. The consistency of RCM with the azeotropic data can be verified by a theoretical test, expressed by the following relation based on topological arguments:

$$4(N_3 - S_3) + 2(N_2 - S_2) + (N_1 - S_1) = 1 \tag{1.25}$$

where N_i and S_i are the number of nodes and saddles, respectively, involving exactly i species from the ternary mixtures. For example, in Figure 1.11a: $N_3 = 0$, $S_3 = 0$, $N_2 = 0$, $S_2 = 1$, $N_1 = 3$, $S_1 = 0$, hence

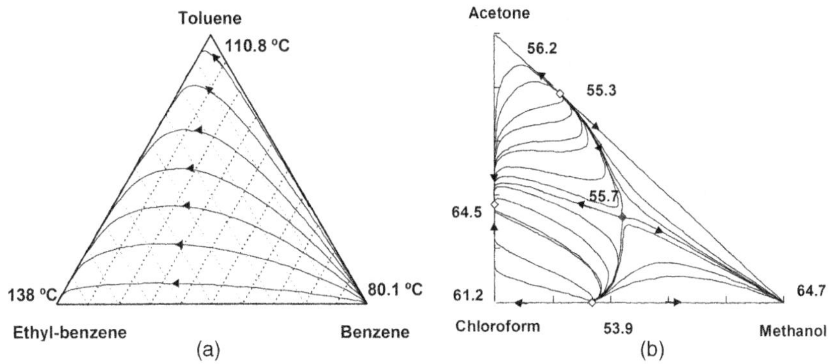

Figure 1.11 RCM for (a) benzene/toluene/ethyl-benzene and (b) acetone/chloroform/methanol

$4(0-0)+2(0-1)+(3-0)=1$. Many different residue curve maps are possible when azeotropes are present (Stichlmair and Fair, 1998; Doherty and Malone, 2001; Petlyuk, 2004).

Residue curve maps have extremely useful applications, such as testing of the consistency of experimental azeotropic data; predicting the order and content of the cuts in batch distillation; checking whether a given mixture is separable by distillation, identification of entrainers, prediction of attainable compositions, and qualitative prediction of composition profile shape; identifying the limiting separation achievable by distillation, and synthesizing separation sequences combining distillation with other methods (Doherty and Malone, 2001; Dimian, 2003; Luyben, 2006).

Note that thermodynamic data is of utmost importance in obtaining reliable RCMs. The adequacy of models and the accuracy of interaction parameters must always be checked. WILSON is very accurate for homogeneous mixtures, while UNIQUAC and NRTL are sufficiently accurate in many cases. An additional advantage is that these models can be applied for both VLE and LLE. Specifying systematically VLLE as an option for flash calculation avoids unreliable azeotrope prediction. UNIFAC should be used only for exploratory purposes, while different sources of equilibrium data should be tested. Nevertheless, when a detailed validated thermodynamic model is not available, pure component and binary azeotropic data often are sufficient to sketch the main characteristics of a RCM. Thereby, it is possible to produce a conceptual design of a separation sequence at an early stage in a project based on very limited information. Notably, nowadays, the generation of a RCM is a standard feature within any process simulator.

Alternatively, a RCM can be represented as a right-angled triangle, a so-called *right-angle triangle diagram,* which is more practical for sketching separation sequences. The following convention is adopted:

A. Pure components: lowest-boiler is on top, intermediate-boiler on bottom-left, and highest-boiler on bottom-right.
B. Azeotropes. A binary azeotrope is represented by a number, which is:
0: no azeotrope,
1: binary minimum-boiling azeotrope, node;
2: binary minimum-boiling azeotrope, saddle;
3: binary maximum-boiling azeotrope, node;
4: binary maximum-boiling azeotrope, saddle.
Ternary azeotrope: m (minimum), M (maximum), or S (intermediate).

For the azeotropic mixture acetone/chloroform/methanol shown in Figure 1.11b the class is 311-S. The first digit represents the maximum-boiling azeotrope acetone/chloroform, the second the minimum-boiling azeotrope chloroform/methanol, the third the minimum-boiling azeotrope acetone/methanol, while the last letter S signifies the ternary saddle azeotrope (Doherty and Malone, 2001; Dimian, 2003).

As explained earlier, all residue curves start at the lightest component and move toward the heaviest component. In this sense they are similar to the compositions in a distillation column. The light components go out the top, and the heavy components go out the bottom. This similarity is very useful for the analysis of distillation systems.

The generation of residue curves can be described mathematically by a dynamic molar balance of the liquid in the vessel M_{liq} and two dynamic component balances for components A and B. The rate of vapor withdrawal (V) is given in moles per unit time:

$$\frac{\mathrm{d}M_{\mathrm{liq}}}{\mathrm{d}t} = -V \tag{1.26}$$

$$\frac{\mathrm{d}(M_{\mathrm{liq}}x_j)}{\mathrm{d}t} = -Vy_j \tag{1.27}$$

The values of x_j and y_j are related by the VLE of the system. Expanding the second equation and substituting the first one gives:

$$M_{\mathrm{liq}}\frac{\mathrm{d}x_j}{\mathrm{d}t} + x_j\frac{\mathrm{d}M_{\mathrm{liq}}}{\mathrm{d}t} = -Vy_j \tag{1.28}$$

$$M_{\text{liq}}\frac{\mathrm{d}x_j}{\mathrm{d}t} + x_j(-V) = -Vy_j \tag{1.29}$$

$$\left(\frac{M_{\text{liq}}}{V}\right)\frac{\mathrm{d}x_j}{\mathrm{d}t} = x_j - y_j \tag{1.30}$$

$$\frac{\mathrm{d}x_j}{\mathrm{d}\theta} = x_j - y_j \tag{1.31}$$

where θ is a dimensionless time variable. The last equation models how the compositions change during the generation of a residue curve. As described next, a similar equation expresses the tray-to-tray liquid compositions in a column under total reflux conditions. This relationship allows the use of residue curves to assess what separations are feasible or infeasible in a given system.

Consider the upper section of a distillation column as shown in Figure 1.12. The column is cut at tray n, at which the passing vapor and liquid streams have compositions $y_{n,j}$ and $x_{n+1,j}$, and the flow rates are V_n and L_{n+1}, respectively. The distillate flow rate and composition are

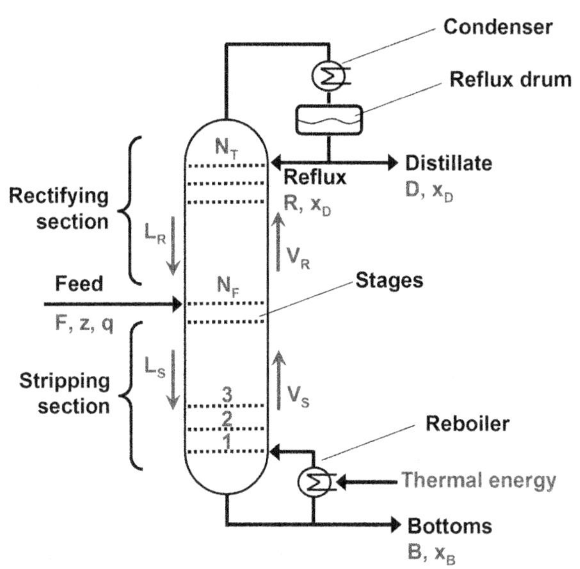

Figure 1.12 Schematics of a distillation column

D and $x_{D,j}$. The steady-state component balance is given by:

$$V_n y_{n,j} = L_{n+1} x_{n+1,j} + D x_{D,j} \tag{1.32}$$

Under total reflux conditions, D is equal to zero and L_{n+1} is equal to V_n. Therefore, $y_{n,j}$ is equal to $x_{n+1,j}$. Now, let us define a continuous variable h as the distance from the top of the column down to any tray. The discrete changes in liquid composition from tray to tray can be approximated by the following differential equation:

$$\frac{dx_j}{dh} \approx x_{n,j} - x_{n+1,j} \tag{1.33}$$

At total reflux this equation becomes:

$$\frac{dx_j}{dh} = x_{n,j} - y_{n,j} \tag{1.34}$$

Note that this is in fact the same equation as developed for residue curves. The significance of this similarity is that the residue curves approximate the column profiles. Therefore, a feasible separation in a column must satisfy two conditions (Dimian, 2003; Luyben, 2006):

1. The distillate compositions and the bottoms compositions must lie near a residue curve.
2. They must lie on a straight line through the feed composition point.

1.8 ANALYSIS OF DISTILLATION COLUMNS

Figure 1.12 illustrates the schematics of a distillation column consisting of an upper (rectifying) section and a lower (stripping) section, with N_T as the total number of stages. Note that the ideal distillation stage is a device that meets three criteria (Kister, 1992a):

1. It operates in steady state and has a liquid product and a vapor product.
2. All vapor and liquid entering the stage are intimately contacted and perfectly mixed.
3. The total vapor leaving the stage is in equilibrium with the total liquid leaving the stage.

The concept of *stage efficiency* is used to account for the non-ideality of a stage. The number of ideal stages is equal to the number of non-ideal stages multiplied by the stage efficiency. The non-ideality may lower or enhance the separation—if it enhances the separation, the stage efficiency can exceed 100%. Vapor leaving a distillation stage is richer than the feed in the more volatile components. Liquid leaving the stage is richer than the feed in the less volatile components. To improve the separation, multiple stages are used. *Stripping stages* concentrate the less volatile components in a liquid stream. A vapor recycle vaporizes (strips) the more volatile components from the liquid. To generate the vapor recycle to the column, heat is supplied to vaporize a portion of the bottom stage liquid—this vapor recycle is termed as *boil-up*. *Rectifying stages* concentrate the more volatile components in a vapor stream. A liquid recycle condenses the less volatile components from the rising vapor. To generate the liquid recycle, cooling is applied to condense a portion of the overhead vapor—the liquid recycle is termed *reflux*.

The stripping and rectifying stages can be combined into a single separation process with internal recycle (Figure 1.12), termed *distillation* or *fractionation*. In a single feed distillation column, the stages above the feed are rectifying and those below it are stripping. In multifeed columns, the more precise functional criterion below is used to distinguish the rectifying from stripping sections. The stripping section has a net down flow of material. The vapor serves only as a recycle stream to remove lights from the liquid. Therefore, the quantity of liquid exceeds the quantity of vapor in the stripping section. The converse applies in the rectifying section. This section has a net up flow of material, and the quantity of vapor exceeds the quantity of liquid (Kister, 1992b).

In a multicomponent distillation of j components there are $j-1$ component balances and $j-1$ equations describing the equilibrium relationship. They form the so called MESH equations:

$$\text{Mass balance}: F_n + V_{n+1} + L_{n-1} = V_n + L_n \tag{1.35}$$

$$\text{Component balance}: F_n z_n + V_{n+1} y_{n+1} + L_{n-1} x_{n-1} = V_n y_n + L_n x_n \tag{1.36}$$

Energy balance :

$$\Delta H_n + F_n H_{F,n} + V_{n+1} H_{V,n+1} + L_{n-1} H_{L,n-1} = V_n H_{V,n} + L_n H_{L,n} \tag{1.37}$$

$$\text{Equilibrium relationship}: y_n = K x_n \tag{1.38}$$

These equations apply to each stage. A rigorous solution simultaneously solves these equations for each stage and each component. The equations can be simplified and solved by analytical shortcut procedures or graphically—as described later. The graphical procedures are also applied to introduce and illustrate several key distillation concepts.

1.8.1 Degrees of Freedom Analysis

The design of a distillation column involves many parameters, such as product compositions, product flow rates, operating pressure, total number of trays, feed tray location, reflux ratio (RR), reboiler heat input, condenser heat removal, column diameter, and column height. Not all of these variables are independent, so a degrees of freedom (DoF) analysis is useful in pinning down exactly how many independent variables can (and must) be specified to completely define the system. A rigorous DoF analysis involves counting the number of variables in the system and subtracting the number of independent equations that describe the system. For a multicomponent, multistage column this can involve hundreds, if not thousands, of variables and equations. Any error in counting is grossly amplified because we are taking the difference between two very large numbers.

The normal situation in distillation design is that the feed conditions are given: flow rate F (mol h^{-1}), composition z_j (mole fraction of component j), temperature T_{F}, and pressure P_{F}. The desired compositions of the product streams are also typically known. Considering a two-product column, the normal specifications are to set the heavy-key impurity in the distillate $x_{D,\mathrm{HK}}$ and the light-key impurity in the bottoms $x_{B,\mathrm{LK}}$. The design problem is to establish the operating pressure P, the total number of trays N_{T}, and the feed tray location N_{F} that produces the desired product purities. All the other parameters are then fixed. Therefore, the number of design degrees of freedom is five: $x_{D,\mathrm{HK}}$, $x_{B,\mathrm{LK}}$, P, N_{T}, and N_{F}. Consequently, if the desired product purities and the pressure are given there are only two degrees of freedom: N_{T} and N_{F}. To emphasize this point, the five variables that could be specified might be the distillate flow rate D, reflux ratio $RR = R/D$, P, N_{T}, and N_{F}. In this case the product compositions cannot be specified but depend on the distillate flow rate and reflux ratio selected (Luyben, 2006). The next sections provide some of the ways used to establish reasonable values of some of the parameters such as the number of stages or the reflux ratio.

1.8.2 McCabe–Thiele Method

The McCabe–Thiele method is a graphical approach that shows very nicely in a graphical form the effects of VLE, reflux ratio, and number of trays (McCabe, Smith, and Harriot, 2005). Although it is limited to binary systems, the effects of parameters can be extended to multicomponent systems. The basic effects can be summarized as follows:

- The easier the separation, the fewer trays are required and the lower the required reflux ratio (also translated into lower energy requirements).
- The higher the desired product purities, the more trays are required—but the required reflux ratio does not increase significantly as the product purities increase.
- There is an engineering trade-off between the number of trays and the reflux ratio. An infinite number of columns can be designed that produce exactly the same products but have different heights, diameters, and energy requirements. Hence, selecting the optimum column involves issues of both steady-state economics and dynamic controllability.
- The minimum values of the number of trays ($N_{\min}$) and of the reflux ratio ($RR_{\min}$) required for a given separation.

All of these items can be visually demonstrated using the McCabe–Thiele method. The distillation column considered is shown in Figure 1.12, with the various flows and composition indicated. Assuming that the feed molar flow rate F and composition z are given, if the product compositions are specified, the molar flow rates of the two products D and B can be immediately calculated from the overall total molar balance and the overall component balance on the light component:

$$F = D + B \tag{1.39}$$

$$Fz = Dx_D + Bx_B \tag{1.40}$$

$$D = F\left(\frac{z - x_B}{x_D - x_B}\right) \tag{1.41}$$

For the moment let us assume that the pressure has been specified, so the VLE is fixed. Let us also assume that the reflux ratio has been specified, so the reflux flow rate can be calculated $R = RRD$. The equimolar overflow assumption is usually made in the McCabe–Thiele method. The liquid and vapor flow rates are assumed to be constant in a given section of the column. For example, the liquid flow rate in the rectifying section L_R is equal to the reflux flow rate R. From an overall balance around the top of the column, the vapor flow rate in the rectifying section V_R is equal to the reflux plus the distillate ($V_R = R + D$). This method uses a xy diagram whose coordinates are the mole fraction of the light component in the liquid x and the mole fraction of the light component in the vapor phase y. The diagonal (45° line) is plotted and so is the VLE curve for the selected pressure. The specified product compositions x_D and x_B are located on the diagonal, as described next.

Figure 1.13 illustrates the construction of the rectifying operating line (ROL), while Figure 1.14 shows the construction of the stripping operating line (SOL). The ROL is a straight line with a slope equal to the ratio of the liquid and vapor flow rates in the rectifying section:

$$\text{Slope ROL} = \frac{L_R}{V_R} = \frac{R}{R+D} = \frac{RR}{1+RR} \tag{1.42}$$

The line intersects the diagonal at the distillate composition x_D, and so it is easy to construct—as shown in Figure 1.15. The proof of this ROL construction can be derived by looking at the top of the column, as shown in Figure 1.13 (Luyben, 2006).

The liquid and vapor flow rates in the stripping section (L_S and V_S) can be calculated if the thermal condition of the feed is known. Since the temperature, pressure, and composition of the feed are given, the fraction of the feed that is liquid—defined as the variable q—can be calculated

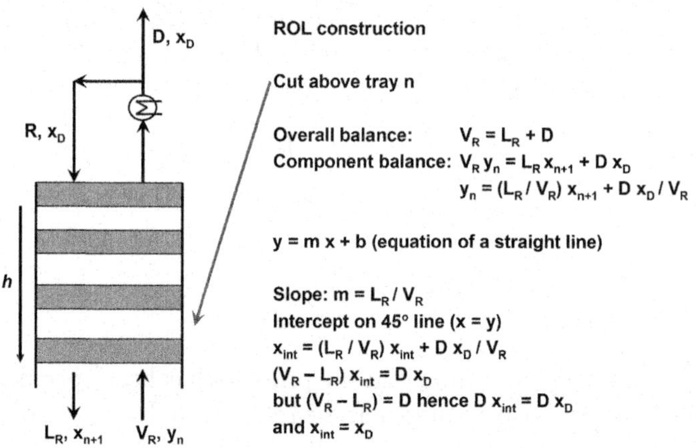

Figure 1.13 Construction of a rectifying operating line (ROL)

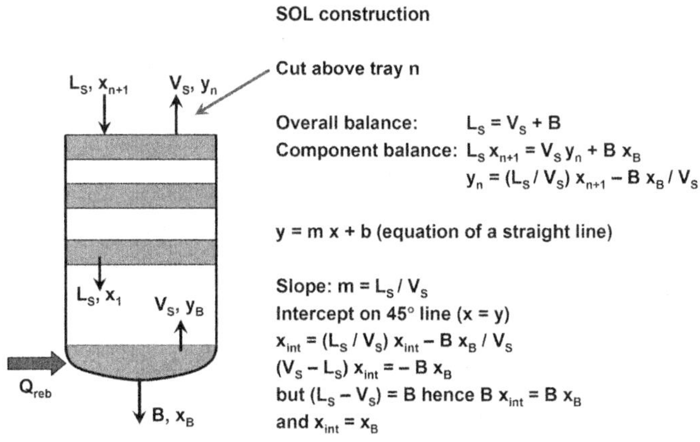

Figure 1.14 Construction of a stripping operating line (SOL)

from an isothermal flash calculation. Knowing q, the liquid and vapor flow rates in the stripping section can be calculated. If the feed is saturated liquid, $q = 1$ and if the feed is saturated vapor, $q = 0$:

$$q = \frac{L_S - L_R}{F} \tag{1.43}$$

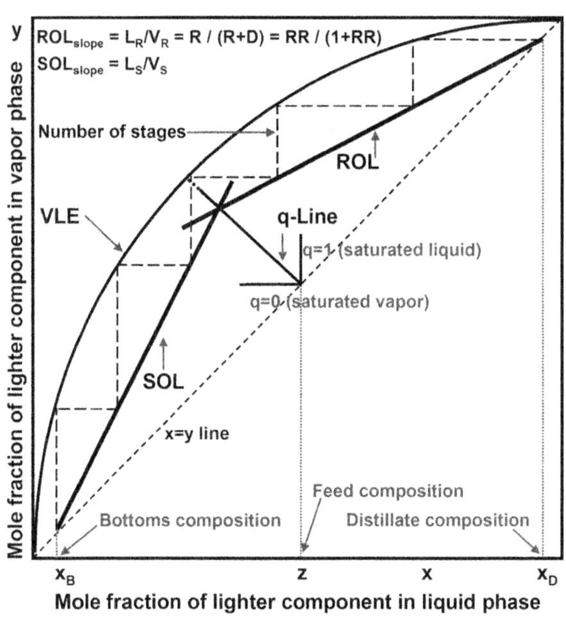

Figure 1.15 McCabe–Thiele method—operating lines and number of stages

$$L_S = qF + L_R \tag{1.44}$$

$$V_S = L_S - B \tag{1.45}$$

The SOL is a straight line with slope L_S/V_S that intersects the diagonal at the bottoms composition x_B. Proof of this construction can be derived by looking at the bottom of the column, as shown in Figure 1.14 (Luyben, 2006). Figure 1.15 shows both operating lines: ROL and SOL. Note that there is a relationship between the intersection point of the two operating lines and feed conditions. As shown in Figure 1.15, a straight line can be drawn from the location of the feed composition z on the diagonal to this intersection point. As proven hereafter, the slope of this line (known as the *q-line*) is a function of only the thermal condition of the feed—defined by parameter q. The slope is $-q/(1-q)$, which makes the construction of the McCabe–Thiele diagram very simple:

- Locate the three compositions on the diagonal (45° line): z, x_D, x_B;
- draw the ROL from the x_D point with a slope of $RR/(1+RR)$;
- draw the q-line from the z point with a slope of $-q/(1-q)$;
- draw the SOL from the x_B point to the intersection of the q-line and the ROL.

The equations of the rectifying and stripping operating lines are given below in terms of the point of intersection of the two lines at y_{int} and x_{int}:

$$\text{ROL}: \quad y_{int} = \left(\frac{L_R}{V_R}\right)x_{int} + \frac{Dx_D}{V_R} \tag{1.46}$$

$$\text{SOL}: \quad y_{int} = \left(\frac{L_S}{V_S}\right)x_{int} - \frac{Bx_B}{V_S} \tag{1.47}$$

Subtracting the two equations gives:

$$(V_R - V_S)y_{int} = (L_R - L_S)x_{int} + (Dx_D + Bx_B) \tag{1.48}$$

The last term on the right-hand side is actually $Fz = Dx_D + Bx_B$. Using the definition of $q = (L_S - L_R)/F$ leads to:

$$(V_R - V_S) = (1-q)F \tag{1.49}$$

$$(L_R - L_S) = -qF \tag{1.50}$$

Substituting these relationships into the previous equation gives:

$$(1-q)Fy_{int} = -qFx_{int} + Fz \tag{1.51}$$

$$y_{int} = \left(\frac{-q}{1-q}\right)x_{int} + \left(\frac{z}{1-q}\right) \tag{1.52}$$

This is in fact the equation of a straight line, with the slope $-q/(1-q)$. The q line is vertical for saturated liquid feed ($q = 1$), and it is horizontal for saturated vapor feed ($q = 0$). On the diagonal, this holds true: $x_{int} = y_{int}$. Consequently, it can be demonstrated that the q line intersects the diagonal (45° line) at the feed composition z:

$$(1-q)x_{45} = -qx_{45} + z \tag{1.53}$$

$$x_{45} = z \tag{1.54}$$

The number of trays is determined by moving vertically from the x_B point on the diagonal to the VLE line—this is in fact the composition of the vapor y_B leaving the partial reboiler. Moving horizontally over to the SOL, this step represents the partial reboiler. The value of x on the SOL is the composition of liquid x_1 leaving tray 1 (when numbering from the bottom of the column up). This stepping is repeated, moving vertically to y_1 and horizontally to x_2. Stepping continues until the intersection of the operating lines is crossed—this is the feed tray. The horizontal line is extended then to the ROL. Continuing to step until the x_D value is crossed gives the total number of trays.

The minimum number of trays for a specified separation corresponds to total reflux operation. If the column is run under total reflux conditions, the distillate flow rate is zero. Therefore, the reflux ratio is infinite, and the slope of the operating lines is unity—this is the 45° line. Thus the minimum number of trays can be determined by simply stepping up between the diagonal (45° line) and the VLE curve, as illustrated in Figure 1.16. The minimum reflux ratio for a specified separation corresponds to having an infinite number of trays. This usually occurs when the operating lines and the q-line intersect exactly on the VLE curve—this is a *pinch condition,* as it would take an infinite number of trays to move past this point. This is also illustrated in Figure 1.16. The minimum reflux ratio is calculated from the slope of this limiting

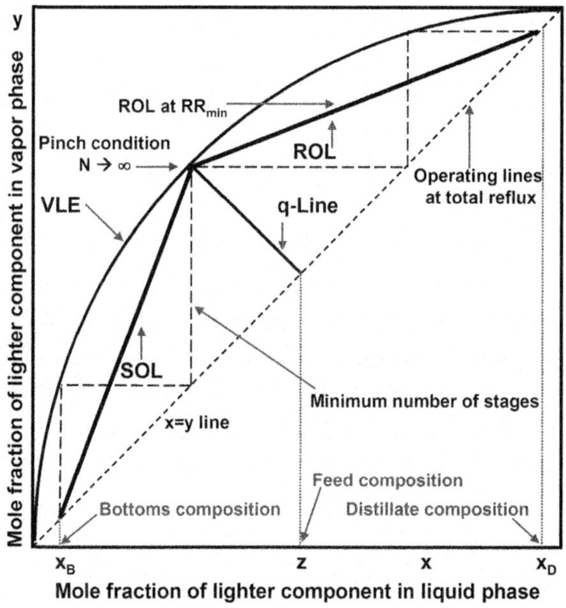

Figure 1.16 McCabe–Thiele method—minimum reflux and minimum number of stages

operating line. The McCabe–Thiele method can be conveniently tested online at: http://www.cheric.org/education/eduaids/distill/McCabe.html (last accessed on 24 December 2012).

Based on the McCabe–Thiele diagram, several observations can be made—which can be applied to all types of separations and distillation columns, not just a binary distillation:

- The further the VLE curve is from the diagonal, the smaller the slope of the rectifying operation line (ROL), meaning a smaller reflux ratio and thus lower energy requirements. A large VLE curve corresponds to large relative volatilities and an easy separation.
- The easier the separation, the fewer trays it takes to make a given separation.
- The higher the product purities, the more trays it takes to make a given separation.
- Increasing product purities does not have a significant effect on the required reflux ratio.
- Increasing the liquid to vapor ratio in a section of a column increases the separation that occurs in that section.

1.8.3 Approximate Multicomponent Methods

Several simple approximate methods were developed for analyzing multicomponent systems, long before the availability of computers for rigorous analysis—the best known being the Fenske–Underwood–Gilliland (FUG) shortcut method. These methods are still quite useful for getting quick estimates of the size of a column (number of trays) and the energy requirements (e.g., reflux ratios and the corresponding vapor boil-up and reboiler heat input):

Fenske equation for minimum number of trays: The minimum number of trays corresponds to total reflux operation (an infinite reflux ratio). The Fenske equation relates the compositions at the two ends of a column to the number of stages in the column under this limiting condition:

$$N_{\mathrm{min}} + 1 = \frac{\log\left[\left(\frac{x_{D,\mathrm{LK}}}{x_{D,\mathrm{HK}}}\right)\left(\frac{x_{B,\mathrm{HK}}}{x_{B,\mathrm{LK}}}\right)\right]}{\log(\alpha_{\mathrm{LK,HK}})} \tag{1.55}$$

where N_{min} is the minimum number of stages required, $x_{D,\mathrm{LK}}$ is the mole fraction of the light-key (LK) component at the top of the column, $x_{D,\mathrm{HK}}$ is the mole fraction of the heavy-key (HK) component at the top of the column, $x_{D,\mathrm{HK}}$ is the mole fraction of the heavy-key component at the bottom of the column, $x_{B,\mathrm{LK}}$ is the mole fraction of the light-key component at the bottom of the column, and $\alpha_{\mathrm{LK,HK}}$ is the relative volatility between the LK and HK components. This equation is applicable to multicomponent systems, but it assumes a constant relative volatility between the two components considered.

Underwood equations for minimum reflux ratio: The Underwood equations can be used to calculate the minimum reflux ratio in a multicomponent system if the relative volatilities of the components are constant. The equations are as follows:

$$\sum_{j=1}^{N_C} \frac{\alpha_j z_j}{\alpha_j - \theta} = 1 - q \tag{1.56}$$

$$\sum_{j=1}^{N_C} \frac{\alpha_j x_{D,j}}{\alpha_j - \theta} = 1 + RR_{\mathrm{min}} \tag{1.57}$$

The following variables are specified: feed composition z_j (mole fractions $j=1, \ldots N_C$), the desired distillate composition $x_{D,j}$ ($j=1, \ldots N_C$), and the feed thermal condition q. The relative volatilities α_j ($j=1, \ldots N_C$) of the multicomponent mixture are known from the VLE. The first equation contains one unknown parameter θ. However, expanding the summation of N_C terms and multiplying through by all the denominator terms $(\alpha_j-\theta)$ gives a polynomial in θ whose order is N_C, which means that there are N_C roots of this polynomial. One of these roots lies between the two relative volatility values α_{LK} and α_{HK}. This is found using some iterative solution method. It is substituted into the second equation, which can then be solved explicitly for the minimum reflux ratio.

Gilliland correlation. An empirical correlation can be used to calculate the final number of stages N from the values calculated through the Fenske and Underwood equations (N_{min}, RR, RR_{min}). Gilliland noted that he could empirically relate the number of stages (N) at a finite reflux ratio (RR) to the minimum number of stages (N_{min}) and the minimum reflux ratio (RR_{min}). The procedure uses a diagram plotting $(RR-RR_{min})/(RR+1)$ on the x-axis and $(N-N_{min})/(N+1)$ on the y-axis. One enters the diagram with the abscissa value, which is known, and then it reads the ordinate of the corresponding point on the Gilliland curve. The only unknown of the ordinate is the total number of stages (N).

Kirkbride equation. This is an empirical equation used to determine the number of stages in the rectifying (N_R) and stripping (N_S) sections, and therefore the feed stage location:

$$\frac{N_R}{N_S}=\left[\left(\frac{x_{F,HK}}{x_{F,LK}}\right)\left(\frac{x_{B,LK}}{x_{D,HK}}\right)^2\frac{B}{D}\right]^{0.206} \tag{1.58}$$

1.9 CONCLUDING REMARKS

The basics of the vapor–liquid phase equilibrium (VLE) reviewed here play a key role, as a very good understanding of VLE is indispensable in the design and control of any distillation system. Most of the key concepts are used extensively throughout this book. Moreover, several practical methods for analyzing distillation columns have been presented. Graphical methods provide valuable insight into how various design and operating parameters affect separations in distillation, while the residue curve map (RCM) representation allows the designer to identify the feasible separations.

REFERENCES

Aspen Technology (2010a) Aspen Plus: User Guide - Volume 1 & 2.

Aspen Technology (2010b) Aspen Physical Property System - Physical Property Models.

Carlson, E.C. (1996) Don't gamble with physical properties for simulations. *Chemical Engineering Progress*, 35–46.

Dimian, A.C. (2003) *Integrated Design and Simulation of Chemical Processes*, Elsevier, Amsterdam.

Doherty, M.F. and Malone, M.F. (2001) *Conceptual Design of Distillation Systems*, McGraw-Hill, New York.

Forbes, R.J. (1970) *A Short History of the Art of Distillation*, E. J. Brill, Leiden.

Gmehling, J., Onken, U., Arlt, W. *et al.* (1993) *Vapor-Liquid Equilibrium Data Collection*, DECHEMA, Frankfurt/Main.

Harmsen, G.J. (2010) Process intensification in the petrochemicals industry: Drivers and hurdles for commercial implementation. *Chemical Engineering and Processing*, *49*, 70–73.

Kister, H.Z. (1992a) *Distillation – Design*, McGraw-Hill, New York.

Kister, H.Z. (1992b) *Distillation – Operation*, McGraw-Hill, New York.

Lei, Z., Chen, B., and Ding, Z. (2005) *Special Distillation Processes*, Elsevier, Amsterdam.

Luyben, W.L. (2006) *Distillation Design and Control using Aspen Simulation*, John Wiley & Sons, Inc., Hoboken.

Luyben, W.L. (2011) *Principles and Case Studies of Simultaneous Design*, John Wiley & Sons, Inc., Hoboken.

Luyben, W.L. and Yu, C.C. (2008) *Reactive Distillation Design and Control*, John Wiley & Sons, Inc., Hoboken.

McCabe, W.L., Smith, J.C., and Harriot, P. (2005) *Unit Operations of Chemical Engineering*, McGraw-Hill, New York.

Mujtaba, I.M. (2004) *Batch Distillation – Design and Operation*, Imperial College Press, London.

Perry, R.H. and Green, D.W. (eds) (1997) *Perry's Chemical Engineers' Handbook*, McGraw-Hill, New York.

Petlyuk, F.B. (2004) *Distillation Theory and its Application to Optimal Design of Separation Units*, Cambridge University Press, Cambridge.

Schmidt-Traub, H. and Gorak, A. (2006) *Integrated Reaction and Separation Operations*, Springer, New York.

Seader, J.D. and Henley, E.J. (1998) *Separation Process Principles*, John Wiley & Sons, Inc., New York.

Sharma, N. and Singh, K. (2010) Control of reactive distillation column - A review. *International Journal of Chemical Reactor Engineering*, *8*, R5.

Stichlmair, J.G. and Fair, J.R. (1998) *Distillation - Principles and Practice*, Wiley-VCH Verlag GmbH, Weinheim.

Sundmacher, K. and Kienle, A. (eds) (2003) *Reactive Distillation: Status and Future Directions*, Wiley-VCH Verlag GmbH, Weinheim.

Sundmacher, K., Kienle, A., and Seidel-Morgenstern, A. (eds) (2005) *Integrated Chemical Processes: Synthesis, Operation, Analysis, and Control*, Wiley-VCH Verlag GmbH, Weinheim.

Taylor, R. and Krishna, R. (1993) *Multi-component Mass Transfer*, John Wiley & Sons, Inc., New York.

2

Design, Control and Economics of Distillation

2.1 INTRODUCTION

This chapter provides the basic principles of design, control, sizing and economic estimates. Distillation is the most important thermal separation method used in the chemical process industry. Considering its many well-known benefits, distillation is and will certainly remain the separation method of choice in the chemical industry for the coming decades – with over 40 000 columns in operation worldwide. However, despite its flexibility and widespread use, one important drawback is the considerable energy requirements, as distillation can make up more than 50% of plant operating cost (Taylor, Krishna, and Kooijman, 2003). Consequently, tremendous research efforts are focusing on energy-efficient distillation. As chemical components have a wide range of boiling points, there are also a wide variety of distillation columns, operating from vacuum conditions (10 mbar) to high pressure (20 bar) and from cryogenic temperature levels conditions up to 350 °C (Luyben, 2011).

Compared to other unit operations, distillation columns are rather complex to design as there are many design variables: operating pressure, total number of stages, feed-stage location, reflux ratio (or reboiler duty), composition specifications for products, and many other variables for more complex distillation columns (e.g., side-draw location, liquid and vapor split in the case of Kaibel and dividing-wall columns). Despite its complexity, several heuristic rules have been developed over recent decades that are quite effective and simplify tremendously the problems

Advanced Distillation Technologies: Design, Control and Applications, First Edition.
Anton Alexandru Kiss.

of optimal design. Conversely, the control of distillation columns is fairly easy and somewhat less critical as distillation towers exhibit well-behaved and stable dynamic responses. Although temperature runaways do not occur in distillation – in contrast to reactors exhibiting nonlinear behavior – tight control of the product streams is important to product-quality issues. The multivariable nature of distillation introduces a certain measure of complexity compared to a single temperature control in a chemical reactor (Luyben, 2011).

The following sections present several approaches to finding the optimal design of distillation columns, as well as a consistent set of methods and relationships for sizing distillation equipment, estimating equipment cost, and specifying energy costs at different temperature levels. Note that the equations, correlations, and parameter values are collected from several very useful books: Douglas (1988), Turton *et al.* (2003), Dimian (2003), and Luyben (2011).

2.2 DESIGN PRINCIPLES

The optimal design of a distillation column involves the classic trade-off between the size of the column (i.e., number of stages) and energy requirements (i.e., reflux ratio or reboiler duty). As the total number of stages is increased, the capital investment in the column shell increases due to the increased height. On the other hand, the energy requirements in the reboiler decrease – this reducing the energy and equipment costs for the heat exchangers (reboiler and condenser). Moreover, the required diameter of the column is also reduced as the vapor rates passing through the column are lower. However, the operating pressure of the column must be specified before quantitatively exploring this trade-off between size and energy. The standard distillation design problem is to find the optimal column given the following specified variables (Luyben, 2011):

- Feed flow rate, composition, and thermal conditions (temperature and pressure);
- distillate composition in terms of heavy-key component impurity;
- bottoms composition in terms of light-key component impurity.

The variables that must be determined so as to satisfy some economic objective functions are the operating pressure, total number of stages, and feed-stage location. Once the best values of these design optimization variables are found, all the other dependent variables of the column are practically fixed (Luyben, 2006, 2011). The temperatures, compositions, and flow rates for all stages are now known, as well as the heat duties in the reboiler and condenser.

2.2.1 Operating Pressure

Selecting the pressure is a rather simple process. Almost all chemical vapor–liquid separations become easier (e.g., have higher relative volatilities) as temperature decreases – with some notable exceptions, as explained later. As a result, a low pressure is preferred in practice in order to provide low temperatures. Nevertheless, lower column pressures also give lower reflux-drum and condenser temperatures. While temperatures down to about 40–50 °C can be achieved by using inexpensive cooling water, operating at temperatures lower than this range would require the use of chilled water or expensive refrigerant for the heat removal in the condenser. Consequently, this sets a lower limit to the operating pressure that is actually economically feasible (Luyben, 2011).

Chemical systems that exhibit the opposite dependence of volatility on temperature (and pressure) include, for example, the water/acetic acid system. In such systems, the selection of pressure is more of a balance between the amount of energy used and its cost. The temperature at the base of the column increases with increasing pressure and, thus, requires a more expensive heat source at higher temperature levels (e.g., hot oil or DOWTHERM™ instead of steam). In addition, a different constraint may occur in some separations, such as for instance a maximum temperature limitation above which undesired phenomena can occur: thermal decomposition, coke formation, and unwanted reactions such as polymerization of one or more components. An example relevant at the industrial scale is the separation of ethyl-benzene and styrene, where the styrene polymerization limits the range of usable temperatures in the distillation column (Luyben, 2011).

In a distillation column, the top temperature is the lowest, while the base temperature is the highest. Consequently, the pressure must be set in such a way that the temperature in the base of the column does not exceed the temperature limit – this being linked to bottoms composition. The pressure drop along the column (e.g., through the trays or packing internals) sets the pressure of the reflux-drum, which is lower than the base pressure of the distillation column. The temperature of the reflux-drum is then actually determined by the distillate composition at the top pressure. Note that expensive refrigeration must be used if this temperature is below that achievable by using cooling water. Such conditions mean that the separation is very expensive in terms of energy, as expensive heat is provided in the reboiler and then removed in the refrigerated condenser (Kister, 1992a, 1992b; Luyben, 2011).

Note that the internals of a distillation column are supposed to provide a very good vapor–liquid contact, without causing an excessive pressure-drop across a column section – for example, structured packing offering a pressure drop of only a few mbar per theoretical stage (Kister, 1992a). This issue is very important as a high pressure drop would mean that more energy is required to drive the vapors up the distillation column. Separations limited by a maximum temperature often require vacuum operation. Such vacuum pressures lead to very small vapor densities that translate further into large diameters for the distillation columns. Moreover, additional equipment – such as vacuum pumps or steam-jet ejectors – is needed to remove the inert components from the condenser (Kister, 1992a, Luyben, 2006, 2011).

2.2.2 Heuristic Optimization

After fixing the pressure of the distillation, the number of stages must be also determined. Heuristic rules can be conveniently applied to obtain the first estimates of the optimal column design:

- Use twice the minimum number of trays ($N_T = 2N_{min}$);
- use 20% more reflux as the minimum reflux ratio ($RR = 1.2RR_{min}$).

These limiting conditions can be found analytically in systems with constant relative volatilities by using the classical Fenske and Underwood equations. However, they can also be easily found in any real system by using a process simulator, such as Aspen Plus. Using a simulator, one must specify the feed conditions, while keeping the compositions of the distillate and bottoms products at their specified values – for example, by using Design Spec and Vary functions in Aspen Plus. Basically, the number of stages is specified at an initial (reasonable, high) value, giving a certain required reflux ratio. Then, the number of trays is gradually increased until the required reflux ratio stops declining any further – this is in fact the minimum reflux ratio (RR_{min}). Then, the number of trays is reduced until the required reflux starts to increase drastically – this is the minimum number of stages (N_{min}). For convenience, Figure 2.1 provides a graphical representation of the dependence of the number of theoretical

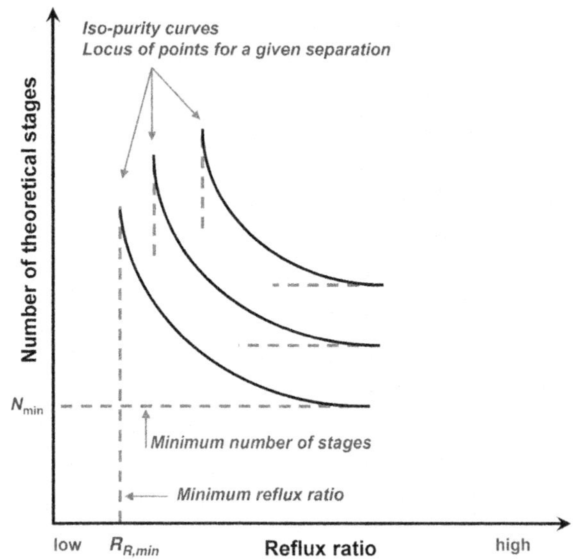

Figure 2.1 Number of theoretical stages versus reflux ratio

stages required for a certain separation as a function of the reflux ratio. Note the existence of a minimum reflux ratio (RR_{min}) and a minimum number of stages (N_{min}) for any desired separation target – the locus of points for a given separation or the so called iso-purity curves.

2.2.3 Rigorous Optimization

The optimal column design can be also obtained using rigorous optimization – by changing the total number of stages over a certain range of values. While keeping the composition of the distillate and bottoms products at their specified values, the required reboiler and condenser duty and the height and diameter of the column are calculated for each case. Then, the feed-stage location is varied for each choice of total number of stages until the location leading to the minimum energy required is found. The heat-transfer rates are then used to calculate the heat-transfer areas required for the reboiler and condenser – using reasonable heat-transfer coefficients (U) and driving forces (ΔT). All this information is then used to calculate the capital cost of the column shell and the additional heat exchangers (reboiler and condenser). Note that the energy cost is largely dominated by the reboiler duty – as long as no refrigeration is used in the condenser. The total annual cost (TAC) –

combining the total investment cost (TIC) and the total operating cost (TOC) – is a simple but effective economic objective function that can be applied to each distillation column. The TAC represents the sum of the annual cost of energy plus the annual cost of capital (e.g., TIC divided by a payback time). Several examples of such calculations are given in case studies described in subsequent chapters of this book.

2.2.4 Feed Preheating

In most cases, distillation columns are not alone in a process and so one may want to consider using hot process streams (e.g., with a temperature higher than that of the distillation feed temperature) to preheat the feed stream in an additional (or existing) heat exchanger. Note that a hotter feed results in lower reboiler duty and higher heat removal in the condenser. Since the latter is typically inexpensive, the capital investment in an additional heat exchanger is very often justified. Moreover, as the bottoms stream from a distillation column is usually hotter than the feed stream, the bottoms product is frequently used for preheating the feed – except when the bottoms product is feeding another downstream distillation column (Kister, 1992a, Luyben, 2011).

Another alternative worth considering is preheating the feed stream using steam. However, this option should be used only for steam at lower pressure than that used in the reboiler – leading to favorable economics, as the lower-pressure steam is less expensive. Clearly, designing a system with a feed preheater that uses the same pressure steam as that used in the reboiler is not economically feasible as the overall investment and energy cost is higher – the heat added in the reboiler has more effect on the separation than the heat added in a preheater (Luyben, 2011).

2.2.5 Intermediate Reboilers and Condensers

Conventional operation of distillation requires adding all the heat at the base of the column, where the temperature is the highest and the most expensive heat source is needed. An alternative worth considering is using an intermediate reboiler located higher up the distillation column, where the temperature levels are lower – hence a lower-temperature, less expensive heat source can be used.

Similarly, instead of removing all the heat at the top of the column – where the temperature is the lowest and the heat is discharged into cooling water – one may use an intermediate condenser (pump-around) located further down in the distillation column where the temperature levels are higher. By doing so, the heat can be recovered at a higher temperature level and transferred into a process stream that needs to be heated (via heat-exchange networks). The use of intermediate condensers and reboilers is frequently seen in columns with a large temperature span across the column, such as petroleum fractionators (Luyben, 2011).

2.2.6 Heat Integration

The thermodynamic efficiency of distillation is very poor, as almost all the energy added in the reboiler is removed in the condenser. However, the overall efficiency can be greatly improved by using multiple-effect distillation (MED) columns. For example, multiple-effect evaporators are widely used to remove water from sea-water or brine in salt production. A series of evaporators are designed with the process stream flowing from one to another. Water is boiled off in each unit by operating the successive evaporation stages at lower and lower pressures. In the lowest pressure unit, the water vapor is condensed using cooling water in a heat exchanger. The heat supplied to this last evaporation stage is actually the water vapor boiled off in the upstream evaporation stage, provided at a higher temperature so that it can be condensed in the jacket of the final stage evaporator. This cascading of evaporation stages, using the vapor from an upstream stage as the heat input to the next stage, is continued with progressively higher temperatures in each stage. The first evaporation stage gets its heat from high-pressure steam. Theoretically, if the steam required to remove a given amount of water by conventional operation is F_S then the steam required to remove the same amount of water in a multiple-stage evaporator (N stages) is significantly reduced: F_S/N (Luyben, 2011; Kiss *et al.*, 2012).

The same basic idea can be applied to distillation columns, if it is possible to operate one column at high and another column at lower temperatures/pressures. The overhead vapor from the high-temperature column is condensed in a condenser/reboiler heat exchanger that generates vapor in a low-temperature column. The energy requirements can be reduced significantly in many separations by the use of heat integration (Kister, 1992a; Kister, 1992b; Luyben, 2011).

The most common situation is when there are two columns making different separations in which the reflux-drum temperature of one column is sufficiently higher than the base temperature of the second column so as to provide reasonable differential temperature driving force for heat transfer without requiring an excessively large heat-transfer area. The condenser heat duty in the high-temperature column and the reboiler duty in the low-temperature column should be similar in magnitude. Any difference in the duties can be easily handled by using either an auxiliary reboiler in the low-temperature column or an auxiliary condenser in the high-temperature column.

In many applications, a two-column heat-integrated system is used instead of a single column for the same separation job. There are several configurations of this type. The feed can be split and fed into the two columns, one operating at high pressure and the other operating at low pressure. The low-pressure column has a water-cooled condenser, while the high-pressure column has a steam-heated reboiler. The pressures are adjusted such that a 20–30 K temperature difference is achieved between the reflux-drum temperature in the high-pressure column and the base temperature in the low-pressure column. The feed split can be adjusted so that the condenser duty in the high-pressure column is exactly equal to the reboiler duty in the low-pressure column. Specification distillate and bottoms products are removed from both columns (Luyben, 2011).

Another common configuration sends the entire feed stream to one of the columns. The distillate from this column is about half of the lighter component. The bottoms stream is fed to the second column, which takes the rest of the lighter component out the top and the entire heavier component out the bottom. Either the first or the second column can be the high-pressure column with the other column the low-pressure column. Heat integration can be in the direction of flow (first column at high pressure) or in the reverse direction to the process flow (second column at high pressure). For this type of heat-integrated system to be economical, the difference in boiling point between the light and heavy components cannot be large. Otherwise, the pressure difference necessary to achieve the condenser and reboiler temperatures would be large, which would result in a high temperature in the base of the high-pressure column and, thereby, require an expensive heat source. The high pressure would reduce also the relative volatilities in most systems, thus leading to increased energy requirements.

2.3 BASICS OF DISTILLATION CONTROL

This section provides the basics for the development of effective distillation control structures. A simple binary distillation column is considered, for which the number of stages and the feed-stage location are fixed and the

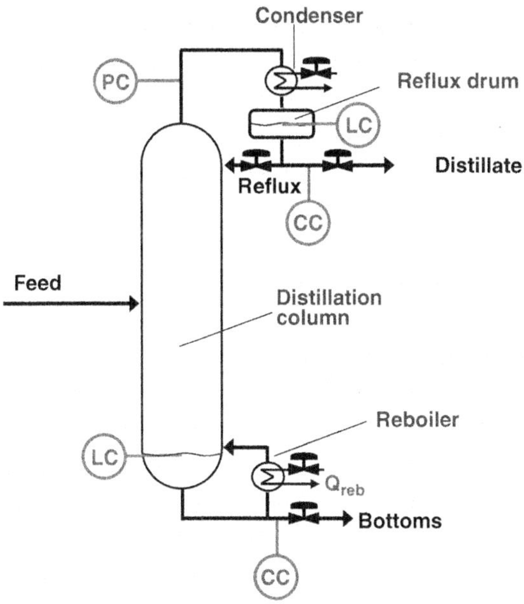

Figure 2.2 Controlled variables and manipulated streams in a binary distillation column

feed flow rate is assumed to be set by an upstream unit. Note that a distillation column control system has three main objectives: (i) to set stable conditions for column operation, (ii) to regulate conditions in the column so that the products always meet the required specifications, and (iii) to achieve the first two objectives most efficiently (e.g., maximum product recovery or/and minimum energy requirements). Figure 2.2 shows the variables typically controlled in a distillation column, including pressure, bottom level, reflux drum level, and top and bottom product compositions. These variables can be classified as follows (Kister, 1992a; Kister, 1992b):

- *Single-loop variables* (e.g., pressure, levels). These are controlled to achieve the first objective, that is, setting stable conditions for operation. The set points are established solely by stability considerations, regardless of product specifications. Controlling the pressures and levels regulates material accumulation in the column. Keeping the levels constant prevents liquid accumulation, while keeping the pressure constant prevents vapor accumulation. Unless accumulation or depletion is prevented, a continuous system cannot operate at steady state and cannot be stable.

- *Unit objective variables* (e.g., top and bottom compositions). These are regulated to achieve the second objective, that is, meeting product specifications. The set points at which these are controlled are determined by product purity considerations alone. Composition controls can be direct (i.e., using composition measurements of the product streams) or indirect (i.e., using a physical property representative of product composition). Typical physical properties used are refractive index, density, vapor pressure, freezing point, and, most often, the temperature of a sensitive tray.

In addition, there are also manipulated variables (e.g., control valves). A stream is manipulated by varying the opening of its control valve. The stream flow rate is thereby changed in order to control a desired variable (which is also known as a controlled variable). Figure 2.2 shows positions of control valves in a typical distillation system. There are five manipulated variables: top and bottom product flow rates, condensation rate, boil-up rate, and reflux flow rate (Kister, 1992a; Kister, 1992b; Luyben, 2011).

Note that there are three inventory variables that must be controlled: pressure, liquid level in the reflux drum, and liquid level in the base. In total, there are five control valves available: condenser cooling water, reboiler steam, reflux, distillate, and bottoms. Three of these must be used to control the three inventory variables. That leaves two control valves available to control two variables. Ideally, the two variables to control would be the compositions of the distillate (heavy-key impurity) and the bottoms (light-key impurity). These are the variables that were used to design the column in the first place. A dual-composition control structure is the ideal scheme but this is rarely implemented in industry since online composition measurements are typically expensive, require high maintenance, and will probably introduce significant delays in the control loop – especially when gas chromatography (GC) is used. For that reason, many distillation columns are controlled in practice by using inferential temperature measurements that are inexpensive, reliable, and sufficiently fast. It is therefore important to decide if one or two temperatures must be controlled, and to determine the best location for their measurement (Luyben, 2011).

2.3.1 Single-End Control

Many industrial distillation columns use some type of single-end temperature control because of its simplicity and low maintenance cost.

However, this simple structure might not provide effective control for some columns. Even if a single-end control structure is possible, one must decide how to select the other control degree of freedom. The most common choices are to hold a constant reflux-to-feed ratio (*R/F*) or to hold a constant reflux ratio (RR). Importantly, any of these ratio-control structures should be effective – from a steady-state viewpoint – for feed flow rate disturbances. As the feed flow rate changes, all flows ratio change directly and composition and temperatures throughout the column end up at the same values – neglecting, of course, any changes in pressure and tray efficiencies. However, for feed composition changes, the compositions and temperatures in the column change at the new steady state with both product purities held constant. Therefore, the choice between the *RR* and *R/F* structure must be based on feed composition disturbances (Luyben and Luyben, 1997; Luyben, 2011).

2.3.1.1 Selecting the Reflux Ratio or Reflux-to-Feed Ratio

Before making this choice, a series of steady-state simulations should be performed to determine the effects of changes in feed composition on the required changes in *R/F* ratio and *RR* – while holding both products at their specified compositions. If the required changes in the *R/F* are very small, then a single-end control structure with *R/F* held constant may be effective. Similarly, if changes in the *RR* are very small, a single-end control structure with *RR* held constant may be effective. If there are significant changes in both, a dual-end control structure would be better.

2.3.1.2 Selecting Temperature Control Stage Location

An important issue in distillation control is the stage location where temperature is to be controlled in a single-end structure. Many methods are available for making this selection, but a simple and effective approach is to select a stage at a point where there are significant changes in temperature from tray to tray. Another useful method is to find the sensitivity of stage temperatures to changes in a manipulated variable (Luyben, 2011). Let us assume a control structure in which the reflux-drum level is controlled by manipulating the distillate flow rate, the base level is controlled by manipulating bottoms flow rate, and the pressure is controlled by manipulating condenser heat removal (i.e., cooling water flow rate). The two remaining control variables are the reflux and the reboiler heat input. With the reflux flow rate fixed, a small (0.1%) change is made in the reboiler heat input. The resulting changes in tray temperatures from the original steady state are calculated. The tray with the largest changes in temperature may be a good location to use for temperature control. Another useful method is the invariant temperature approach. Here the feed composition is varied over the expected range of values with the distillate/bottoms held at their specified compositions. Temperature profiles are then plotted for each feed composition. If there are trays with temperatures that do not change for different feed compositions, these trays might be good locations to control.

However, the most elegant method for determining the control-tray location is the *singular value decomposition* (SVD), as described by Luyben (2011). The steady-state gains between all the stage temperatures and the two manipulated variables ($\Delta T_n/\Delta R$ and $\Delta T_n/\Delta Q_R$) are calculated, and the $N \times 2$ matrix is decomposed (e.g., using MATLAB® software) into three matrices $\boldsymbol{u}\sigma V^T$. The $\boldsymbol{u}$ vectors are plotted against tray numbers, and the trays with the largest magnitudes of $\boldsymbol{u}$ indicate the locations in the column that can provide effective control (Luyben, 2011).

In columns separating components with low relative volatilities, it is often not possible to use temperatures because there is very little change from tray to tray. Although pressure changes may have more effect than composition, direct composition control is often required in these cases. Control wisdom suggests that it is more effective to control the impurity instead of the purity of a stream as there is more change in the composition of the impure component for a change in the manipulated variable. The steady-state gain is higher, thus offering the potential for tighter control. A process engineer would intuitively select the composition of the product stream as controlled variable. However, if the product purity is high (the impurity level is very small) it may be difficult to obtain an accurate online measurement. The high purity may also result in very nonlinear dynamic responses. An effective and practical method for dealing with this problem is to move away from the end of the column and control the composition on an intermediate tray where the impurity is larger and can be more accurately measured – thus also reducing the nonlinearity problems (Kister, 1992b; Luyben, 2006).

Once the control structure is established, the temperature or composition controller must be tuned in a consistent and repeatable way. Realistic lags or dead times must be inserted in the dynamic simulation loop (e.g., 1 min dead time for temperature controllers, and 3–5 min for composition controllers). Failure to account for dynamic lags or dead times results in the prediction of a controller performance that is unrealistically better than what is attainable in the real plant – where these dynamic elements are always present due to measurement and valve dynamics. For most distillation columns, the ratio between the dead-time and the time constant is rather small. The relay-feedback test is an effective way to determine the ultimate gain and period of a loop. Controller tuning parameters are then easily calculated from the ultimate properties using classic tuning rules. Tyreus–Luyben tuning is quite appropriate for distillation columns, where rapid and oscillatory swings in the manipulated variables can cause major hydraulic problems, such as flooding or weeping (Luyben and Luyben, 1997).

2.3.2 Dual-End Control

If single-end control is inadequate – as indicated by the control analysis discussed previously – then two temperature controllers, or two composition controllers, or even a combination of one temperature controller and one composition controller may be required. In this situation, one must be a cautious about the controller tuning, as the two loops inherently interact with each other. When one product is more important than the other it is best to tune less aggressively the less important loop first (e.g., critically damped) and then to tune tightly the important loop. In this manner, the multivariable control problem is dynamically decoupled by providing relatively fast closed-loop dynamics for the important loop and considerably slower closed-loop dynamics for the less important loop. Therefore, the coupling effects of the less important loop are sufficiently slow such that the important loop can absorb them without difficulty. Accordingly, variability in the important loop can be maintained consistently at a relatively low level (Riggs, 2001). This approach approximates the performance of single composition control without allowing the less important product composition to suffer large offsets from the set point. When the control of both products is of similar importance, both control loops must be tuned together – for example, both control loops must be detuned equally to the point where the effects of coupling are at an acceptable level (Riggs, 2001).

Note that tuning one loop with the other on manual, and then doing the same with the other loop might lead to an unstable system when both control loops are put on automatic (Luyben, 2011). An effective and practical way to handle this problem is to use sequential tuning. The fastest loop is tuned first with the other controller on manual – usually the loop with reboiler heat input as the manipulated variable. Changes in the vapor rates travel up through the column rapidly, so changing reboiler heat input quickly affects the temperature and compositions on all trays. On the other hand, changes in the reflux affect conditions in the column slowly because of the liquid hydraulic lags. Typically, it takes 3–6 s per tray to produce a change in the liquid inflow/outflow on a tray. So a change in the reflux flow rate at the top of a 40-tray column may take 2–4 min to show up on the bottom trays. Once the fast loop is tuned, keep it on automatic and tune the next fastest loop.

There is a variety of choices for the manipulated variables (including L, D, L/D, V, B, V/B, B/L, and D/V) that can be paired so there is a large number of possible configuration choices, although only few of them of practical significance. Typically, the distillate purity is controlled using L, D, or L/D, while V, B, or V/B is used to control the bottoms purity – thus there are nine possible combinations. Figure 2.3a illustrates the LV configuration, in which the reflux controls the overhead composition and the flow controller on the reboiler duty regulates the bottoms composition. Accordingly, the distillate flow rate (D) is used to control the level in the reflux drum, and the bottoms flow rate (B) to control the

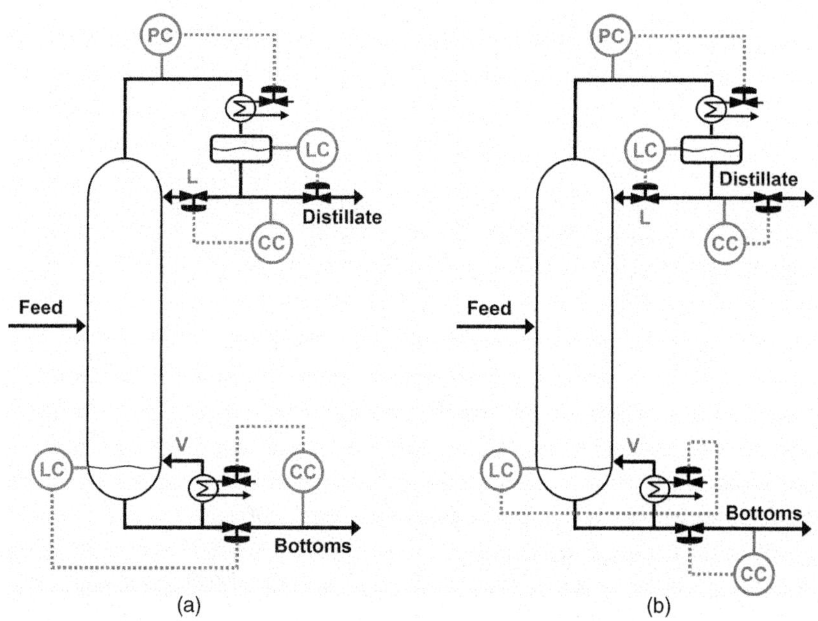

Figure 2.3 Dual-end control of a binary distillation column: (a) LV and (b) DB configurations

reboiler level. Note that for single control structures either *L* or only *V* is used for the composition control. Similarly, Figure 2.3b shows the *DB* configuration, where *D* is adjusted to control its purity, and *B* is manipulated to control its purity – thus leaving the reflux (*L*) to control the level in the reflux drum, and the boil-up vapor (*V*) for the reboiler level (Kister, 1992b; Riggs, 2001).

The configurations that use *D* or *B* as a manipulated variable for composition control are referred to as *material balance* (MB) *configurations* as they use the overall material balance for the column to adjust the product compositions. In fact, *DB* is also known as the *super material balance configuration.* The other configurations that do not manipulate *D* or *B* for composition control are known as *energy balance configurations* since they directly adjust the vapor/liquid traffic in the column for composition control. In addition, the $L/D + V/B$ configuration is known as the *double ratio configuration.*

The major factors affecting the control performance of a particular configuration are coupling, sensitivity to disturbances, and the response time for changes in the manipulated variables. Although highly susceptible to coupling, the most commonly used configuration (*LV*) is the easiest to implement, it provides good dynamic response, and is the least sensitive to feed composition disturbances. However, the *DB* configuration has advantages for certain high-purity columns if the levels are tuned tightly, but it is non-self-regulating and, hence, open-loop unstable. The double ratio configuration is typically the least affected by coupling and has a good dynamic response, but it is more difficult to implement and it is rather sensitive to feed composition disturbances (Riggs, 2001).

Although it is not possible to choose a priori the optimal control configuration, there are some guidelines that can drastically reduce the chances of selecting a poor configuration for a particular column:

- For high reflux ratio cases ($RR > 5–8$), configurations that use the material balance as manipulated variables (*D* and/or *B*) or ratios (*L*/*D* and *V*/*B*) are preferred: *LB* or *L*,*V*/*B* when the top distillate is more important, and *DV* or *L*, *D*/*V* when the bottom product is more important.
- For lower reflux ratio cases ($RR < 3–5$), configurations that use the energy balance as manipulated variables (*L* and/or *V*) or ratios are preferred: *LV* or *L*, *V*/*B* when the distillate product is more important, and *LV* or *L*, *D*/*V* when the bottom product is more important.

Notably, the single-end control is much easier to implement, tune, and maintain than dual-end control. The choice between single- and dual-end control is based on the tradeoff between the additional cost associated with dual-end control (e.g., increased controller maintenance, cost of analyzers) and its economic benefits (e.g., increased product recovery, reduced utility costs). While single-end control is widely used in the chemical process industry, dual-end control is typically preferred for large refinery columns, and columns that produce high volumes of chemical intermediates – mainly because of their high energy usage (Riggs, 2001).

2.3.3 Alternative Control Structures

The control structure considered so far (LV) – assuming that the reflux-drum level is controlled by distillate flow rate and the base level is controlled by bottoms flow rate – is the most widely used for several reasons. The major reason is that it provides attenuation of flow disturbance down through a series of process units since proportional-only level control provides gradual, non-oscillatory variations in the product streams feeding downstream units. In a multiunit process this can provide stable operation with reduced variability in product quality.

However, there are also situations where other structures should be used. The most frequently occurring case is when the reflux ratio is high. If $RR > 3$ then the reflux-drum level should be controlled by the reflux flow rate, and not by the distillate flow rate. This switch in the level control loop is needed because any small change in the vapor coming overhead in the column will require a large change in the distillate flow rate since it is four-times smaller than the vapor (e.g., $V = R + D$). When the reflux is controlling the reflux-drum level, the distillate flow rate can be set in different ways. If it is used to control a temperature or composition, the tuning of the reflux-drum level controller should be proportional only, with a gain of 5 instead of the normal value of 2. The reflux should change fairly quickly when the distillate is changed, because reflux is really what is affecting the temperature or composition. If the distillate is manipulated to hold a constant RR, the reflux flow rate is measured and sent to a multiplier with a constant that is the reciprocal of the desired RR. The output signal from the multiplier is the set-point signal of the distillate flow controller.

An interesting dilemma appears in a high RR column when the single-end analysis indicates that a R/F control structure is better than a RR control structure: How do we solve the conflicting objectives of wanting an *R/F* structure that fixes the reflux for a fixed-feed flow rate, but not wanting to control the reflux-drum level with the small distillate stream? An effective solution is to control the reflux-drum level with the reboiler heat input and control a tray temperature (or composition) near the top of the column with the distillate flow rate. The reboiler heat input affects the pressure in the condenser. If the pressure controller is on automatic, the increase in reboiler heat input will increase the pressure, and the pressure controller will increase the condenser heat removal (increase cooling water flow rate), leading to more vapor being condensed and thus increasing the liquid level in the reflux drum. The type and tuning of the reflux-drum level controller differ from the normal proportional-only controller with a gain of 2. A proportional-integral (PI) controller should be used, and a relay-feedback test must be run to tune this level loop because the dynamics of the column and the condenser pressure controller are nested inside the level loop. Fortunately, these high *RR* columns usually occur when the amount of the light component in the feed is rather small. The resulting temperature profile shows a break near the top of the column that can be effectively controlled by manipulating the distillate flow rate. Note that the reflux-drum level loop is nested inside the temperature loop. The variable that is really affecting temperature is the vapor boil-up, which is changed by the level controller as the level is affected by changes in distillate flow rate. In cases were reaction and distillation are integrated into one unit (e.g., reactive distillation), control is even more complex – this topic is addressed in a later chapter. For more information on RD control the reader is directed to the work of Nagy *et al.* (2007), Luyben and Yu (2008), and Sharma and Singh (2010).

2.3.4 Constraint Control

Some of the most common column constraints include:

- *Maximum reboiler duty:* This can result from various reasons, such as: fouled or plugged heat-exchanger tubes in the reboiler that reduce the heat transfer-rate; an increase in the pressure that reduces the heat transfer driving force (i.e., temperature difference, ΔT); an inadequately sized steam trap that causes condensate to return into the reboiler tubes; an improperly sized control valve on the steam to the reboiler that limits the steam flow; an increase in the column feed rate that requires a reboiler duty that is higher than the designed maximum duty (Riggs, 2001).

- *Maximum reboiler temperature:* For some specific systems, an elevated temperature in the reboiler can promote undesired reactions (e.g., polymerization) to the level that excessive fouling of the reboiler results. A duty constraint can be identified when the steam control valve remains fully open and the requested steam flow is always above the measured flow rate (Riggs, 2001).
- *Maximum condenser duty:* This constraint can be due to: fouled or plugged tubes in the condenser that reduce its heat duty; a higher ambient temperature that decreases the heat transfer driving force (e.g., during hot summer days); an improperly sized coolant flow control valve; an increase in coolant temperature; an increase in the column feed rate that requires a duty that exceeds the designed maximum duty of the condenser. Condenser duty constraints are usually identified when the pressure reaches a certain level, or when the reflux temperature rises to a specific value (Riggs, 2001; Luyben, 2011).
- *Flooding and weeping:* Flooding is the excessive accumulation of liquid inside the column. There are several types of flooding – typically caused by spray entrainment, froth entrainment, downcomer backup or downcomer choke – and Kister (1992a,b) showed that each type results from excessive levels of vapor/liquid traffic in the column. Weeping results when the vapor flow rate is too low to keep the liquid from draining from a tray onto the tray below. The onset of flooding or weeping is usually correlated to the pressure drop across (a part of) the column.

It should be emphasized that the reboiler duty is practically adjusted to respect each of these constraints and prevent the process from violating them. Several approaches can be used for the constraint control of a distillation column (Riggs, 2001; Luyben, 2011):

- Convert from dual-end control to a single-end control structure;
- reduce the product impurity set points for both top and bottom products;
- reduce the column feed rate to maintain the purity of the products.

2.3.5 Multivariable Control

Distillation is a classic example of a (constraint) process that is a multi-input multi-output (MIMO) system sometimes with a quite nonlinear behavior. MIMO systems can greatly benefit from multivariable control, such as model predictive control (MPC). MPC can be successfully applied not only to one column but also to a series of columns (e.g., an entire separation train). Remarkably, the MPC controller can efficiently operate a system of columns, against a large complex set of operative constraints that limit the overall throughput – to maximize the system throughput. Advanced PID (proportional, integral, derivative) constraint controllers applied to this problem require a separate control loop for each combination of constraints, thus resulting in a very large number of separate control configurations. Conversely, an MPC controller can handle directly the full range of constraints combinations with a single controller. Moreover, an MPC controller is easier to maintain than a custom-built advanced PID controller for such a complex system. While in most cases MPC can provide significant control improvements over PID control for a single column, these improvements pale in comparison to the economic advantages offered by applying MPC to large-scale distillation processes (Riggs, 2001; Agachi *et al.*, 2006).

2.4 ECONOMIC EVALUATION

Engineering economics are applied at many levels, ranging from very rough (order-of-magnitude) estimates to the reasonably precise estimates needed to get the money to proceed with buying equipment and constructing the plant. The most precise estimates strive to achieve an accuracy of ±10%. However, long before the project reaches the stage of actually committing to the expenditure of hundreds of millions of dollars, a long series of preliminary scouting studies must be made to see if there is any possibility of making money (Dimian, 2003; Luyben, 2011).

At the conceptual design stage, the main objective is to compare alternative flow-sheet designs and operating parameters. In such cases, precise economic estimates of capital and energy costs are not required - being better left off to cost estimators. Approximate numbers are sufficiently good to decide which of the many alternative flow-sheet arrangements is the most economically promising.

At later stages in the project, more precision is required, but one should remember that even though fairly accurate estimates of equipment and energy costs can be generated, the profitability of the process depends heavily on the ability of the marketing department to provide estimates of the cost of raw materials and the potential selling price of the products. Unfortunately, marketing estimates are usually unreliable and, consequently, very simple economic analysis approaches are taken hereafter. There are dozens of books that concentrate on cost estimation, and there are also quite rigorous computer software packages available for cost estimation. AspenTech (2010) provides the Aspen Economic Analyzer (former ICARUS Process Evaluator) and K-Base software that can be used when more precision is required. Alternatively, the CapCost program given in Turton *et al.* (2003) can also be used.

2.4.1 Equipment Sizing

Each major piece of equipment in the process must be properly designed, meaning that its size must be known. The size (e.g., diameter, length) of vessels used as column shells and tanks must be determined. These dimensions depend on the flow rates through the equipment. The heat-transfer area required in heat exchangers (e.g., reboiler, condenser) must be determined based on realistic estimates of overall heat-transfer coefficients, heat-transfer rates, and differential temperature driving forces. The capital cost of a compressor (e.g., used in heat pumps assisted distillation) depends directly on the work required to achieve the compression ratio and throughput specified (Dimian, 2003; Luyben, 2011).

2.4.1.1 Column Shells

Vertical cylindrical vessels are typically used for distillation, absorption, and liquid–liquid extraction. Such columns can have trays or packing as internals. When trays are used, the tray spacing must be specified. To provide enough room for a person to crawl inside the column for repairs and maintenance work, a typical value of tray spacing is 0.6 m. Packed columns are separated into several sections of packing with liquid redistribution every 4–8 m. The height equivalent of a theoretical plate (HETP) must be estimated to determine the total length of the vessel (Kister, 1992a). Commercial process simulators, such as Aspen Plus, can calculate the diameter of the vessel. Knowing the diameter and the height, the capital cost can be found using available cost equations.

Many types of packing and/or trays are suitable as internals, such as, for example Sulzer Mellapak™ (Plus), CY, BX(Plus), I/C/P/R-ring, Pall rings, CASCADE MINIRINGS®, Raschig rings, Raschig Super-Ring-/Pak, INTRALOX® (Ultra), Berl saddles, Nutter rings, hollow fibers, VGPlus™ trays, SUPERFRAC® trays, (wire-mesh-packed) sieve trays, bubble cap trays, or valve trays – see also vendor catalogues.

2.4.1.2 *Liquid Surge Tanks*

The reflux drums and column bases in distillation columns are typically sized to provide about 5 min of liquid holdup when the vessel is half full – based on the total liquid stream entering the tank (e.g., total residence time of 10 min for a full tank).

2.4.1.3 *Flash Tanks*

Two criteria must be evaluated when sizing a flash tank. The diameter can be set by the requirement of 5 min of liquid holdup, or it may be set by a maximum superficial vapor velocity up through the vessel to minimize the liquid entrainment. An F factor of 0.5 (in engineering units) is normally used, along with an aspect ratio of 2 that is very common in practice (Luyben, 2011):

$$F\,\text{factor} = V_{\max}\rho_V = 0.5 \tag{2.1}$$

where $V_{\max}$ is expressed in ft s^{-1} and ρ_V in lb ft^{-3}.

2.4.1.4 *Decanters*

In the case of decanters, more holdup time is required to give the two liquid phases ample time to settle into light and heavy phases (e.g. aqueous and organics phase). Holdup times of 20–30 min are very often used.

2.4.1.5 *Pumps, Valves, and Piping*

Since the work of pumping liquids is typically small and the cost of a pump is usually much less than that the major vessels, pumping cost can be most often neglected at the conceptual design stage. The same is true for valves and piping (Luyben, 2011).

2.4.1.6 *Heat Exchangers*

Heat exchangers of many types are used in chemical processes. Tube-in-shell heat exchangers are the most common, but plate heat exchangers are

Table 2.1 Overall heat transfer coefficients for various systems

System	Heat transfer coefficient		
	$W\,m^{-2}\,K^{-1}$	$kcal\,h^{-1}\,m^{-2}\,K^{-1}$	$Btu\,h^{-1}\,ft^{-2}\,F^{-1}$
Gas–gas	170	150	30
Gas–condensing vapor	280	240	50
Gas–vaporizing liquid	280	240	50
Liquid–liquid	570	490	100
Liquid–condensing vapor	850	730	150
Liquid–vaporizing liquid	850	730	150

also used when fouling is not a problem as they provide more heat-transfer area and lower pressure drop per volume of vessel. Fired furnaces are mainly used when high temperatures are required. Many heat exchangers use utilities (steam or cooling water) to heat or cool process material. Other heat exchangers transfer heat between process streams. To find the capital cost of a heat exchanger, only the heat-transfer area is needed (along with the material of construction). However, to determine the heat-transfer area one needs the heat duty, the overall heat-transfer coefficient, and the differential temperature driving force. The heat duty is obtained directly from the simulation. The overall heat-transfer coefficient depends on the phases on the two sides of the heat exchanger. Table 2.1 provides typical values for the overall heat transfer coefficients (Dimian, 2003; Luyben, 2011).

Note that the differential temperature driving force varies with the temperature. At higher temperatures, the economic optimum differential temperature is also larger. A useful heuristic is to use a ΔT that is around 5% of the absolute temperature (expressed in Kelvin). This provides a reasonable compromise between investment in the heat exchanger area, and the energy costs of high- or low-temperature heat sources or sinks. Therefore, if a heat exchanger operates at around 225 °C (498 K), the minimum approach temperature at the limiting end of the heat exchanger should be about 25 K. In cryogenic service, a small ΔT is used because the low temperature is attained by expensive compression. The condenser of a distillation column using cooling water has a heat sink that comes in at about 30–35 °C and leaves at 40–45 °C. A log-mean temperature difference (LMTD) is used with the two differences between the reflux-drum temperature and inlet/outlet temperatures of the cooling-water (Dimian, 2003; Luyben, 2011):

$$\mathrm{LMTD} = (\Delta T_{\mathrm{A}} - \Delta T_{\mathrm{B}})/\ln(\Delta T_{\mathrm{A}}/\Delta T_{\mathrm{B}}) \qquad (2.2)$$

2.4.1.7 *Compressors*

For distillation, compressors are mostly used in heat pump assisted configurations. The power requirements of a compression system depend on the compression ratio and the suction temperature (which should be kept low). Compressors are driven by steam turbines (using high-pressure steam) or electric motors, but in both cases the energy is high level and expensive. Multistage compression systems (with equal compression ratios) use inter-cooling to follow closely an isothermal compression. Compressor power requirements are obtained from the simulation.

2.4.2 Equipment Cost

Two types of capital equipment costs are presented in cost estimation references. Bare-module cost is the cost of the equipment itself. Installed cost is the cost of the equipment plus all the costs of installing it in the process, e.g. equipment and setting, piping, civil and electrical, structural steel, instrumentation, insulation, paint and manpower (AspenTech, 2010; Luyben, 2011). The equations provided hereafter are for installed costs.

Of course, the total capital cost of a plant includes much expenditure beyond the investment in the major pieces of equipment. These include site preparation, off-site units (boiler house, raw material and product tanks, piping, buildings for staff, maintenance shops). The total investment in a typical plant is about three to four times the cost of the major equipment. In this book, a small three-year payback period is used to account for this factor. An alternative approach would be to increase the capital cost of the equipment by a factor ~3, and use a 9–10 years payback time (Luyben, 2011).

The equations given below are taken from Douglas (1988), Turton *et al.* (2003), and Dimian (2003). Conventional carbon steel materials of construction and reasonable pressure levels are assumed. If these assumptions are not valid for a process, then appropriate correction factors should be used. Correction factors during periods of rapid inflation should also be applied. The Marshall & Swift equipment cost index (M&S) had a value of 1536.5 at the end of 2011. All costs are estimated in US dollars:

Vessels and column shells:

$$\text{Purchased cost (\$)} = (\text{M\&S}/280)(957.9D^{1.066}H^{0.802}F_c) \tag{2.3}$$

$$\text{Installed cost (\$)} = (\text{M\&S}/280)(957.9D^{1.066}H^{0.802})(2.18 + F_c) \tag{2.4}$$

Table 2.2 Material factors (F_m) for pressure vessels

Factor	Shell material			
	Carbon steel	Stainless steel	Monel®	Titanium
F_m clad	1.00	2.25	3.89	4.25
F_m solid	1.00	3.67	6.34	7.89

where both D (diameter) and H (height) are expressed in meters. The cost factor (F_c) takes into account the material (F_m) and pressure (F_p), as follows (with P in bar):

$$F_c = F_m F_p \tag{2.5}$$

$$F_p = 1 + 0.0074(P - 3.48) + 0.00023(P - 3.48)^2 \tag{2.6}$$

Table 2.2 provides the numerical values to be used for the material factor – F_m (Dimian, 2003), while Figure 2.4 provides a convenient graphical illustration of the cost of column shells as a function of their height and diameter - caution to be observed for extreme values.

Column trays:

$$\text{Installed cost (\$)} = (\text{M\&S}/280)97.2D^{1.55}HF_c \tag{2.7}$$

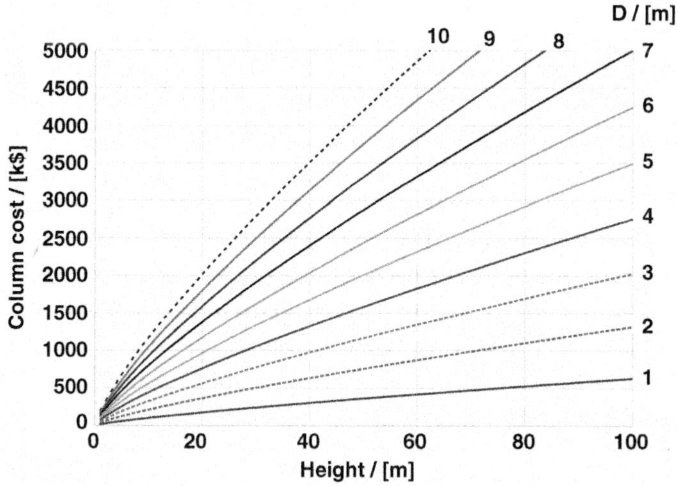

Figure 2.4 Cost of column shells as a function of height and diameter (the latter is indicated on the line)

For standard 24 inch tray spacing the overall correction factor is $F_c = F_t + F_m$. The tray factor (F_t) is 0 for sieve trays, 1.8 for bubble cap, and around 3 for more complex trays. The material factor (F_m) is 1.0 for carbon steel and 1.7 for stainless steel, but it is much larger for high alloys.
Column packing: The cost of random packing (size 50–25 mm) varies in the range €1200–1800 m^{-3} for saddles (ceramic), €500–1400 m^{-3} for Pall rings (plastic), and €1750–3200 m^{-3} for Pall rings (stainless steel). The cost of structured packing per cubic meter depends on the vendor and material of construction. Nevertheless, structured packing is significantly more expensive than random packing, but this is practically offset by its higher efficiency.

Heat exchangers (shell-and tubes):

$$\text{Purchased cost (\$)} = (\text{M\&S}/280)(474.7A^{0.65}F_c) \tag{2.8}$$

$$\text{Installed cost (\$)} = (\text{M\&S}/280)(474.7A^{0.65})(2.29 + F_c) \tag{2.9}$$

The heat exchange area (A) is given in m^2, for a size of $20 < A < 500\,m^2$ per shell. The cost factor is defined as $F_c = F_m(F_d + F_p)$, where F_m, F_d, and F_p are the correction factors for material, design type, and design pressure, respectively – as provided by Table 2.3 (Dimian, 2003).

Compressors (including centrifugal machine, motor drive, base plate, and coupling):

$$\text{Purchased cost (\$)} = (\text{M\&S}/280)(664.1P^{0.82}F_c) \tag{2.10}$$

where P is the brake power expressed in kW, within the range $25 < P < 750$ kW. The correction factor F_c varies with design type as follows: 1 for centrifugal (motor), 1.07 for reciprocating (steam), 1.15 for centrifugal (turbine), and 1.29 and 1.82 for reciprocating (motor and gas engine).

Table 2.3 Material, design, and pressure factors for shell & tubes heat exchangers

Shell/tubes	F_m	Design type	F_d	Design pressure (bar)	F_p
CS/CS	1.00	Kettle reboiler	1.35	<10	0.00
CS/brass	1.30	Floating head	1.00	20	0.10
CS/Monel	2.15	U-tube	0.85	30	0.25
CS/SS	2.81	Fixed-tube sheet	0.80	60	0.52
SS/SS	3.75	—	—	>75	0.55
Monel/Monel	4.25	—	—	—	—
CS/titanium	8.95	—	—	—	—

Table 2.4 Cost factors for turbo blowers

	Discharge pressure drop (bar)		
Cost factor	0.2	0.6	2
Constant C	2282	7271	4821
Exponent m	0.529	0.598	0.493
Q range ($m^3 s^{-1}$)	0.05–5	0.5–15	1–7.5

Turbo blowers:

$$\text{Purchased cost (\$)} = (\text{M\&S}/280)CQ^m \tag{2.11}$$

The capacity Q is expressed in $m^3 s^{-1}$, while the constant C and exponent m depend on the maximum discharge pressure drop as shown in Table 2.4 (Dimian, 2003).

2.4.3 Utilities and Energy Cost

Common process utilities include electricity, steam, refrigerants, compressed air, cooling water, hot water, hot oil, process water, demineralized water, municipal water, and water from other sources (river, lake, sea, or ocean). Waste disposal costs can also be treated like a utility expense. Unlike other expenses (e.g., capital and labor costs), utility prices do not correlate directly with inflationary indexes, as the basic energy costs are independent of capital and labor and they vary unpredictably.

Basically, utility price is linked to both inflation and energy cost. To reflect this dual dependence, a two-factor utility cost equation is needed (Ulrich and Vasudevan, 2006):

$$C_{S,u} = a(CE\,\text{PCI}) + b(C_{S,f}) \tag{2.12}$$

where $C_{S,u}$ is the price of the utility, a and b are utility cost coefficients, $C_{S,f}$ is the price of fuel (\$ GJ^{-1}), and CE PCI (Plant Cost Index) is an inflation parameter for projects in the USA. The utility cost coefficients and more detailed information and examples are provided by Ulrich and Vasudevan (2006).

The cost of energy also depends on the temperature level of the heat source or sink. As high level energy, electricity is the most expensive while low-pressure steam is the least expensive. The following costs should be used as guidelines only, since the actual costs always depend on the site location and the local availability of utilities (Dimian, 2003; Luyben, 2011):

- Electricity ≈ \$16.8 GJ^{-1}

  Heating utilities:
- Low-pressure steam (at 6 bar/87 psia and 160 °C/433 K/320 °F) ≈ \$7.78 GJ^{-1}
- Medium-pressure steam (at 11 bar/160 psia and 184 °C/457 K/363 °F) ≈ \$8.22 GJ^{-1}
- High-pressure steam (at 42 bar/611 psia and 254 °C/527 K/490 °F) ≈ \$9.88 GJ^{-1}

  Cooling utilities:
- Chilled water at 5 °C, returned at 15 °C ≈ \$4.43 GJ^{-1}
- Refrigerant at −20 °C ≈ \$7.89 GJ^{-1}
- Refrigerant at −50 °C ≈ \$13.11 GJ^{-1}

Another very valuable source of information about the cost of utilities and equipment is the book of prices (*DACE-Prijzenboekje*) edited by the Dutch Association of Cost Engineers. This lists the following cost ranges for 2012: 0.06–0.1 €/m^3 natural gas, 0.07–0.14 €/kWh electricity, 25–30 €/ton high-pressure steam, 11–12 / 25–30 €/GJ (heavy/domestic) fuel oil, 0.65–1.75 €/m^3 cold drink-water.

2.4.4 Cost of Chemicals

Estimation of capital and energy costs can usually be made with reasonable accuracy (±20%). Estimation of the costs of raw materials and the selling price of products is usually subject to much more uncertainty. Consumption figures are typically used to express the amount of chemicals and energy required per unit of product (e.g., 1 kg of finished product). Some very useful sources of chemical prices are the following websites (last accessed on 24 December 2012):

- http://www.icis.com/StaticPages/Students.htm
- http://www.icis.com/chemicals/channel-info-chemicals-a-z/
- http://ed.icheme.org/costchem.html

2.5 CONCLUDING REMARKS

This chapter has summarized some of the most important principles of the (optimal) design, control, sizing and economics of distillation – which still remains the dominant separation method used in the chemical process industry. The basic design concepts presented here can be integrated into a complete design of the entire process. Several quantitative examples of the application of these principles are given in the numerous case studies presented in subsequent chapters of this book.

The efficient design and effective control of the separation section of any chemical process are vital to its profitable and safe operation (Dimian, 2003; Luyben, 2011). Use L or V for single-end control configuration, while in the case of dual-end control use an energy balance configuration for low reflux ratio cases and use material balance or ratio configurations for high reflux ratio columns. Usually, the control of one product is much more important than control of the other. In such cases, use L as manipulated variable when the distillate is the most valuable product and V when the bottom product is the most important. The less important product should be controlled as follows: for low reflux cases by using an energy balance variable (L or V) or a ratio knob (L/D or V/B) and for high reflux columns by using D, L/D, B, or V/B. For these cases, it is essential to tune the important loop tightly and tune the less important loop much less aggressively (Riggs, 2001; Luyben, 2011).

Numerous equations are available in the literature to estimate the cost of various equipment types such as column shell, internals, heat exchangers (e.g., reboiler, condenser), decanters, flash and surge tanks, as well as pumps, valves, and piping. In addition, the cost of chemicals and utilities required can be also estimated within a reasonable range, using various data sources. However, these costs should be used as guidelines only, as the actual costs always depend on the site location and the local availability of utilities.

REFERENCES

Agachi, P.S., Nagy, Z.K., Cristea, M.V., and Imre-Lucaci, Á. (2006) Model based control: case studies in process engineering, Wiley-VCH.

Aspen Technology, Aspen Process Economic Analyzer: User guide, 2010.

Dimian, A.C. (2003) *Integrated Design and Simulation of Chemical Processes*, Elsevier, Amsterdam.

Doherty, M.F. and Malone, M.F. (2001) *Conceptual Design of Distillation Systems*, McGraw-Hill, New York.

Douglas, J.M. (1988) *Conceptual Design of Chemical Processes*, McGraw-Hill, New York.

Harmsen, G.J. (2010) Process intensification in the petrochemicals industry: Drivers and hurdles for commercial implementation. *Chemical Engineering and Processing*, *49*, 70–73.

Kiss, A.A., Flores Landaeta, S.J., and Infante Ferreira, C.A. (2012) Towards energy efficient distillation technologies – Making the right choice. *Energy*, *47*, 531–542.

Kister, H.Z. (1992a) *Distillation – Design*, McGraw-Hill.

Kister, H.Z. (1992b) *Distillation – Operation*, McGraw-Hill.

Lei, Z., Chen, B., and Ding, Z. (2005) *Special Distillation Processes*, Elsevier.

Luyben, W.L. (2006) *Distillation Design and Control using Aspen Simulation*, John Wiley & Sons, Inc., Hoboken.

Luyben, W.L. and Yu, C.C. (2008) *Reactive Distillation Design and Control*, John Wiley & Sons, Inc., Hoboken.

Luyben, W.L. (2011) *Principles and Case Studies of Simultaneous Design*, John Wiley & Sons, Inc., Hoboken.

Luyben, W.L. and Luyben, M.L. (1997) Essentials of process control, McGraw-Hill.

Mujtaba, I.M. (2004) *Batch Distillation - Design and Operation*, Imperial College Press, London.

Nagy, Z.K., Klein, R., Kiss, A.A., and Findeisen, R. (2007) Advanced control of a reactive distillation column. *Computer Aided Chemical Engineering*, *24*, 805–810.

Petlyuk, F.B. (2004) *Distillation Theory and its Application to Optimal Design of Separation Units*, Cambridge University Press.

Riggs, J.B. (2001) *Chemical Process Control*, Ferret Publishing.

Schmidt-Traub, H. and Gorak, A. (2006) *Integrated Reaction and Separation Operations*, Springer, New York.

Sharma, N. and Singh, K. (2010) Control of reactive distillation column - A review. *International Journal of Chemical Reactor Engineering*, *8*, R5.

Stichlmair, J.G. and Fair, J.R. (1998) *Distillation - Principles and Practice*, Wiley-VCH Verlag GmbH, Weinheim.

Sundmacher, K. and Kienle, A. (eds) (2003) *Reactive Distillation: Status and Future Directions*, Wiley-VCH Verlag GmbH, Weinheim.

Sundmacher, K., Kienle, A., and Seidel-Morgenstern, A. (eds) (2005) *Integrated Chemical Processes: Synthesis, Operation, Analysis, and Control*, Wiley-VCH Verlag GmbH, Weinheim.

Taylor, R. and Krishna, R. (1993) *Multicomponent Mass Transfer*, John Wiley & Sons, Inc., New York.

Taylor, R., Krishna, R., and Kooijman, H.A. (2003) Real world modeling of distillation. *Chemical Engineering Progress*, *99*, 28–39.

Turton, R., Bailie, R.C., Whiting, W.B., and Shaelwitz, J.A. (2003) *Analysis, Synthesis and Design of Chemical Processes*, 2nd edn, Prentice Hall, Englewood Cliffs.

Ulrich, G.D. and Vasudevan, P.T. (2006) How to estimate utility costs. *Chemical Engineering*, *113*, 66–69.

3

Dividing-Wall Column

3.1 INTRODUCTION

Process intensification represents a dominant trend in the chemical process engineering, mostly due to increasing awareness of the limited energy resources of the modern society (Huang *et al.*, 2007). In addition, the latest achievements in process modeling as well as growing computational power and advanced numerical methods make process intensification possible. Among others, process integration shows great potential that must be fully harnessed by the chemical process industry (Harmsen, 2010; Yildirim, Kiss, and Kenig, 2011).

For the separation of multicomponent mixtures, most often a sequence of distillation columns is applied. At least two columns are necessary when separating three components, such as for instance the direct and indirect sequences shown in Figure 3.1a,b (Yildirim, Kiss, and Kenig, 2011). A more energetically favorable alternative configuration is illustrated in Figure 3.1c (Yildirim, Kiss, and Kenig, 2011). In this so-called Petlyuk configuration, the vapor and liquid streams leaving the first column are directly connected with the second column (Petlyuk, Platonov, and Slavinskii, 1965). Basically, one condenser and reboiler are effectively replaced by thermal (heat) coupling of the prefractionator with the main column, while the required condenser and reboiler are attached to the main column. This integrated configuration is generally known as the *fully thermally coupled distillation column* (FTCDC). Remarkably, this is not limited only to three component mixtures. According to Christiansen, Skogestad, and Liena (1997), a Petlyuk column is in general: " . . . *a column arrangement, separating three or*

Advanced Distillation Technologies: Design, Control and Applications, First Edition.
Anton Alexandru Kiss.

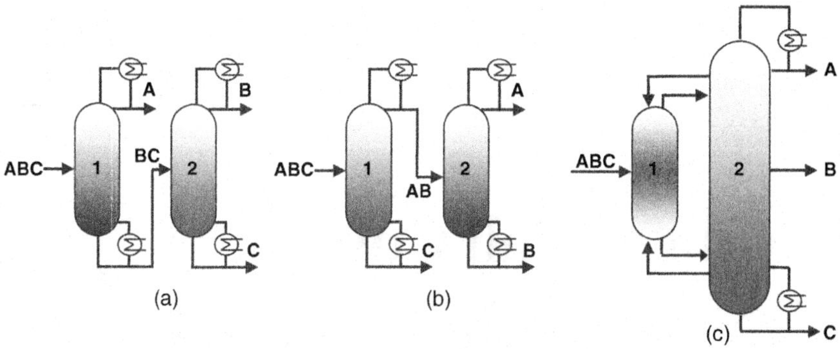

Figure 3.1 Direct sequence (a), indirect sequence (b), and Petlyuk configuration (c) for separating a three-component mixture ABC (A: light, B: middle, and C: heavy boiling component)

more components using a single reboiler and a single condenser, in which any degree of separation (purity) can be obtained by increasing the number of stages, provided the reflux is above a certain minimum value and the separation is thermodynamically feasible."

The prefractionator of the Petlyuk configuration performs a sharp split between A (light boiling component) and C (heavy boiling component), whereas the middle boiler B is distributed naturally between the top and bottom products. A further separation towards high-purity components takes place in the second column. The improvement in thermal efficiency leads to considerable energy savings of about 30% as compared to the direct or indirect sequences (Schultz *et al.*, 2002). Moreover, since only one reboiler and one condenser are used, the capital costs are also reduced.

The main reason for the energy efficiency in a Petlyuk setup is the avoidance of remixing of internal streams, which in arrangements of two columns in series exhibits concentration peaks of middle boiling components either above or below the feed stage. Additional gain comes from the fact that the prefractionator arrangement that distributes the intermediate component between top and bottom allows greater freedom to match the feed composition with a tray in the column to further reduce mixing losses at the feed tray. This mixing as well as remixing of streams with different compositions that occurs at feed points and along the column – inevitably accompanied by the entropy of mixing formation – is an intrinsic source of thermodynamic inefficiency of the separation process occurring in a multicomponent distillation column (Dejanović, Matijašević, and Olujić, 2010).

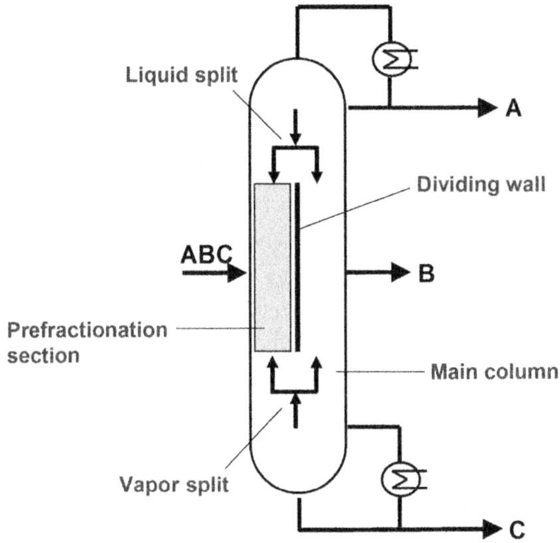

Figure 3.2 Schematics of a dividing-wall column

Further equipment integration and cost savings can be made if the two columns of the Petlyuk configuration are integrated into one shell. This alternative to conventional columns is identified as a dividing-wall column (DWC) and illustrated in Figure 3.2 (Kaibel, 1987; Yildirim, Kiss, and Kenig, 2011). According to Schultz *et al.* (2002), the DWC will become a standard distillation tool in the next 50 years. Unlike conventional columns with a side draw, a DWC can deliver high purity side product, and thus is able to produce three pure products in just a single shell. Kaibel (1987) extended further the basic ideas of separating multicomponent mixtures to four and more products in one shell, including also chemical reactions and connecting DWC units in series.

Note that due to the large number of design parameters it has been virtually impossible to simulate, for many years, the design and operation of DWC units. The first industrial application of a DWC–as packed column–was accomplished in 1985 by BASF SE at Ludwigshafen, Germany (Kaibel, 1987; Parkinson, 2007). Other columns equipped with structured packing followed and presently there are around 70 packed DWC units in operation in BASF plants worldwide–some of notable dimensions, with diameters above 5 m and heights up to 100 m (Yildirim, Kiss, and Kenig, 2011). Internals for nearly all of these columns were delivered by Julius Montz GmbH, and the necessary process and mechanical design know-how was developed in close cooperation between BASF and Montz (Dejanović, Matijašević, and Olujić, 2010). Overall, the number of applications of DWC increased rapidly to more than 100 in 2010 (Harmsen, 2010). With the increasing number of applications, the number of patents has also grown and today nearly half of all patents in this area are owned by BASF and Montz – see the review paper of Dejanović, Matijašević, and Olujić (2010) for a complete list of patents.

The recent review papers of Dejanović, Matijašević, and Olujić (2010), Yildirim, Kiss, and Kenig (2011), and Kiss and Bildea (2011) give a more comprehensive overview of DWC technology, covering the design and theoretical background, specific industrial applications, and dynamics and process control issues.

3.2 DWC CONFIGURATIONS

For a three-component separation, two different DWC configurations can be applied. The first type, shown in Figure 3.3a , was patented by Wright (1949) and is more common (Yildirim, Kiss, and Kenig, 2011). The dividing wall and the feed and side draws are usually placed close to the middle of the column (Asprion and Kaibel, 2010). The second configuration is shown in Figure 3.3b,c. The wall is located either at the lower or at the upper part of the column. This configuration was patented by Monro (1938) and was first applied in 2004 (Kaibel *et al.*, 2006). The column in Figure 3.3b is referred to as a *split shell column with common overhead section and divided bottoms section*, while the column shown in Figure 3.3c is called a *split shell column with divided*

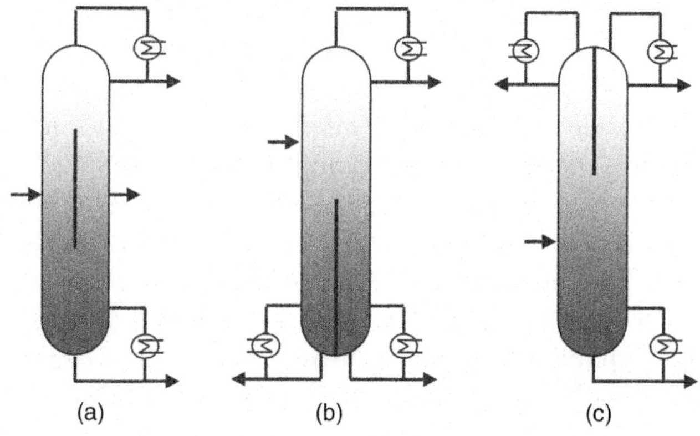

Figure 3.3 Basic types of dividing-wall columns

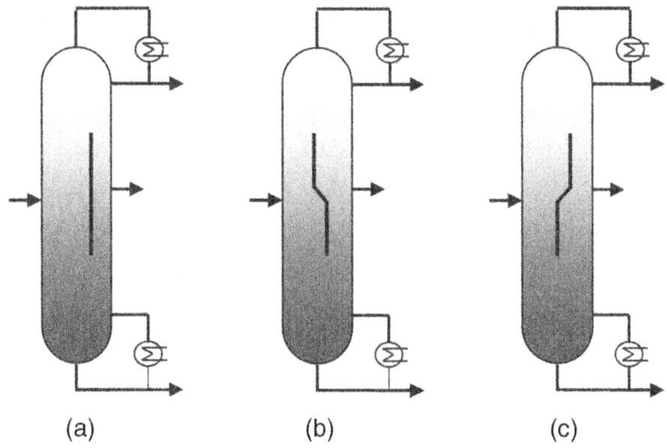

Figure 3.4 Different positions and shape of the dividing wall

overhead section and common bottoms section (Schultz *et al.*, 2006; Yildirim, Kiss, and Kenig, 2011). Moreover, the wall can be shifted from the center towards the column walls (Figure 3.4a), and can have diagonal sections as well (Figure 3.4b,c).

DWC units can also be applied for the separation of more than three components. The number of possible configurations grows accordingly with increasing number of components. Figure 3.5 shows the basic DWC types applied for such separations (Yildirim, Kiss, and Kenig, 2011). In the *Kaibel column* configuration, illustrated in Figure 3.5a, separation is

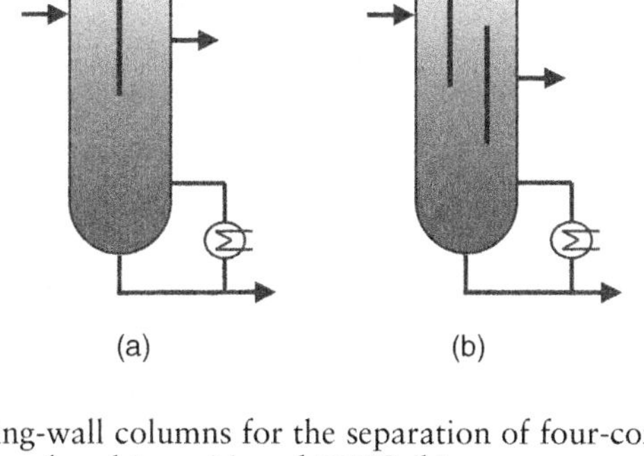

Figure 3.5 Dividing-wall columns for the separation of four-component mixtures: Kaibel column (a) and multi-partitioned DWC (b)

performed in a shell with one dividing wall (Strandberg and Skogestad, 2006) while the two middle-boiling products are accumulated at the right-hand side of the dividing wall (Yildirim, Kiss, and Kenig, 2011). According to Kaibel *et al.* (2006), the configuration with only one wall is less thermally efficient and it can be improved by the application of additional dividing walls, as shown in Figure 3.5b. This multi-partitioned configuration is referred to as a *Sargent arrangement* (Agrawal, 2001a). Despite several theoretical studies carried out so far, no industrial application has yet been reported (Dejanović, Matijašević, and Olujić, 2010).

A further configuration, shown in Figure 3.6a, is known as the *Agrawal arrangement* (Agrawal *et al.*, 2001; Yildirim, Kiss, and Kenig, 2011). In this configuration, the feed enters the middle partition of the DWC. This can easily be understood if the DWC is seen from the top view (Figure 3.6b). Further arrangements for the wall are also conceivable, for instance the configuration shown in Figure 3.7 (Yildirim, Kiss, and Kenig, 2011). Recently, Rong and Turunen (2006) and Rong (2010) has introduced a procedure that allows a quick synthesis of three-product DWC units for a given separation task and has shown possible alternative configurations for the separation of a four-component mixture – depicted in Figure 3.8 (Yildirim, Kiss, and Kenig, 2011) – but these studies do not cover the Kaibel column configuration.

The proper selection of column internals is necessary to achieve efficient heat and mass transfer and, hence, the required purity. Dividing-wall columns can be equipped with trays or with different kinds of packing.

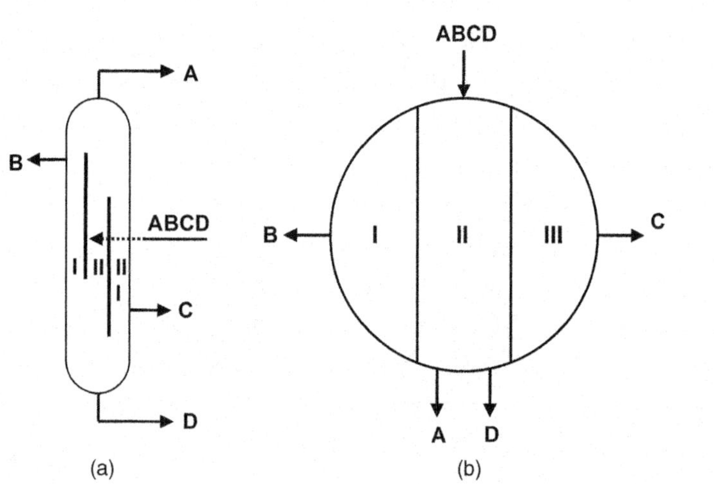

Figure 3.6 Agrawal arrangement for the separation of four components

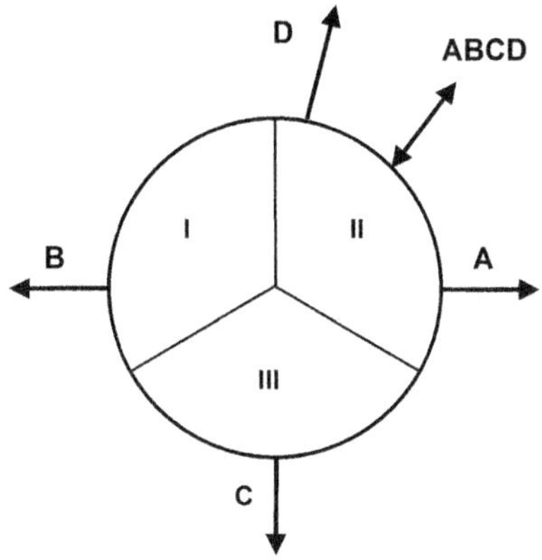

Figure 3.7 Top view of triangular wall structures

Generally, the selection criteria for optimal column internals for DWC are similar to those for conventional distillation columns. The DWC units built and delivered so far by Julius Montz are all equipped with structured packing. This is not really surprising considering that these columns are used mainly for the separation of various heavier chemicals, which are carried out under vacuum conditions (Dejanović, Matijašević, and Olujić, 2010). While BASF exclusively applies packed columns, other companies (e.g., Koch-Glitsch and CEPSA Refinery) install trays (Jobson, 2005). The advances in trays and packing were reported by Olujić *et al.* (2003).

The wall construction is different for tray and packed columns. Generally, tray DWCs are easier to build and the dividing wall that is welded on the column can strengthen the shell stability. The construction of a packed DWC is more complex, with the welding of the wall being especially difficult. During installation, it must be assured that the packing material does not touch the column walls. If the packing does touch the wall, according to Kaibel *et al.* (2006) this would result in excessive liquid flow and negatively affect the separation. Recently, non-welded wall technology was developed in cooperation between Julius Montz GmbH and BASF SE. Using unfixed walls, the column design becomes much simpler. Another advantage of non-welded dividing walls is a faster and more precise installation (Kaibel *et al.*, 2006). Other benefits include fewer manholes and lower weight, since the manufacturing requires less metal. The revamping of conventional columns becomes faster, simpler, and cheaper, too. Remarkably, the implementation of a non-welded partition wall by Montz in the mid-1990s led to a real breakthrough of this technology, first within BASF and then in general (Dejanović, Matijašević,

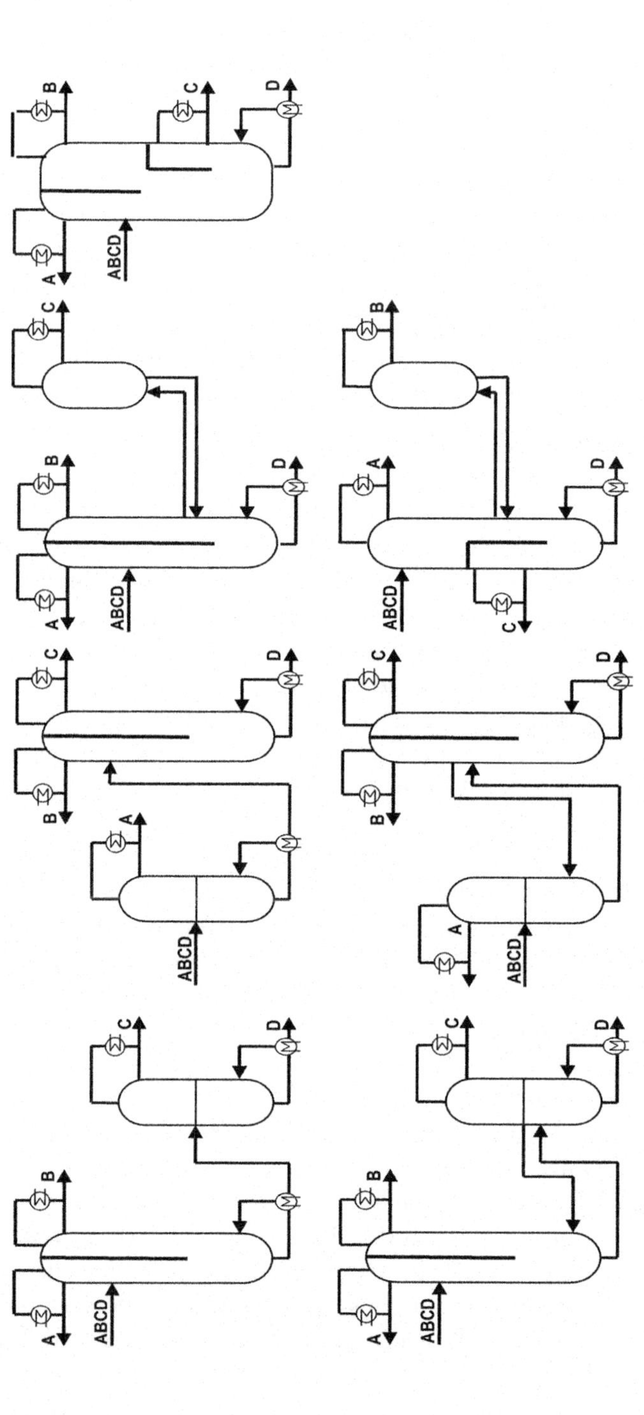

Figure 3.8 Possible DWC configurations for the separation of a four-component mixture according to the synthesis procedure of Rong (2010)

and Olujić, 2010). Adopting a non-welded partition wall enabled a significantly large increase of the application window, with possibilities for revamping existing columns – so this can be considered as a milestone in the development and implementation of DWC technology. This is also indicated by a sudden increase in the number of DWC units delivered by Montz over the ensuing years, reaching over 85 deliveries in 2009 (Dejanović, Matijašević, and Olujić, 2010; Yildirim, Kiss, and Kenig, 2011). More details about the DWC equipment and some key construction issues are provided in subsequent sections of this chapter.

3.3 DESIGN OF DWC

Regarding the design, construction, and performance of DWCs in industry, no public information was available until 2000 when a DWC was for the first time built outside BASF. This was a tray column that led to a wider acceptance of DWC technology. Becker, Godorr, and Kreis (2001) – from Linde AG and Sasol – awakened public awareness by describing the design and construction of the first two-tray DWC units of impressive dimensions (one is 100 m tall, with a diameter of 5.2 m) installed in a Sasol plant in South Africa, with the purpose of recovering valuable petrochemicals from Fischer–Tropsch synthesis products. Others soon joined in, resulting in several publications describing design configurations considered suitable and/or implemented for various refining applications dominated traditionally by tray columns (Dejanović, Matijašević, and Olujić, 2010). During the past decade the number of papers describing the design and modeling (steady-state and dynamics), control and operation of DWCs has flourished (Dejanović, Matijašević, and Olujić, 2010; Yildirim, Kiss, and Kenig, 2011; Kiss and Bildea, 2011).

A theoretical analysis of thermodynamically reversible distillation, carried out by Petlyuk (1965, 2004), concludes that the sequence can be extended to the separation of any number of components:

- In each column section only the components with extreme volatilities (i.e., lightest and heaviest) are taken as key components. This results in $n(n-1)$ sections required for separating an n-component mixture, instead of $2(n-1)$ in the conventional scheme.
- One reboiler and one condenser are sufficient, independent of the number of products.
- All products (n) can be obtained with high purity.

Being thermodynamically equivalent to a Petlyuk column, a DWC can be considered as a single shell realization of the Petlyuk configuration. However, note that a Petlyuk column is not always the best choice, as there are mixtures and conditions that may favor implementation of alternative configurations (Shah, 2002). Although a fully thermally coupled system always has the lowest minimum vapor flow, the energetic optimum strongly depends on the feed composition. Certain conventional arrangements provide greater energy savings for lower contents of middle boiling component in the feed, a symmetric distribution of high and low boiling components, as well as large differences in relative volatilities (Becker, Godorr, and Kreis, 2001; Dejanović, Matijašević, and Olujić, 2010).

The optimal design of a DWC requires adequate models and computer-based simulations. However, commercial process simulators do not include particular subroutines for DWC units. The so-called decomposition method developed by Triantafyllou and Smith (1992) simplifies the design problem. The existing DWC configuration is replaced by a sequence of conventional distillation columns. Figure 3.9 shows some possible decomposition variants (Yildirim, Kiss, and Kenig, 2011).

The literature reveals that there are several design methods available that concern mostly ternary separations. However, they can be relatively easily extended to cover cases with more components. When designing a DWC system for the separation of a three-component feed into three products the number of degrees of freedom (i.e., design parameters) significantly increases as compared to that required when designing conventional configurations of two columns in series – where the two

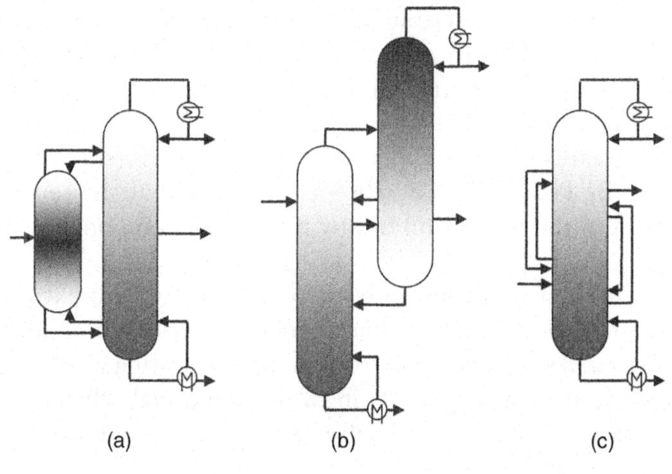

Figure 3.9 Decomposition of DWC into simple column sequences

columns can be optimized independently of each other (Dejanović, Matijašević, and Olujić, 2010). The required design parameters for a DWC are: number of stages in six different sections (e.g., common top and bottom sections, sections above and below the feed stage and side draw, respectively), vapor split ratio, liquid split ratio, reflux ratio, heat input in the reboiler, and the side product flow rate (Figure 3.2).

The initial steps are similar to designing conventional columns: selecting column configuration and operating pressure, combined with an appropriate VLE model. However, the next steps differ considerably and will be explained hereafter: establishing initial configuration, short-cut or detailed design, stage and reflux requirements, optimization, equipment sizing, and process control system.

3.3.1 Heuristic Rules for DWC Design

The design parameters require good initial values and some guidance to ensure convergence of simulations (Dejanović, Matijašević, and Olujić, 2010). The following list of heuristics, as proposed by Becker, Godorr, and Kreis (2001), provides pretty good initial estimates for both short-cut and detailed simulations:

- Design a conventional two-column system as a base case (e.g., in-/direct sequence);
- take the total numbers of stages for DWC as 80% of the total number of stages required for the conventional two-column sequence: $N_{DWC} = 0.8(N_1 + N_2)$;
- place the partition (i.e., dividing wall) in the middle third of the column (e.g., 33–66% H);
- set the internal flow rates in the DWC, as determined by the reboiler or condenser duty, at 70% of the total duties of two conventional columns: $Q_{DWC} = 0.7(Q_1 + Q_2)$;
- use equalized vapor and liquid splits ($r_V = 0.5$, $r_L = 0.5$) as initial values.

However, this is only enough to make a good start with the simulation as it requires a lot of tuning to achieve the optimal DWC design. Carrying out rigorous DWC simulations requires a certain level of experience, and they are computationally demanding, depending on the configuration chosen to represent the DWC (e.g., 1–4 column models) and the modeling approach (short-cut or detailed).

3.3.2 Approximate Design Methods

The shortcut method suggested by Triantafyllou and Smith (1992) can be applied for the initial design of a DWC. This method is based on the Fenske–Underwood–Gilliland–Kirkbride model (FUGK) – a quite well-known combination of models used to establish the minimum number of equilibrium stages (Fenske), minimum reflux (Underwood), stage requirement at chosen operating reflux ratio (Gilliland), and the feed stage (Kirkbride) for a given separation. The basic assumptions of this model are constant relative volatility and constant molar flows. The authors approximated a DWC configuration with a three-column model (e.g., distributed or sloppy sequence). The prefractionator section is modeled as a column with a partial condenser and a partial reboiler in which the sloppy (non-sharp) separation takes place. Assuming VL equilibrium in the condenser and reboiler of the prefractionator, compositions of coupling streams are calculated by performing simple flash calculations (Dejanović, Matijašević, and Olujić, 2010). The top and bottom columns are modeled as a side stream column, and are linked by equalizing vapor traffic at the bottom of the top column and the top of the bottom column, respectively. The reflux ratio of the prefractionator is adjusted until its number of stages equals the number of stages between the feed locations in the top and bottom of the main column. The recoveries in the prefractionator column are then optimized by a simple procedure so as to minimize either the vapor flow (i.e., minimum energy use) or the number of stages (i.e., minimum investment cost), for a given reflux ratio. The reflux ratios of the main column and prefractionator are then optimized to minimize the total cost (Dejanović, Matijašević, and Olujić, 2010). A similar short-cut method was used by Muralikrishna, Madhavan, and Shah (2002), who proposed a useful visualization tool to represent all feasible designs on a single plot, which can also be utilized for a simple optimization procedure. The calculations are relatively simple and can be used for mixtures of any number of components (Mueller *et al.*, 2007; Dejanović, Matijašević, and Olujić, 2010).

Amminudin and Smith (2001) and Amminudin *et al.* (2001) pointed out that using the Kirkbride equation to find the thermal coupling locations can lead to errors when transferred to rigorous simulations. They proposed a semi-rigorous design method based on an equilibrium stage composition concept previously applied for the synthesis of azeotropic distillation sequences (Amminudin *et al.*, 2001). The proposed design procedure starts from a defined composition of products, and works backward to determine the required design parameters to achieve them. This and other similar procedures for product feasibility determination can be extended to the separation of more than three components, providing that the mixture shows ideal behavior (i.e., the volatility order does not change with the compositions). Amminudin *et al.* (2001) claim that their method provides a more accurate base-case design than the standard FUGK based short-cut methods. However, their method is somewhat computationally more demanding and a literature survey shows that the method initially developed by Triantafyllou and Smith (1992) is still preferred because of its simplicity.

Sotudeh and Shahraki (2007, 2008) proposed a method based on Underwood equations only, as they consider inadequate the use of Fenske equation for the minimum number of stages in DWC design. The main reason for this is because the Fenske equation is based on an assumption of equal compositions of liquid and vapor streams at top and bottom of the prefractionator, which for DWCs is clearly not the case (Dejanović, Matijašević, and Olujić, 2010). According to the case study presented in their paper, this method provided a more economical design than the method of Muralikrishna, Madhavan, and Shah (2002).

Rangaiah, Ooi, and Premkumar (2009) have developed a procedure for the quick design of DWC using AspenTech HYSYS®. This tool is based on the equilibrium stage model, which assumes that the streams leaving a column segment (stage) are in thermodynamic equilibrium. This is actually an idealized description that requires experimental parameters such as efficiency factors for tray columns or HETP values for packed columns. A more consistent alternative is given by the rate-based approach that directly takes into account the process kinetics and specifics of column internals. In this case, the rates of mass and heat transfer between the liquid and vapor phases are determined explicitly (Mueller and Kenig, 2007). For the rate-based model, adequate correlations describing pressure drop, liquid hold-up, and mass and heat transfer coefficients are necessary (Mueller and Kenig, 2007).

3.3.3 $V_{\min}$ Diagram Method

Another simple – yet interesting and effective – approach to the conceptual design of DWC was suggested by Halvorsen and Skogestad (2003a–c, 2011). This method is based on a graphical treatment of the minimum energy represented by normalized vapor flow, as a function of the feed distribution. It is based on Underwood's equations, using the assumptions of constant molar flows, infinite number of stages, and constant relative volatilities. This obviously means that the method is applicable only to zeotropic mixtures. The required input parameters are the feed composition, feed quality expressed by the liquid fraction, K-values, and the desired product purities or recoveries. The Underwood equations are then used to determine the minimum vapor ($V_{\min}$) and liquid flows needed to perform all the binary separations of a specified feed mixture. These are in fact product splits occurring in each section of the column, assuming an infinite number of theoretical stages. In practice, the infinite number of stages can be approximated by setting the number of stages for each simulation to at least $4N_{\min}$ as calculated by the Fenske equation. This value is also suggested by Halvorsen (2001) and has been confirmed by rigorous simulations showing that no decrease in the reboiler heat duty can be achieved by further increasing the number of stages (Dejanović, Matijašević, and Olujić, 2010, 2011).

The $V_{\min}$ diagram conveniently shows the vapor and liquid traffic needed in every column section, which can directly serve as the basis for column design. The basic claim of the $V_{\min}$ method is that the minimum vapor flow required for the separation of n components feed into n pure products in any arrangement corresponds to that required for the most difficult binary split (shown as the highest peak in the $V_{\min}$ diagram) – getting all the other separations "for free." The number of stages can be preliminary considered as twice the minimum number, as determined using the Fenske equation. The $V_{\min}$ diagram plots the vapor flow rate above the feed (V/F) versus the net flow of the top product (D/F) per unit of feed. For each given pair (D/F, V/F) all the other properties are completely determined, such as all component recoveries and product compositions. The feed enthalpy condition is given by the liquid fraction (q) in the feed stream (Dejanović *et al.*, 2011).

Figure 3.10 plots the $V_{\min}$ diagram for an equimolar ternary system [benzene–toluene–xylene (BTX) or ABC]. The $V_{\min}$ diagram shows how the feed components of a ternary feed (ABC) are distributed to the top and bottom products in a simple two-product "infinite stage" distillation column as a function of the operating point (D/F, V/F). For values of V/F above the upper boundary following the three peaks in the diagram ($0,0$-P_{AB}-P_{AC}-P_{BC}-$1,0$), the column is over-fractionating, meaning that valuable energy is wasted. Note that the point located at $x,y = 1,0$ in Figure 3.10 is more generally defined as $1,(1-q)$ but in this particular case a saturated liquid feed stream was considered ($q = 1$).

The values at the peaks (P_{AB}, P_{AC}, P_{BC}) give the vapor flow for the corresponding sharp neighbor component splits. The knots are $V_{\min}$ for the so-called *preferred splits*, where a sharp split between two key-

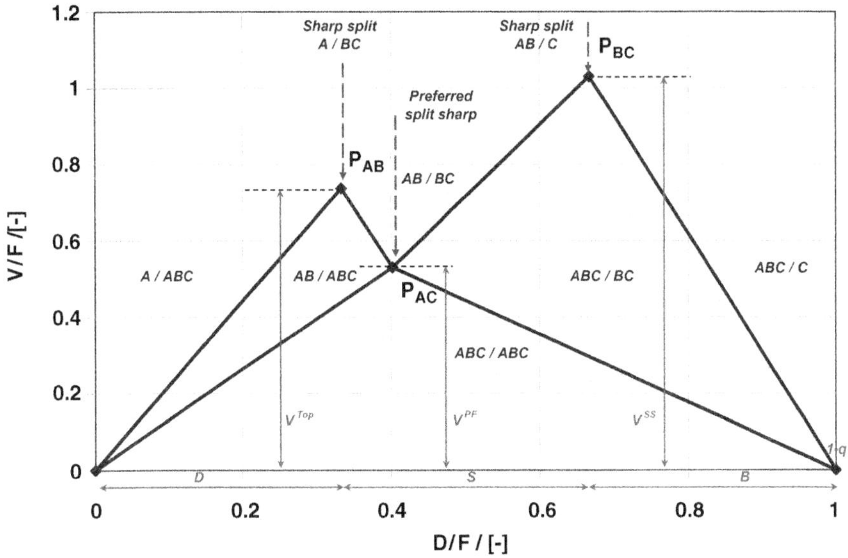

Figure 3.10 V_{min} diagram of a BTX (benzene–toluene–xylene) ternary system (equimolar mixture)

components is specified, while allowing intermediate components to be distributed. Only the sharp split between each possible pair of key components must be solved to find the diagram for a multicomponent feed. For the example shown in Figure 3.10, only three points are needed: P_{AB}: sharp A/B split, P_{BC}: sharp B/C split, and P_{AC}: sharp A/C split. P_{AC} is the *preferred split* that is the minimum energy operating point for a sharp separation between the heavy and light keys while the intermediate distributes to both column ends. At any operating point at or above the V-shaped P_{AB}-P_{AC}-P_{BC}, a sharp A/C split is obtained but with higher energy than that required at the exact point P_{AC} (Halvorsen and Skogestad, 2011). Note that the minimum energy required for the ternary separation in a DWC corresponds to the highest peak (i.e., P_{BC} in Figure 3.10).

The V_{min} method provides an excellent visual tool that can provide good initial parameters for rigorous design in a simple manner, regardless of the number of components or column configuration. All the necessary information on internal flow rates of vapor and liquid in different column sections can be easily calculated from the V_{min} diagram, as described in previous papers (Halvorsen and Skogestad, 2003a–c, 2011; Dejanović *et al.*, 2011). The calculated internal vapor and liquid flow rates, vapor and liquid splits, reflux and boil-up ratio, as well as the product flow rates can be used as very reliable initial estimates of parameters for the rigorous model (Dejanović *et al.*, 2011).

The V_{min} method is available as a software package that can be used to gain insight into the actual requirements for separating a multicomponent feed mixture into products (Dejanović *et al.*, 2011). However, the V_{min} diagram can also be developed by performing a series of binary split calculations using rigorous distillation models from commercial process simulators such as Aspen Plus® or ChemCAD (Halvorsen and Skogestad, 2003a–c; Dejanović *et al.*, 2011; Kiss *et al.*, 2013). Moreover, with basic programming skills, the V_{min} method can be easily implemented in MATLAB® or Microsoft® Excel, and then be used for fast screening of alternative configurations and to identify energy saving potential.

3.3.4 Optimal Design of a DWC

Dejanović *et al.* (2011) proposed a calculation procedure for an effective method for establishing the stage and reflux requirements for a three-product DWC, which can be relatively easily extended to other DWC applications as well (e.g., azeotropic, extractive, and reactive distillation in a DWC). The procedure can be summarized as follows – while Figure 3.11 illustrates the proposed algorithm of the detailed model (Dejanović *et al.*, 2011):

1. Define the separation task: feed composition, feed state, and recoveries of key components.
2. Calculate the reboiler vapor flow and distillate flow rate for every possible binary split of key components, for an infinite number of stages (e.g., $N > 4N_{min}$ when using Fenske equation).
3. Construct V_{min} diagram and calculate the product, internal liquid, and vapor flow rates.
4. Initialize rigorous simulations using values from V_{min} diagram. Change the number of stages per column section – as well as the number of reactive trays (N_{RD}), reactants ratio (Rx_R), solvent feed stage (N_S), and solvent to feed ratio (SFR) when applicable.
5. Adjust the vapor and liquid split ratios until the reboiler duty (Q_R) is minimized, while maintaining the purities of all products.
6. Repeat the calculations by re-initialization, while gradually reducing the number of stages in each section until $N(RR+1)$ is minimized.

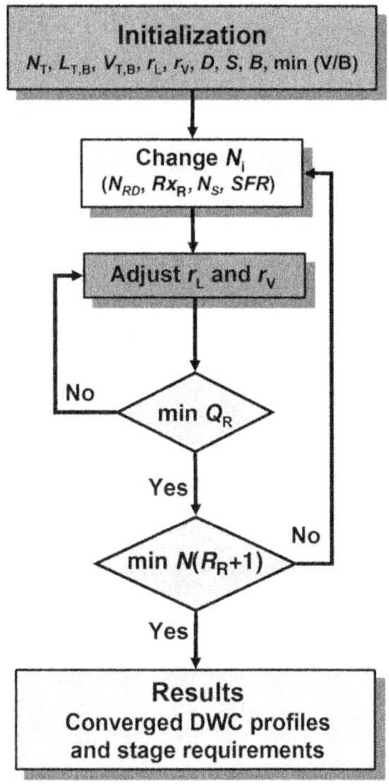

Figure 3.11 Algorithm of detailed DWC modeling for optimal design configuration

3.4 MODELING OF A DWC

Commercial process simulators such as Aspen Plus, HYSYS, or ChemCAD still do not include DWC as a distinct model, not even in its simplest form – although the MultiFrac unit from Aspen Plus can be used to simulate Petlyuk configurations that are thermodynamically equivalent to a DWC, if the heat transfer across the wall is negligible. In practice, a DWC unit can be modeled as a sequence of simple column sections. Several approaches can be used, ranging from one to four column models, each having specific advantages and disadvantages (Dejanović, Matijašević, and Olujić, 2010). As a rule of a thumb, one-column pump-around models have better convergence properties than models with more columns. While the four-column model is rather difficult to initialize and exhibits slower convergence, it is much easier to evaluate the results and perform appropriate sizing – considering that each column section is represented individually (Dejanović, Matijašević, and Olujić, 2010).

In any case, the final design of a DWC must be based on results obtained by rigorous simulations carried out using established stage-by-stage models to provide the required design parameters and a detailed performance simulation of a DWC system. Owing to the absence of off-the-shelf models for DWC, there are several ways of using a process simulator to simulate a DWC.

3.4.1 Pump-Around Model

As illustrated by Figure 3.9c, a DWC for ternary separations can be represented as a single column in which various sections of the DWC are situated vertically above one another. The traffic of vapor and liquid within the model are regulated using liquid pump-around streams and vapor bypasses in such way as to imitate a DWC (Dejanović, Matijašević, and Olujić, 2010). It is computationally simpler and easier to initiate than other arrangements, since it involves simulating only one column – conveniently carried out in any process simulator. However, it was reported that it can lead to convergence problems as the entire vapor and liquid are drawn off in two points of the column, and nothing remains to flow to the next tray (Becker, Godorr, and Kreis, 2001; Dejanović, Matijašević, and Olujić, 2010).

3.4.2 Two Columns Sequence Model

Several two-column sequences are possible, which are thermodynamically equivalent to a DWC, such as the ones shown in Figure 3.9a,b. Two-column configurations are generally easy to set-up and offer a bit more flexibility than the pump-around model. There is no practical difference in convergence time or results obtained by various two-column configurations. The sequence shown in Figure 3.9a is usually preferred, simply because it represents the real-life sequence of a prefractionator and main column, and the results can be more easily inspected. When initializing such sequences, the following procedure is recommended by Dejanović, Matijašević, and Olujić (2010):

- Specify the composition of the coupling streams, and the flows from the prefractionator to the main column in the following way:
  - Set the vapor stream composition using the feed stream composition with all impurities (heavy components) set to zero. Keeping

    the same composition, set the total flow rate as determined by the short-cut model. Set the temperature as the dew-point temperature at the operating pressure.
  - Set the liquid stream composition using the feed stream composition with all impurities (light components) set to zero. Keeping the same composition, set the total flow rate as determined by the short-cut model. Set the temperature as the boiling point temperature at the operating pressure.
- Run the main column once as standalone, using the boil-up ratio calculated by the short-cut method and highest reflux ratio from the direct sequence, for reboiler and condenser specifications.
- Run the prefractionator column once, as standalone unit.
- Run the entire sequence together, using these converged profiles and the composition of the coupling streams and setting the desired distillate and bottom product purities for the condenser and reboiler specifications, respectively.

Note that, compared to the pump-around model, a downside of all multi-column sequences is that the overall product recoveries cannot be used for the condenser and reboiler specifications (e.g. internal design spec of the column) because the coupling streams also count as feed streams, and are used for recovery calculations.

3.4.3 Four Columns Sequence Model

This model uses a sequence of four columns representing the sections of a DWC as shown in Figure 3.2: top common rectifying section, feed side section (prefractionator), side stream section, and bottom common stripping section. Such a configuration reflects best the actual situation, but it is the most difficult to initialize – as more interconnecting streams need to be estimated – and it is also the slowest to converge. The general initialization procedure is similar to the one described for two-column models and, hence it is not repeated here. Remarkably, the four-column configuration allows for maximum flexibility with regard to specifications for different column sections, as well as vapor and liquid splits. Consequently, it is considered to be the most suitable configuration for dynamic simulations (Dejanović, Matijašević, and Olujić, 2010) and it is actually recommended to set up such a model before exporting a steady-state simulation to a dynamic one (e.g., exporting from Aspen Plus to Aspen Dynamics).

3.4.4 Simultaneous Models

Equation-based models that enable simultaneous flow-sheet solving could most likely provide better convergence properties and speed, as compared to sequential approach of commercial process simulators. Such an approach – based on a three-column model similar to the one used for a shortcut calculation that is simultaneously solved by matrix inversion – has better convergence properties than the usual sequential models, as reported by Becker, Godorr, and Kreis (2001) and Dejanović, Matijašević, and Olujić (2010).

3.4.5 Simulation of a Four-Product DWC

The simplest four-product DWC configuration is the Kaibel column (Figure 3.5a). Such a single longitudinal partition wall configuration can be simulated as a single pump-around column, three- or four-column sequence, with the same considerations and procedures as described in previous sections. The key difference is the two additional degrees of freedom – second side product flow rate and the number of stages between the two side products – that make this model somewhat more difficult to converge. An alternative configuration was introduced by Christiansen, Skogestad, and Liena (1997) and consists of adding a horizontal wall in between the two side draw stages. This configuration can be simulated as a sloppy three-column sequence because there is only heat integration between the bottom and top part of the main column (Dejanović, Matijašević, and Olujić, 2010).

The most complex configuration (Figure 3.5b) is a multi-partitioned DWC separating the middle part of the column into three parallel sections – usually referred to as a *fully extended Petlyuk sequence*. However, in this case a three-column (in parallel) model appears to be the most appropriate choice although it is still difficult to handle. A similar procedure can be used for the initial composition and flow estimates of 12 coupling streams, as described for three-product sequence.

3.4.6 Optimization Methods

A common difficulty associated with detailed simulation is related to the estimation of the number of stages in each section. Since the number of stages is an integer variable, column optimization falls into a class of mixed integer non-linear programming problems (MINLPs). This cannot be done easily within commercially available process simulators, as, for example, only the sequential quadratic programming (SQP) method is available in Aspen Plus. To overcome this drawback, usually an external optimization routine is required, coupled with process simulators capabilities.

One rigorous design method – proposed by Dünnebier and Pantelides (1999) – is based on the models of detailed column superstructures for a given separation task, followed by a mathematical optimization procedure. The tools used were local optimization code (CONOPT) coupled with the gPROMS process modeling tool. Their method can simultaneously provide design parameter determination and performance simulation. However, as the authors admitted, the problem could not be solved to global optimality. Instead the proposed superstructures were used only as a means to generate a number of promising candidate structures that were then optimized with fixed structure.

Wenzel and Röhm (2003) have presented a method for simultaneous design and optimization using an external optimization routine. The authors say that the steady-state design of a DWC is the only remaining difficulty that lies in the way of its wider acceptance. They used Aspen Plus software in conjunction with an external optimization routine based on an evolutionary algorithm to minimize total annualized costs. A similar approach was reported by Kiss *et al.* (2012), who connected rigorous Aspen Plus simulations with simulated annealing optimization in MATLAB to obtain the optimal design of a reactive DWC. For an overview of applications of mathematical programming methods to distillation system design the review paper of Grossmann, Aguirre, and Barttfeld (2005) is recommended.

3.5 DWC EQUIPMENT

The dividing wall can be made of one or several metal plates, splitting the column into two sides. The simplest configuration has one metal plate welded along the inner walls of the column. This technique was adopted for the first implementation of a DWC. At that time, welding was the only way to avoid infiltrations, while providing additional strength to the column. Only small tolerances in column shell are allowed, as the welds must be accurate enough. It also can be thermally isolated in cases where the temperature difference across the wall is over 40 °C (Asprion and Kaibel, 2010). The use of a totally welded wall has been reported by Montz (Asprion and Kaibel, 2010), Linde (Dejanović, Matijašević and Olujic, 2010), Koch-Glitsch, and Sulzer Chemtech (Sander, 2007). Engineering know-how was in the meantime also developed and implemented successfully by Linde, Uhde, and UOP.

The strength added by the welded wall leads to high mechanical or thermal stress, while reducing the flexibility of the column design. Montz and TU Delft proved that vapor permeation did not affect the column performance as liquid leaks did (Dejanović, Matijašević, and Olujić, 2010). Thus, Montz propose using metal segments that provide a tight liquid seal by small deformations. BASF came up with a similar design using springs. Both solutions allow off-centered or even tilted walls and feature higher design tolerances for the column shell. The wall segments can be introduced through manholes, giving more space for tray or packing installation. Figure 3.12 shows the Montz proprietary system (Asprion and Kaibel, 2010). Since Sulzer has acquired technology rights from Montz, they could also commercialize this technology, but this needs to be confirmed independently.

Moreover, partially welded walls solutions have been adopted in industry and experimental setups. A thin layer of PTFE (Teflon®) in the wall sides can deal with unevenness in the column. The wall could be supported with a specially designed plate attached to the column shell (Barroso-Muñoz *et al.*, 2010a, 2010b). Similarly, PTFE gaskets seal-off the dividing wall, supported by bolt-bars welded to the column body. For certain applications, critical parts like feed or side draw sections could be welded – but this requires previous mechanical studies. Koch-Glitsch reports the use of this solution on a xylene column from Exxon-Mobil (Slade, Stober, and Simpson, 2006). Other companies, such as UOP, KBR (formerly Kellogg Brown & Root), and Udhe GmbH have built DWC with undisclosed wall designs (Dejanović, Matijašević, and Olujić, 2010). The wall positioning has an enormous influence on the vapor split. As the vapor

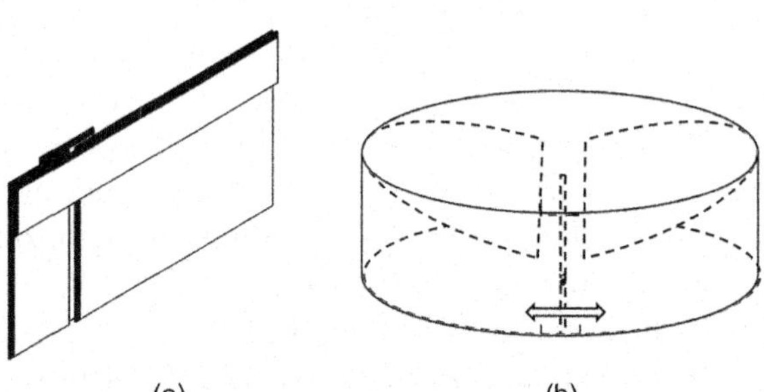

Figure 3.12 Montz proprietary non-welded wall system (a) and reflux splitter implemented by Sulzer (b)

rises freely, the pressure drop on both sides of the wall determines the vapor distribution within the two sections. Therefore, the wall design should provide the required vapor split during normal operation. In off-design conditions, controlling the liquid split allows maintenance of an acceptable vapor split (Slade, Stober, and Simpson, 2006). Vapor velocities reported in DWC can be up to 1.1 $m\,s^{-1}$, while liquid velocities reach 0.0057 $m\,s^{-1}$ (Barroso-Muñoz *et al.*, 2010a, 2010b).

Choosing the appropriate wall solution depends on the effect of vapor leaks during normal operation. The Montz proprietary non-welded system is expected to be more expensive, owing to its novelty and additional benefits. Welded walls could be less expensive, but they are not suitable for every application. The most cost-effective solution in non-demanding applications is the non-welded wall systems with PTFE seals, as this avoids paying for proprietary technology while dealing with certain column imperfections (Dejanović, Matijašević, and Olujić, 2010).

3.5.1 Liquid/Reflux Splitter

The liquid splitter defines the liquid load for each DWC section, controlling the column operation. Some DWC control schemes vary the liquid split fraction to determine the purity of the side product and minimize the energy use. Thus, it is important to have a reliable system for smooth operation. Based on their experience, Montz and BASF offer the most advanced solution (Asprion and Kaibel, 2010). It consists of a case with a dividing body, creating three chambers: feed, reflux, and pass-out. Positioning the dividing body directs the liquid to both sides of the column. The system can be driven by a pneumatic motor or by magnetic means for pressure-tight design. Montz provides corrosion-proof materials or treatments for susceptible parts (Asprion and Kaibel, 2010).

A simpler design employs pipes, valves, and gravity to split the liquid flow across the wall. Koch-Glitsch uses a chimney tray to collect the liquid above the wall. Then, pipes split it in equal proportion for both sides of the wall. The liquid control is exerted though a valve in a by-pass induced by gravity. Koch-Glitsch tested this design prior to its implementation in revamped xylenes columns at Exxon-Mobil (Slade, Stober, and Simpson, 2006). Other patented systems use two pipes to collect an uneven amount of liquid, splitting it across the walls by gravity. Again, control is exerted by valves. Different intake heights allow more liquid to flow towards one section when a valve in the other pipe is closed. Figure 3.13 reveals two possible configurations: with and without a siphon (Zuber,

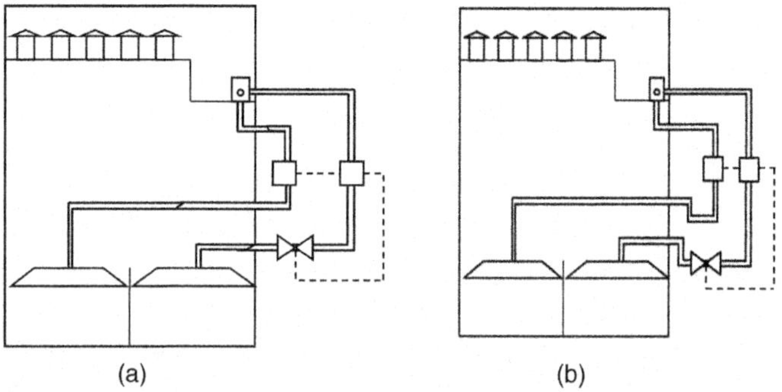

Figure 3.13 Possible configurations for pipes reflux splitter solutions: without siphon (a) and with siphon (b)

Gottlieb-Bachmann, and Ausner, 2008). Note that an additional control loop is needed, as the liquid level in the downcomer is not fixed. This might lead to poor accuracy and response time. Overall, the reflux splitter based on pipes features unlikely vacuum applicability.

Some experimental solutions include basic setups. For instance, liquid can be drawn to a side tank and then fed to both column sections by gravity. As this also requires separate pipes and control valves its industrial applicability is limited and the associated cost for implementation is relatively high. Additionally, a second control loop is required to prevent the side-tank flooding (Barroso-Muñoz *et al.*, 2010a, 2010b). Moreover, Sulzer employs the solution depicted in Figure 3.12b. A dividing body is located at the middle of the column, with the downcomers from the upper trays just above it (Sander, 2007). The dividing body is actuated mechanically or magnetically, depending on the complexity required. The design from Sulzer seems to be less prone to fouling than that from Montz – at the expense of a higher pressure drop. According to Sulzer, the reflux splitter performs well (Sander, 2007), but no industrial implementation has been reported so far.

Choosing one solution depends on the accuracy required for the liquid split, as well as the column operating conditions. For instance, the pipes reflux splitters are an option for standard processes conditions, probably presenting problems in fouling, sensitive, or vacuum applications. The Montz system might be more expensive, but it features vacuum capability and more accurate and tailor-made performance. As for the dividing wall, neither UOP nor KBR or Udhe disclose their reflux splitter solutions. Given its relationship with BASF, Linde might be using the Montz reflux splitter but no literature reports point this out (Asprion and Kaibel, 2010).

3.5.2 Column Internals

The internal mass transfer equipment for a DWC is similar to the ones for regular columns. Support beams are smaller as the column section is splitted. Moreover, the type of dividing wall is not limiting. Both welded and not-welded walls can operate correctly with trays and packing (Asprion and Kaibel, 2010). For packing columns, liquid bypassing can be an issue, as the DWC features more wall area than a conventional column. In the past, heating the walls was considered as a possible solution (Dejanović, Matijašević, and Olujić, 2010). However, the use of heavy-duty wall-wipers is an elegant proposal. In the Montz wall design, wall-wipers provide support for both the packing and the wall. The so-called self-centering packing avoids the wall contact (Asprion and Kaibel, 2010).

As mentioned before, the Montz wall design also facilitates installation of the systems, as the wall and packing are split into sections that pass through manholes. Montz claims that its removable man-way sections make installation simpler and faster. Other solutions also incorporate man-way removable segments that provide easy maintenance but increase the risk of leaks. Moreover, conventional construction techniques can also be applied with DWCs. For instance, Koch-Glitsch uses its FLEXILOCK® trays technology in DWCs (Slade, Stober, and Simpson, 2006), while similar non-bolted equipment is available from Sulzer Chemtech (Dejanović, Matijašević, and Olujić, 2010).

Trays are less sensitive to liquid bypassing through the wall. Thus, boxed downcomers have been applied, but not for all DWC units. In principle, any kind of tray can be installed in a DWC, but most applications so far use two-pass sieve trays. The use of multi-pass trays has not been reported yet. The cross-flow two-pass sieve trays depicted in Figure 3.14 have been patented, but their application was not reported. The downcomers can be parallel or perpendicular to the wall. The DWC of Sasol features the first mode, while Koch-Glitsch prefers the second mode (Dejanović, Matijašević, and Olujić, 2010; Slade, Stober, and Simpson, 2006). It is worth noting that the tray rings provide additional stiffness to the wall installation.

3.5.3 Equipment Sizing

As dividing-wall columns can be equipped with trays or random or structured packing, the same knowledge applies as for conventional distillation columns. However, peculiarities related to essential details

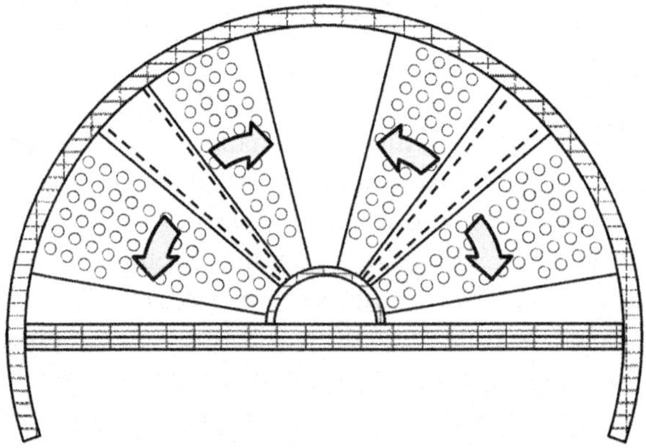

Figure 3.14 Cross-flow two-pass sieve trays

associated with dimensioning of these columns are not described in the open literature.

In an overview paper on DWC technology, Shah (2002) suggests using the sizing and costing procedures from Aspen Plus for conventional distillation columns. This was also applied by other authors in more recent studies (Errico *et al.*, 2009; Rangaiah, Ooi, and Premkumar, 2009). The authors considered the separated sections as parallel cylindrical columns with a given cross sectional area, which are further sized as regular columns and capital costs are determined accordingly using published correlations, to allow generation of total annual cost (TAC) as a basis for comparisons of alternatives. This academic approach can provide a good indication of the potential benefits. The actual know-how used in industry is still not in the public domain as it belongs to only a few companies that have successfully implemented the DWC technology in practice (Dejanović, Matijašević, and Olujić, 2010).

Notably, the basis for sizing a DWC is a converged column profile obtained by rigorous simulation, such that the liquid and vapor flow rates and properties are known for every single stage. In the conventional design, the sizing has to ensure stable column performance with regard to the liquid and vapor loads. For DWCs, the sizing has one additional objective: to ensure the desired vapor split across the wall. Unlike the liquid split that can be set precisely by an external device, the vapor split ratio is normally self-adjusting, depending on the amount of flow resistance imposed by the column internals in conjunction with liquid load – and corresponds to that needed to obtain the same pressure drop on both sides of the partition wall (Dejanović, Matijašević, and Olujić, 2010).

For structured packing columns a good indication of the mass transfer efficiency and the pressure drop can be obtained using the Delft model proposed by Olujić *et al.* (2004). The most distinctive feature of this model is that it does not require any packing-specific empirical parameter, just the accurate dimensions of the corrugated sheets used to build a packing element. This allows maximum flexibility regarding variations in specific geometric area, corrugation inclination angle, as well as bends at lower or both ends as encountered in high-performance structured packing. This is important because it allows options to arrange packed beds on opposite sides of the partition wall to ensure – at fixed liquid splits – the proper pressure drop (equal on both sides), hence the proper split of vapor flows to occur just below the partition wall (Dejanović, Matijašević, and Olujić, 2010).

Note that in demanding separations requiring installation of multiple beds of packing the quantity of the installed liquid collectors and distributors can be considerable, contributing significantly to the pressure drop across the column. However, this can be predicted with good confidence using a simple method proposed recently by Rix and Olujić (2008). A particular feature of this method, of importance for designing packed DWC units, is the possibility of tuning the pressure drop of the liquid redistribution sections. This method allows establishment of the free area that will generate the desired amount of pressure drop, as a means of ensuring the required vapor flow rates in the sections separated by the dividing wall (Rix and Olujić, 2008; Dejanović, Matijašević, and Olujić, 2010).

In the absence of a dedicated method for tray columns, the best approach is to make preliminary hydraulic design of a DWC as a combination of several sieve tray columns (Dejanović, Matijašević, and Olujić, 2010), utilizing a well-proven method elaborated in detail by Stichlmair and Fair (1998).

Earlier DWC designs were limited because the wall had to be placed in the middle, dividing the column into two sections of same shape and equal cross section size. This was due to the wall being welded to the column shell, and placing it in the middle was required to minimize the mechanical stress. However, the research of TU Delft and J. Montz showed that the partition wall does not need to be welded, because a fast equalization of pressures on two sides of the partition wall minimizes the occurrence and durability of mechanical leaks of both phases from one side to the other (Dejanović, Matijašević, and Olujić, 2010). Certainly, a pronounced temperature gradient across the partition wall could induce a vapor leakage by natural convection, but this can be avoided by adopting adequate practical provisions. The non-welded wall was recognized as an important breakthrough in DWC technology because it allows the easy revamp of existing columns with less installation time and less investment costs, while the ability to place the wall off-center extends the range of applications with respect to composition, relative volatilities, and thermodynamic condition of mixture feeds. This has been implemented with success in many columns equipped with structured packing, and several patents are described in the literature (Dejanović, Matijašević, and Olujić, 2010). Moreover, adopting non-welded walls enables the installation of multiple walls, which effectively means placing more conventional columns into only one shell. So far, this has greatly improved the area of packed DWC application, and it is expected to be widely used in tray applications as well (Dejanović, Matijašević, and Olujić, 2010).

Consequently, each section of the column can be sized using commercially available software (e.g., Aspen Plus, HYSYS, ChemCAD, PRO/II) for the particular liquid and vapor loads. In this way an equivalent cylindrical cross section area needed to accommodate the required vapor and liquid loads can be easily calculated. The latitudinal position of the wall is then set in such way that each column section has an equivalent cross-section area to that calculated by individual sizing. This approach was also used and reported in more recent publications about DWC technology (Dejanović, Matijašević, and Olujić, 2010; Kiss and Ignat, 2012; Kiss and Suszwalak, 2012; Kiss *et al.*, 2013).

The number of stages corresponding to the height of the partition wall can differ on the two sides of the wall and this represents an additional design issue. Extending the wall to include more stages may help in one respect but it will certainly be at the cost of a shift in composition on the other side – which might become detrimental to the overall performance (Dejanović, Matijašević, and Olujić, 2010).

3.5.4 Constructional Aspects

Basically, there is no technical difference between DWC units and conventional columns, providing that the internals designed for a DWC comply with the increased process and mechanical design requirements. Regarding the column dimensions, the diameters of industrial DWC units cover a wide range: 0.3–4 m for units equipped with structured packing, and up to above 6 m for tray columns. Owing to their robust construction, tray columns are in fact easier to deal with. Moreover, the dividing wall can also be used to increase the mechanical strength of the construction. For example, the Sasol columns contain two-pass sieve trays with flow paths in parallel to the partition wall, while the tray rings ensure additional mechanical stiffening of the partition wall (Becker, Godorr, and Kreis, 2001). Importantly, by shortening (halving) the distances between opposite walls the dimensions of support beams of the trays can be kept smaller (Dejanović, Matijašević, and Olujić, 2010).

Remarkably, the welding procedure is still the most critical fabrication step – particularly when structured packing is employed – because the partition wall must remain absolutely flat. In addition, a DWC with a welded wall implies two half-cylindrical columns of equal cross sectional area – meaning that in case of unequal loads one side will always be oversized and, hence, operating at a vapor load below the optimal value. This implies less capital saving than in case of having a proper cross section area on both sides of the partition wall. As previously mentioned, the use of a non-welded wall solves this problem at its source in the most appropriate way (Dejanović, Matijašević, and Olujić, 2010).

The temperature difference across the wall may become a serious concern when separations of wide boiling mixtures are considered, with proper insulation being essential to maintain the desired separation efficiency and ensure the proper functioning of a DWC. An effective solution for large temperature gradients in packed columns – that can be in opposite directions in the upper and lower sections due to composition changes imposed by the separation – is a partition wall made of two metal sheets separated by a gas space in between, as recommended in a BASF patent by Kaibel, Stroezel, and Pfeffinger (1998). This insulating partition wall can be sealed (preferred) or flushed with an inert gas. For tray columns, a similar but simpler solution was proposed in a UOP patent by Stacey (2003) – a separate vertical separation isolation wall is placed at a short distance from the partition wall to thermally isolate downcomers from the partition wall, including downcomers on both sides of the wall.

In an Air Products and Chemicals patent by Kovak (2007), the radial-crossflow trays for a DWC with one downcomer are placed centrally and two downcomers are placed adjacent to partition wall, with liquid flowing over two oppositely oriented active areas on each subsequent tray alternately towards and from the partition wall. In another patent by Agrawal (2001b) an uncommon feature is the introduction of an additional horizontal partition wall in the central zone of the column, which allows the obtainment of two side products, a bottom product from the upper part and an overhead product from the lower part of the column. In this way, three distillation columns are effectively placed within one shell, and two columns placed above each other on the side draw side and separated by a horizontal partition wall, each provided with its own reboiler and condenser, respectively.

The introduction of partially vaporized or overheated liquid feeds requires special provisions to allow the separation and subsequent smooth delivery of vapor and liquid to the packing beds above and below, respectively. Jansen *et al.* (1996) from Montz have designed an effective liquid flashing box containing a demister at the vapor outlet side that can be fixed on the partition wall opposite to the inlet of the column feed (Dejanović, Matijašević, and Olujić, 2010).

Note that, owing to the additional partition wall, a packed DWC contains more wall area than a conventional column. Therefore, bypassing of both liquid and vapor flows is a serious design concern, as this is detrimental to the separation efficiency. Direct contact of packing elements with the column walls must be avoided to prevent excessive wall flow of liquid – this is usually accounted for by providing the packing with effective wall wipers. Particularly in the case of pronounced unevenness (i.e., poor column diameter tolerances), the wipers need to be sufficiently long and strong, while great care needs to be taken to adjust them properly during installation (Dejanović, Matijašević, and Olujić, 2010). For packed DWC units it is essential to have a good working system of wall wipers and to adjust them properly during installation. Several patents by BASF and Montz propose various constructive solutions (Kaibel, 1999; Kaibel, Stroezel, and Pfeffinger, 1999; Rust and Kaibel, 2005; Jansen *et al.*, 2004).

The flexibility provided by a non-welded wall enabled a tailor-made approach to assembling DWCs, to meet special process requirements. Moreover, the non-welded wall makes feasible building dividing-wall columns for the separation of four and more component mixtures into four or more pure products. However, the first (Kaibel) column designed for this purpose employs only one partition wall with different positions in upper and lower section, which is thermodynamically less efficient than a complex multiple partition wall configuration – as shown in Figure 3.5b. The number of packed beds is double in complex configurations, thus implying also additional space for the necessary liquid redistribution sections and the need to assemble at least three separate partition walls in different locations – which clearly introduces installation related challenges. In this case and process, the degree of complexity is indeed much larger, and so mechanical design and construction methods need to be well established prior to making an attempt in this direction. In addition, a reliable sizing procedure is required to allow cost estimates and enable a fair comparison of alternatives.

Notably, currently, the mechanical design and installation have reached a stage in which the costs associated with manufacturing and installation of rather complex internal structures are only 20% higher than that of a conventional column, while the time needed for installation is roughly the same (Dejanović *et al.*, 2010, 2011).

3.6 CASE STUDY: SEPARATION OF AROMATICS

The separation of aromatics in a platforming process delivers five product streams that are obtained from a conventional direct sequence of four distillation columns. This case study investigates new separation alternatives using combinations of dividing-wall columns (DWC) as well as conventional and Kaibel distillation columns. Remarkably, these new separation sequences using intensified distillation technology are able to separate all five products (lights, benzene, toluene, xylene, and heavies) at high purity levels, using only two distillation columns as main equipment.

Hereby, we use reported compositions but each product cut was lumped into its key component (Dejanović *et al.*, 2011). Accordingly, the feed stream contains 12.32 wt% lights (e.g., *n*-pentane), 20.42 wt% benzene, 26.87 wt% toluene, 27.36 wt% xylene, and 13.03 wt% heavies (e.g., trimethyl-benzene) – for convenience noted here as L, B, T, X, and H, respectively. For the alternative separation schemes investigated here the feed basis is 12 500 kg h^{-1} (equivalent to a production rate of 100 ktpy) and the target product purity for each product cut is min. 99.5 wt%. Steady-state simulations were carried out in AspenTech Aspen Plus using the rigorous RADFRAC unit enhanced with the RateSep (rate-based) model. Owing to the nature of the components involved in the separation, NRTL–Redlich–Kwong was selected as the most adequate property model (Kiss and Rewagad, 2011).

All the classic and novel alternatives described hereafter were optimized in terms of minimal energy demand using the sensitivity analysis tool combined with the sequential quadratic programming (SQP) method available in Aspen Plus (AspenTech, 2010). Several optimization variables are used: total number of stages, feed-stage location, side-stream location, location and length of the dividing wall, reflux ratio, and liquid and vapor split (Kiss and Suszwalak, 2012).

Figure 3.15 illustrates the possible distillations sequences, including the naming convention, namely, grouping the initials of the product cuts obtained in each column (Kiss *et al.*, 2013). To choose the most energy efficient alternative configurations based on DWC and Kaibel columns, we use the concept of minimum vapor flow (V_{min}). Each point of the V_{min} diagram (D/F, V/F) – corresponding to a sharp split between

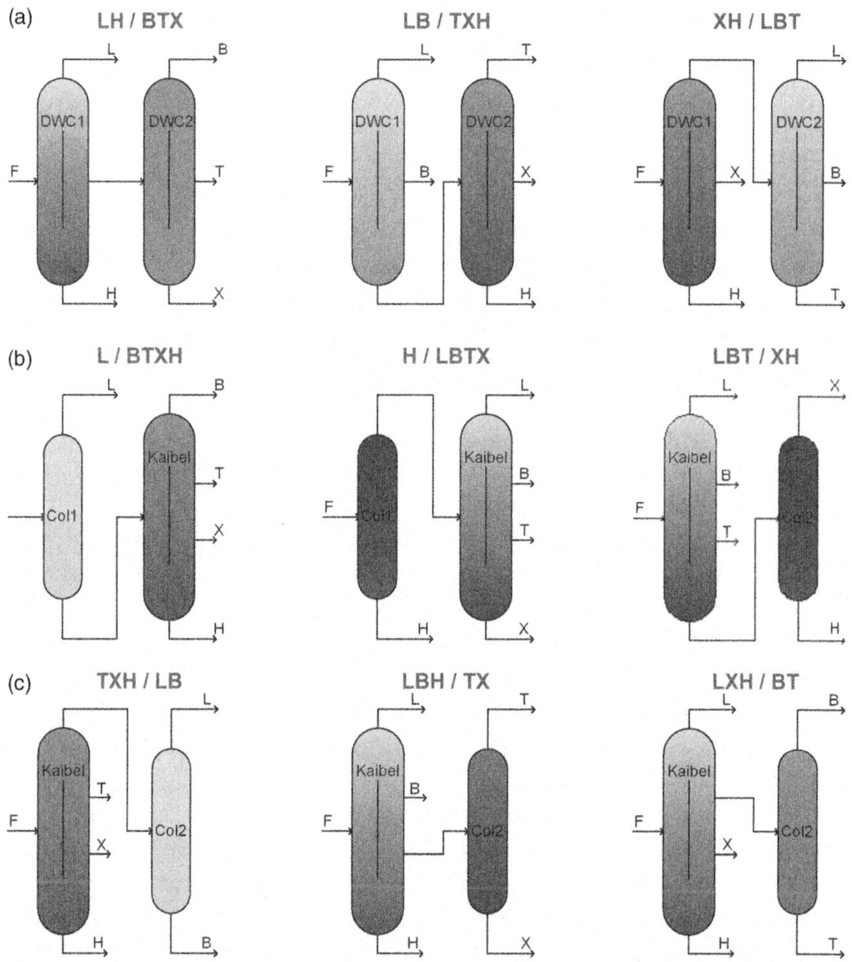

Figure 3.15 Alternative configurations based on two DWC units (a) and combinations of conventional and Kaibel columns (b) and (c)

two components of the feed mixture – was computed by using a series of rigorous binary distillation calculations as described in the literature (Halvorsen and Skogestad, 2011). All rigorous simulations were performed using the Aspen Plus process simulator and NRTL–Redlich–Kwong property model. Sections with an infinite number of stages were assumed (e.g., in practice, four times the minimum number of stages is sufficient) and a maximum recovery of 0.1% was specified for the light key component in the bottom stream, and the heavy key component in the distillate stream, respectively. The minimum vapor flow required for the separation in a DWC is given by the highest peak in the V_{min} diagram, as long as the prefractionator performs the easiest separation (i.e., lower peak in the diagram).

However, this method for plotting V_{min} diagrams is not so straightforward for a Kaibel column because – compared to a DWC – the prefractionator is not operated at the preferred split (i.e., easiest separation), but it performs a sharp split between two key components (Dejanović *et al.*, 2011; Halvorsen and Skogestad, 2011; Ghadrdan, Halvorsen, and Skogestad, 2011). Thus the minimum vapor flow at the top of the prefractionator is actually higher as compared to a DWC unit. Nevertheless, Dejanović *et al.* (2011) describe another method to calculate the key points of the V_{min} diagram for a Kaibel column by performing simulations using a three-column configuration (e.g., distributed sequence).

Table 3.1 lists the energy requirements for all sequences considered, while Figure 3.16 plots the V_{min} diagrams for most relevant sequences (Kiss *et al.*, 2013). Note that in the plotted V_{min} diagrams only the important points corresponding to the desired sharp splits in each configuration are shown. Based on the results provided by the V_{min} diagrams (Table 3.1) the LB/TXH and L/BTXH configurations are indicated as the best solutions in terms of minimum energy. These results should not be really surprising, as plotting the V_{min} diagrams for the two-DWC configurations one can see that the T/X and X/H are the most difficult separations (i.e., largest peaks), thus requiring higher energy amounts for a sharp split. Since the minimum energy requirements in a DWC are given by the highest peak in the V_{min} diagram, in the case of a

Table 3.1 Ideal energy requirements for each distillation sequence, based on V_{min} diagrams

	V/F							
	Col1	Col2	Col3	Col4	DWC1	DWC2	Kaibel	Total
Direct sequence	0.3194	0.8065	1.2552	1.7518	—	—	—	4.13
LH/BTX	—	—	—	—	1.3909	1.2778	—	2.67
LB/TXH	—	—	—	—	0.7508	1.5011	—	2.25
XH/LBT	—	—	—	—	1.2843	1.0667	—	2.35
L/BTXH	0.3307	—	—	—	—	—	1.7524	2.08
H/LBTX	1.2812	—	—	—	—	—	1.4940	2.78
LBT/XH	—	1.7508	—	—	—	—	1.2550	3.01
TXH/LB	—	0.6743	—	—	—	—	1.7518	2.43
LBH/TX	—	1.4311	—	—	—	—	1.5033	2.93
LXH/BT	—	1.1895	—	—	—	—	1.7515	2.94

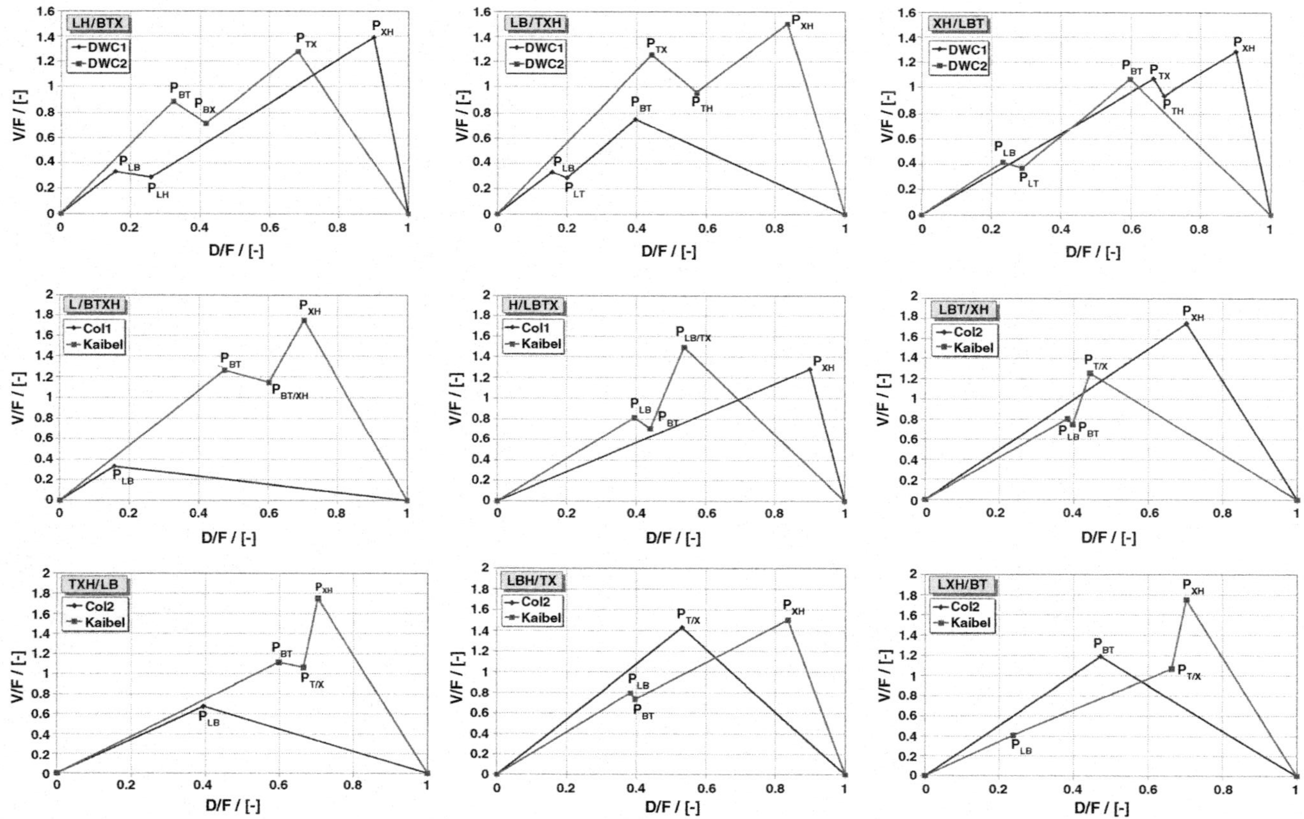

Figure 3.16 $V_{\min}$ diagrams of all configurations based on two DWC units and combinations of conventional and Kaibel columns

multicomponent feed mixture, it is essential to group the difficult separations between two components in the same column to achieve the maximum energy efficiency. The same statement is also valid for the Kaibel configurations. Compared to conventional heuristic rules, V_{min} diagrams provide very accurate results in the case of selecting the most energy efficient thermally coupled separation sequences.

A sequence of two RADFRAC units was used to model each DWC or Kaibel column since no off-the-shelf DWC unit is available in current commercial process simulators. This configuration is thermodynamically equivalent to a DWC, as long as the temperature difference on both sides of the wall indicates that there is no heat transfer between the two sides. The condenser pressure and the stage pressure drop were again set to 1 bar and 10 mbar, respectively. The SQP optimization method and the effective sensitivity analysis tool from Aspen Plus® were also used in the optimization of the DWC and Kaibel columns. The objective of the optimization is to minimize the total reboiler duty required, as follows:

$$\begin{aligned}&\text{Min}(Q) = f(N_T, N_F, N_{SS}, N_{DWS}, N_{DWC}, V, RR, r_V, r_L)\\&\text{Subject to}\, \boldsymbol{y}_m \geq \boldsymbol{x}_m\end{aligned} \tag{3.1}$$

where the optimization parameters used here are total number of stages (N_T), feed location (N_F), side-draw stage (N_{SS}), wall size (N_{DWS}) and location (N_{DWC}), boil-up rate (V), reflux ratio (RR), and liquid and vapor split (r_L and r_V), while $\boldsymbol{y}_m$ and $\boldsymbol{x}_m$ are the vectors of the obtained and required purities for the m products. Note that in order to determine the optimal ratio between the energy cost and the number of stages an additional objective function was used, Min N_T $(RR + 1)$, which approximates very well the minimum total annual cost (TAC) of a conventional distillation column, according to the procedure described by Dejanović *et al.* (2011).

Table 3.2 provides the resulting optimal design parameters for the two-DWC system (LB/TXH), while Figure 3.17 shows the temperature and the composition profiles (Kiss *et al.*, 2013). Benzene distributes, in principle, towards the top and bottom of the dividing-wall, but it accumulates on the other side and subsequently it is collected as side

Table 3.2 Design parameters for the optimal two-DWC system LB/THX

	Design parameter									
	N	N_{PF}	N_{top}	N_{btm}	N_{feed}	N_{side}	RR	r_L	r_V	Q_{reb} (kW)
DWC1	37	18	7	12	17	25	9.1	0.09	0.24	1725
DWC2	51	28	9	14	23	28	4.1	0.28	0.47	1734

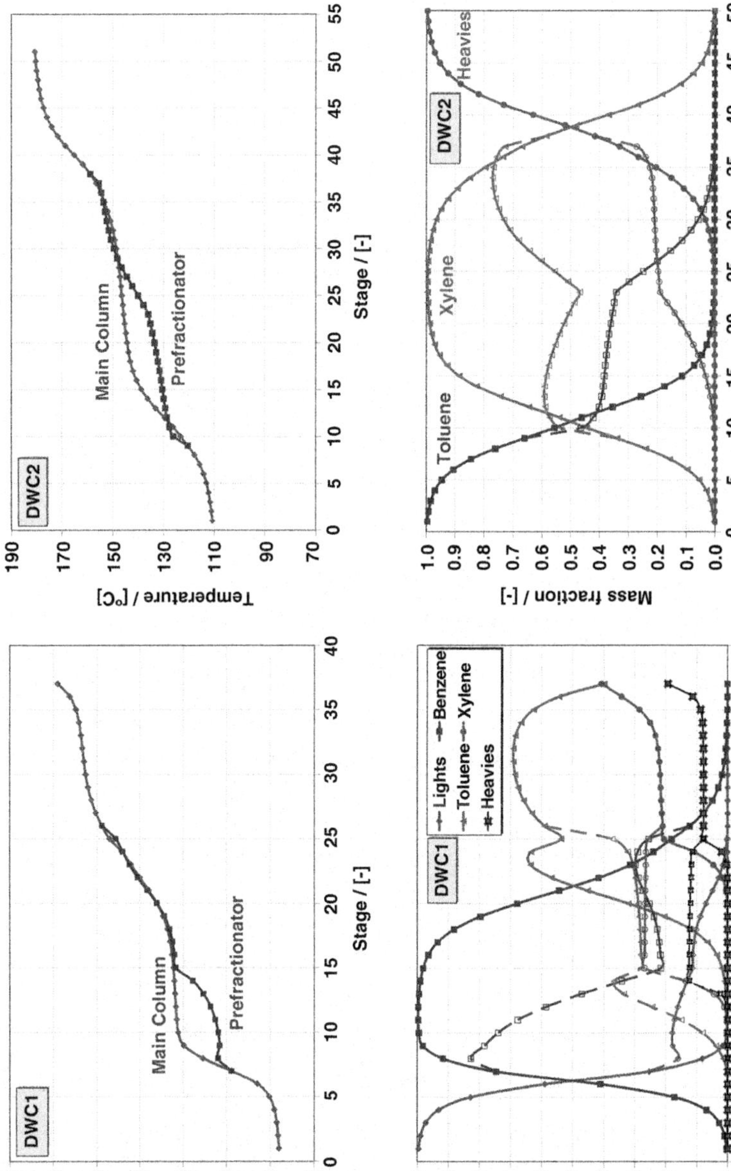

Figure 3.17 Temperature and composition profiles along the two DWCs of the LB/TXH system

stream. On the other hand, DWC2 clearly exploits all the advantages of the wall, delivering high purity toluene at the top, xylene as side stream and heavies at the bottom. Compared to the conventional sequence, the two-DWC system achieves 14% energy savings and a notable reduction of 20% in the total number of stages required.

The proven efficiency of the DWC is also featured by a Kaibel column, as they both share the same structural design. Thus, the modeling and optimization processes for the Kaibel column in the L/BTXH system are similar to those described for a DWC unit (Yildirim, Kiss, and Kenig, 2011; Kiss and Suszwalak, 2012). Figure 3.18 (Kiss *et al.*, 2013) illustrates the temperature and composition profiles along the optimally designed Kaibel column consisting of 65 stages, with a 39-stage prefractionator located between stages 4 and 10, feed on stage 28, toluene and xylene side-draw from stage 20 and 36, liquid and vapor split of 0.44 and 0.64 respectively, reflux ratio 8.5, and reboiler duty of 2761 kW. Note that the overall results for the Kaibel system also include those for the first column of the direct sequence – Col1 (Kiss *et al.*, 2013). The L/BTXH scheme achieves an important 17% increase in energy efficiency compared to the conventional sequence, while requiring 24% less total number of stages.

As clearly illustrated by Table 3.3 (Kiss *et al.*, 2013), the novel proposed schemes are less expensive to operate than the conventional direct sequence. Additionally, higher energy savings are possible when the conventional system is not an optimal one – as is typically the case in existing industrial plants. Notably, the novel separation schemes achieve only marginal investment savings when compared to the conventional sequence. One reason is that the internals installation costs for DWC or Kaibel columns was considered to be 20% higher than those of their conventional counterparts, due to greater construction complexity (Dejanović *et al.*, 2011). Moreover, although the overall number of stages is lower, and the DWC and Kaibel columns have increased diameter and height. Consequently, the proposed distillation schemes are more appropriate for building new plants although they could also be considered for revamping projects and thus re-using existing columns.

3.7 CONCLUDING REMARKS

Ever since its first industrial application, DWC has moved from a conceptual to a proven technology, steadily growing in number and size of applications (Dejanović, Matijašević, and Olujić, 2010). Numerous industrial applications are known today, mainly concerning separations of

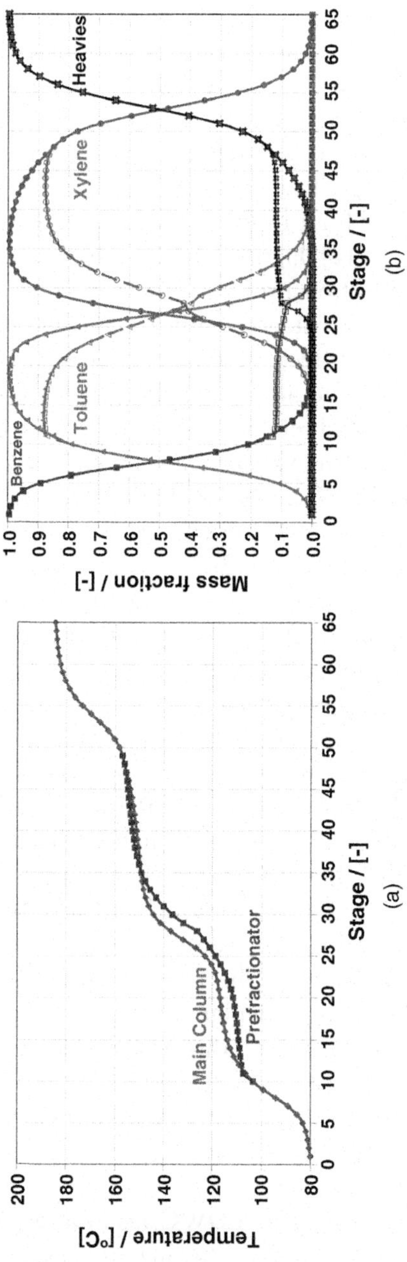

Figure 3.18 Temperature (a) and composition (b) profiles along the Kaibel column of the L/BTXH system

Table 3.3 Comparison of the studied systems in terms of key performance parameters (100 ktpy rate)

Parameter (units)	Direct sequence	LB/THX	L/BTHX
Total energy requirements (kW)	4030	3460	3346
Total number of stages	111	88	84
Lights recovery (%)	99.9	99.9	99.9
Benzene recovery (%)	99.6	99.6	99.5
Toluene recovery (%)	99.6	99.5	99.5
Xylene recovery (%)	99.5	99.5	99.6
Heavies recovery (%)	99.2	99.3	99.5
Total investment cost (TIC) (k$)	1906	1733	1886
Total operating cost (TOC) (k$ y^{-1})	939	741	683
Total annual cost (TAC) (k$ y^{-1})	1129	914	872
CO_2 emissions (kg h^{-1})	564	484	468

ternary mixtures. The development and implementation efforts focus nowadays on the separation of more than three components or applications of extractive, azeotropic, and reactive distillation in a DWC (Yildirim, Kiss, and Kenig, 2011).

To maximize the potential energy savings, multi-partitioned DWC configurations could be employed, but these are still at the stage of theoretical studies and so far no attempt at practical or experimental validation has been reported (Dejanović, Matijašević, and Olujić, 2010). Moreover, these configurations might be demanding or expensive, meaning that viable designs will not necessarily be the most efficient ones. Nevertheless, the recent best practice shows that adopting the non-welded partition wall design approach minimizes the difficulties associated with the construction of complex columns, at least for those equipped with structured packing (Dejanović, Matijašević, and Olujić, 2010; Yildirim, Kiss, and Kenig, 2011).

Despite its many advantages, one should keep in mind that DWC has some limitations as well. The complete separation sequence carried out in a DWC is operated at the same pressure, which can reduce its cost effectiveness and practical applicability. For example, one cannot combine, in a DWC unit, a distillation taking place at vacuum with one at ambient or elevated pressure. Another potential drawback is that a DWC unit is larger (diameter and height) than any single column of a conventional (e.g., direct or indirect) separation sequence – which in certain cases can render the application of DWC impractical. In addition, although DWC provides significant energy savings, the energy required has to be supplied and rejected at the highest (e.g., in reboiler) and the lowest (e.g., in condenser) temperature levels. This can reduce the overall economy of the column, since more expensive utilities must be used (Dejanović, Matijašević, and Olujić, 2010). Moreover, the application of heat pumps – transferring heat from the condenser to reboiler – is drastically reduced. Although more difficult to control than conventional columns, recent studies and the industrial experience indicate that the control of DWC units is in fact satisfactory (Kiss and Bildea, 2011). Dynamic simulations can be used additionally to provide insight into the dynamic behavior of the DWC system, and give valuable guidance for choosing the appropriate control strategy.

In terms of DWC design, there are sufficient short-cut and detailed methods described in the literature. Short-cut designs are based mostly on the Fenske–Underwood–Gilliland–Kirkbride equations and were proven suitable for *ideal mixtures*. However, detailed methods are required for final designs in any case. The employment of V_{min} diagrams is especially useful when setting the stage for rigorous simulations, as these can provide good initial values for the liquid and vapor splits. Regrettably, the available process simulators still do not include a DWC as a separate model, although the MultiFrac unit of Aspen Plus comes very close (e.g., it is usable for the thermodynamic equivalent Petlyuk configuration). As an alternative, one can use various sequences of conventional column modules to replicate a particular DWC configuration. However, there are certain difficulties regarding the initialization and convergence of these decomposed models. Experience shows that generally it is more beneficial to use models with as little as possible column sections to reduce the convergence time and increase stability and robustness of the simulation (Dejanović, Matijašević, and Olujić, 2010).

Dimensioning of DWC is still the proprietary knowledge of very few industrial companies. This is important not only from a hydraulic design viewpoint but also from a process control point of view. The main reason is that the vapor split is established spontaneously depending on the flow resistances on each side of the wall. This means that vapor split is effectively set during the dimensioning phase and it cannot be changed later. The pressure drop is a natural means of establishing cross sectional vapor distribution. Therefore, it is of great importance to have reliable predictive models that are versatile enough to account properly for the adjustments needed in the geometry of packing (e.g., corrugation dimensions and angles) or/and trays (e.g., free area, weir height) in order to arrive at the required vapor flows that will ensure – in combination with fixed liquid flows – an equal pressure drop in each of the parallel sections of the DWC. A real challenge in this respect is designing an internal configuration for a four and more products column with multiple partitions, which appears to be so demanding that it has not been attempted yet in practice – although there has been some progress in this direction (Dejanović, Matijašević, and Olujić, 2010; Yildirim, Kiss, and Kenig, 2011).

Experimental research on DWCs is presently aimed at acquiring knowledge that supports further development and implementation of extractive, azeotropic, and reactive distillation applications. For example, a reactive DWC can be seen as an ultimate efficiency approach due to the integration of reactive distillation with DWC technology. However, relevant research is still required to provide the necessary experimental evidence to further develop and validate advanced predictive models. Regarding the number and variety of industrial applications, DWC can already be considered a success story concerning process intensification in distillation. Being a genuine representative of sustainable distilling, DWC will certainly develop into a standard type of distillation column in the near future (Dejanović, Matijašević, and Olujić, 2010; Yildirim, Kiss, and Kenig, 2011).

REFERENCES

Agrawal, R. (2001a) Processes for multicomponent separation, US Patent No. 6286335.

Agrawal, R. (2001b) Process for separation of multicomponent fluids using a multizone distillation column, describing a dividing wall column configuration with additional horizontal partition on product side, with side reboiler and side condenser above and below the horizontal partition wall, respectively, US Patent No. 6250106 B1.

Amminudin, K.A., Smith, R., Thong, D.Y.C., and Towler, G.P. (2001) Design and optimization of fully thermally coupled distillation columns. Part 1: Preliminary design and optimization methodology. *Chemical Engineering Research and Design*, *79*, 701–715.

Amminudin, K.A. and Smith, R. (2001) Design and optimization of fully thermally coupled distillation columns. Part 2: Application of dividing wall columns in retrofit. *Chemical Engineering Research and Design*, *79*, 716–724.

Aspen Technology, Aspen Plus: User guide - Volume 1 & 2, 2010.

Asprion, N. and Kaibel, G. (2010) Dividing wall columns: Fundamentals and recent advances. *Chemical Engineering and Processing: Process Intensification*, *49*, 139–146.

Barroso-Muñoz, F.O., Hernandez, S., Hernandez-Escoto, H. *et al.* (2010a) Experimental study on pressure drops in a dividing wall distillation column. *Chemical Engineering and Processing*, *49*, 177–182.

Barroso-Muñoz, F.O., Hernandez, S., Segovia-Hernandez, J.G. *et al.* (2010b) Implementation and operation of a dividing wall distillation column. *Chemical Engineering Technology*, *34* (5), 746–750.

Becker, H., Godorr, S., and Kreis, H. (2001) Partitioned distillation columns - why, when & how. *Journal of Chemical Engineering*, *108*, 68–74.

Christiansen, A.C., Skogestad, S., and Liena, L. (1997) Complex distillation arrangements: Extending the Petlyuk ideas. *Computers & Chemical Engineering*, *21*, 237–242.

Dejanović, I., Matijašević, Lj., and Olujić, Z. (2010) Dividing wall column - A breakthrough towards sustainable distilling. *Chemical Engineering and Processing: Process Intensification*, *49*, 559–580.

Dejanović, I., Matijašević, L., Halvorsen, I.J. *et al.* (2011) Designing four-product dividing wall columns for separation of a multicomponent aromatic mixture. *Chemical Engineering Research and Design*, *89*, 1155–1167.

Dejanović, I., Matijašević, Lj., and Olujić, Z. (2011) An effective method for establishing the stage and reflux requirement of three-product dividing wall columns. *Chemical and Biochemical Engineering Quarterly*, *25*, 147–157.

Dünnebier, G. and Pantelides, C.C. (1999) Optimal design of thermally coupled distillation columns. *Industrial and Engineering Chemistry Research*, *38*, 162–176.

Errico, M., Tola, G., Rong, B.-G. *et al.* (2009) Energy saving and capital cost evaluation in distillation column sequences with a divided wall column. *Chemical Engineering Research and Design*, *87*, 1649–1657.

Ghadrdan, M., Halvorsen, I.J., and Skogestad, S. (2011) Optimal operation of Kaibel distillation columns. *Chemical Engineering Research and Design*, *89*, 1382–1391.

Grossmann, I.E., Aguirre, P.A., and Barttfeld, M. (2005) Optimal synthesis of complex distillation columns using rigorous models. *Computers and Chemical Engineering*, *29*, 1203–1215.

Halvorsen, I.J. (2001) Minimum energy requirements in complex distillation arrangements, PhD Thesis, Norwegian University of Science and Technology, Trondheim.

Halvorsen, I.J. and Skogestad, S. (2003a) Minimum energy consumption in multicomponent distillation. 1. Vmin diagram for a two-product column. *Industrial and Engineering Chemistry Research*, *42*, 596–604.

Halvorsen, I.J. and Skogestad, S. (2003b) Minimum energy consumption in multicomponent distillation. 2. Three-product Petlyuk arrangements. *Industrial and Engineering Chemistry Research*, *42*, 605–615.

Halvorsen, I.J. and Skogestad, S. (2003c) Minimum energy consumption in multicomponent distillation. 3. More than three products and generalized Petlyuk arrangements. *Industrial and Engineering Chemistry Research*, *42*, 616–629.

Halvorsen, I.J. and Skogestad, S. (2011) Energy efficient distillation. *Journal of Natural Gas Science and Engineering*, *3*, 571–580.

Harmsen, J. (2010) Process intensification in the petrochemicals industry: Drivers and hurdles for commercial implementation. *Chemical Engineering and Processing*, *49*, 70–73.

Huang, K., Wang, S.-J., Shan, L. *et al.* (2007) Seeking synergistic effect - A key principle in process intensification. *Separation and Purification Technology*, *57*, 111–120.

Jansen, H., Leben, J., Rietfort, T., and Zich, E. (1996) Column for overheated liquids. US Patent No. 5580425.

Jansen, H., Leben, J., Rietfort, T., and Zich, E. (2004) Column partition, describing segments with plug and clamp connectors, for assembling a partition wall without fixing it to column walls, US Patent No. 6770173 B1.

Jobson, M. (2005) Dividing wall distillation comes of age. *The Chemical Engineer*, *766*, 30–31.

Kaibel, G. (1987) Distillation columns with longitudinal subdivision. *Chemie Ingenieur Technik*, *59*, 533–533.

Kaibel, G., Stroezel, M., and Pfeffinger, J. (1998) Distillation column for separating a liquid into a plurality of pure fractions, US Patent No. 5785819.

Kaibel, G. (1999) Distillative separation of mixtures and apparatus for this purpose, with different provisions to ensure certain liquid load in the wall zone, US Patent No. 5897748.

Kaibel, G., Stroezel, M., and Rheude, U. (1999) Dividing wall column for continuous fractionation of multicomponent mixtures by distillation, with a detachable partition wall and elastic wall wipers that eliminate liquid and vapour bypassing and fix the wall, US Patent No. 5914012.

Kaibel, B., Jansen, H., Zich, E., and Olujić, Ž. (2006) Unfixed dividing wall technology for packed and tray distillation columns, in *Distillation & Absorption Symposium 2006*, IChemE, Rugby, pp. 252–266.

Kiss, A.A. and Bildea, C.S. (2011) A control perspective on process intensification in dividing-wall columns. *Chemical Engineering and Processing: Process Intensification*, *50*, 281–292.

Kiss, A.A. and Rewagad, R.R. (2011) Energy efficient control of a BTX dividing-wall column. *Computers & Chemical Engineering*, *35*, 2896–2904.

Kiss, A.A., Segovia-Hernandez, J.G., Bildea, C.S. *et al.* (2012) Reactive DWC leading the way to FAME and fortune. *Fuel*, *95*, 352–359.

Kiss, A.A. and Suszwalak, D.J.-P.C. (2012) Enhanced bioethanol dehydration by extractive and azeotropic distillation in dividing-wall columns. *Separation & Purification Technology*, *86*, 70–78.

Kiss, A.A. and Ignat, R.M. (2012) Enhanced methanol recovery and glycerol separation in biodiesel production – DWC makes it happen. *Applied Energy*, *99*, 146–153.

Kiss, A.A., Ignat, R.M., Flores Landaeta, S.J., and de Haan, A.B. (2013) Intensified process for aromatics separation powered by Kaibel and dividing-wall columns. *Chemical Engineering and Processing: Process Intensification*, article in press. Doi: 10.1016/j.cep.2012.06.010.

Kiss, A.A. and Suszwalak, D.J.-P.C. (2012) Innovative dimethyl ether synthesis in a reactive dividing-wall column. *Computers & Chemical Engineering*, *38*, 74–81.

Kovak, K.W. (2007) Radial-crossflow distillation trays for divided wall column applications, US Patent No. 7234691 B2.

Monro, D.A. (1938) Fractionating apparatus and method of fractionation, US Patent No. 2134882.

Mueller, I. and Kenig, E.Y. (2007) Reactive distillation in a dividing wall column – rate-based modeling and simulation. *Industrial & Engineering Chemistry Research*, *46*, 3709–3719.

Mueller, I., Pech, C., Bhatia, D., and Kenig, E.Y. (2007) Rate-based analysis of reactive distillation sequences with different degrees of integration. *Chemical Engineering Science*, *62*, 7327–7335.

Muralikrishna, K., Madhavan, V.K.P., and Shah, S.S. (2002) Development of dividing wall distillation column design space for a specified separation. *Chemical Engineering Research and Design*, *80*, 155–166.

Olujić, Ž., Kaibel, B., Jansen, H. *et al.* (2003) Distillation column internals/configurations for process intensification. *Chemical and Biochemical Engineering Quarterly*, *17*, 301–309.

Olujić, Ž., Behrens, M., Colli, L., and Paglianti, A. (2004) Predicting the efficiency of corrugated sheet structured packings with large specific surface area. *Chemical and Biochemical Engineering Quarterly*, *18*, 89–96.

Parkinson, G. (2007) Dividing-wall columns find greater appeal. *Chemical Engineering Progress*, *103*, 8–11.

Petlyuk, F.B., Platonov, V.M., and Slavinskii, D.M. (1965) Thermodynamically optimal method for separating multicomponent mixtures. *International Chemical Engineering*, *5*, 555–561.

Petlyuk, F.B. (2004) *Distillation Theory and its Application to Optimal Design of Separation Units*, Cambridge University Press, Cambridge.

Rangaiah, G.P., Ooi, E.L., and Premkumar, R. (2009) A simplified procedure for quick design of dividing-wall columns for industrial applications. *Chemical Product and Process Modeling*, *4* (1), 1934–2659.

Rix, A. and Olujić, Ž. (2008) Pressure drop of internals for packed columns. *Chemical Engineering and Processing*, *47*, 1520–1529.

Rong, B.G. and Turunen, I. (2006) A new method for synthesis of thermodynamically equivalent structures for Petlyuk arrangements. *Chemical Engineering Research & Design*, *84*, 1095–1116.

Rong, B.G. (2010) Systematic synthesis of dividing-wall columns for multicomponent distillations, in *Distillation & Absorption 2010* (eds A.B.de Haan, A. Gorak, and H. Kooijman), IChemE, Rugby, pp. 85–90.

Rust, H. and Kaibel, G. (2005) Dividing wall column, with (non-welded) partition wall constructed partly of an elastic material pressing against the column wall to avoid leakage of liquid and/or vapour to the other side of the partition wall, US Patent No. 6958111 B2.

Sander, S. (2007) Increased purity at reduced cost. *Sulzer Technical Review*, *1*, 19–21.

Schultz, M.A., Stewart, D.G., Harris, J.M. *et al.* (2002) Reduce costs with dividing-wall columns. *Chemical Engineering Progress*, *98*, 64–71.

Schultz, M.A., O'Brien, D.E., Hoehn, R.K. *et al.* (2006) Innovative flowschemes using dividing wall columns. *Computer Aided Chemical Engineering*, *21*, 695–700.

Shah, P.B. (2002) Squeeze more out of complex columns. *Chemical Engineering Progress*, *98*, 46–55.

Slade, B., Stober, B., and Simpson, D. (2006) Dividing wall column revamp optimizes mixed xylenes production. *IChemE Symposium Series*, *152*, 252.

Stichlmair, J.G. and Fair, J.R. (1998) *Distillation - Principles and Practice*, Wiley-VCH Verlag GmbH, Weinheim.

Strandberg, J. and Skogestad, S. (2006) Stabilizing operation of a 4-product integrated Kaibel column, in *Distillation & Absorption 2006*, IChemE, Rugby, pp. 638–647.

Sotudeh, N. and Shahraki, B.H. (2007) A method for the design of divided wall columns. *Chemical Engineering and Technology*, *30*, 1284–1291.

Sotudeh, N. and Shahraki, B.H. (2008) Extension of a method for the design of divided wall columns. *Chemical Engineering and Technology*, *31*, 83–86.

Stacey, P.C. (2003) Dividing wall column fractionation tray, where heat transfer through the vertical partition wall is mitigated by providing a separate vertical isolation wall between the downcomer and the dividing wall, US Patent No. 6645350 B1.

Triantafyllou, C. and Smith, R. (1992) The design and optimisation of fully thermally coupled distillation columns, *Transactions IChemE, Part A*, *70*, 118–132.

Wenzel, S. and Röhm, H. (2003) Design of complex distillation columns by overall-cost optimization. *Chemical Engineering and Technology*, *27*, 484–490.

Wright, R.O. (1949) Fractionation apparatus, US Patent No. 2471134.

Yildirim, O., Kiss, A.A., and Kenig, E.Y. (2011) Dividing wall columns in chemical process industry: A review on current activities. *Separation and Purification Technology*, *80*, 403–417.

Zuber, L., Gottlieb-Bachmann, C., and Ausner, I. (2008) Reflux divider for a column having portions for the transfer of material arranged in parallel, US Patent Application No. 2008/0251127 A1.

4

Optimal Operation and Control of DWC

4.1 INTRODUCTION

The advantages of using dividing-wall distillation columns for ternary separation are the main drivers for commercial implementation—especially for ternary mixtures where the middle-boiling component is in the largest amount (Dejanović, Matijašević, and Olujić, 2010). However, there are also major hurdles (Harmsen, 2010), such as the concern that the benefits of dividing wall columns (DWCs) are obtained at the cost of lack of controllability and, consequently, flexibility in operation. But what is in fact this controllability property and why is it so important? Basically, the controllability of a system denotes the ability to reject the expected disturbances and to move the system to new operating points using only certain admissible manipulations (Luyben and Luyben, 1997). Practically, this means that a DWC should be able to deliver on-spec products despite common transitory regimes arising due to planned changes or unexpected disturbances (Kiss and Bildea, 2011).

Although a large amount of the existing literature focuses on the control of binary distillation columns, there are only a limited number of studies on the control of a DWC. A critical overview is made here of the most important DWC control studies made to date. Notably, various authors have selected different ternary chemical systems to be separated and have explored many control structures (the selection and pairing of manipulated and controlled variables) with different control objectives and different control algorithms, from proportional integral derivative

Advanced Distillation Technologies: Design, Control and Applications, First Edition.
Anton Alexandru Kiss.

(PID) to model predictive control (MPC). Some authors control only two compositions, while others control three or even four compositions. Several papers look at inferential temperature control instead of—or in combination with—composition control. Nevertheless, the main conclusion of the review paper of Kiss and Bildea (2011) is that the dividing-wall distillation column has good controllability properties, providing that an appropriate control structure is implemented.

The approach we take here is to present previously reported work on the controllability of DWC, by following the historical development of different control structures. For each paper analyzed hereafter, the novelty and key details of the approach are presented and—when available—the results of the dynamic simulations proving the performance of the controllers are also discussed and/or shown. Two case studies that are industrially relevant are also described in this chapter, namely the separation of the ternary mixtures pentane–hexane–heptane and benzene–toluene–xylene, respectively.

4.2 DEGREES OF FREEDOM ANALYSIS

In the following sections, the first subscript in the concentration notation denotes the stream (D, S, or B), while the second subscript refers to the components. The components are labeled 1 (or A), 2 (or B), and 3 (or C), for the light, intermediate and heavy, respectively. Figure 4.1 illustrates the schematics of a generic DWC divided into six sections (Kiss and Bildea, 2011). In addition, Table 4.1 provides the flow relationships in the DWC based on the mole balance at the interface between sections (Kiss and Bildea, 2011). Remarkably, the specification of L, V, S, r_L, and r_V is sufficient to determine all the flow rates in a DWC, in all sections.

The dividing-wall distillation system has several degrees of freedom. Most of the degrees of freedom are usual for a typical distillation column with side stream, namely, the product rates (distillate D, side stream S, bottoms B), the condenser duty (Q_c) and the reboiler duty (Q_r)—or equivalently, the liquid reflux (L) and the vapor boil-up rate (V). These variables can be also combined; for example, instead of the reflux rate (L), the reflux ratio (RR or $R=L/D$) could be manipulated.

The additional degree of freedom arises from the flow of liquid between the two sections of the column: at the top of the divided-wall section the liquid coming down from the rectifying section can be split, in a controlled manner, between the two sides of the wall by using a total liquid trap-out tray and sending part of the total liquid to the

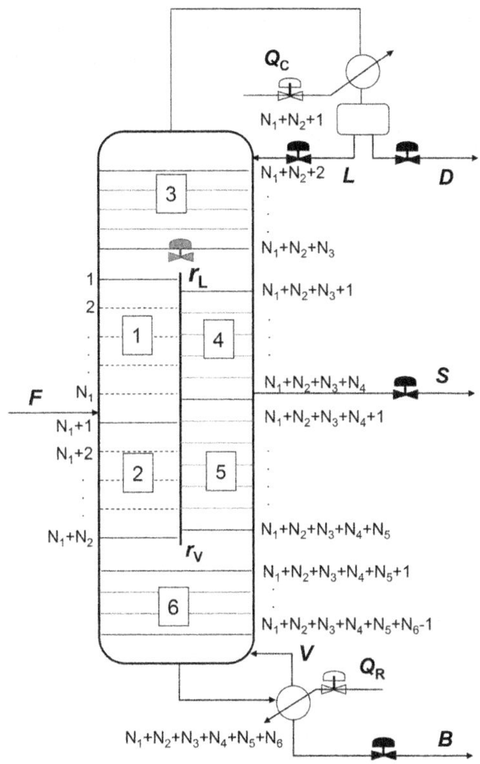

Figure 4.1 Schematics of a dividing-wall column (DWC), including notation of streams

prefractionator side and the rest to the side stream side. Thus this internal liquid split (r_L) is available for control purposes. Note that at the bottom of the dividing-wall section the vapor flow is split proportionally to the cross-sectional area of each side and the hydrodynamic conditions (e.g., flow resistance). The cross-sectional area of each side is fixed by the physical location of the wall, and this is already set at the design stage and, hence, it cannot be changed later on, during operation. Because the location of the wall fixes how the vapor flow splits between the two sides of the column, the vapor split (r_V) variable is not generally adjustable during operation for control purposes.

Typically, the distillate (D) and bottoms flow rates (B) are used to maintain liquid levels in the reflux drum and column base, respectively. Moreover, the condenser duty (Q_c) controls the pressure. Therefore, the remaining degrees of freedom left (L, S, Q_r, r_L) can be used to control four variables. Note that in some control structures the roles of the flow rates

Table 4.1 Flow relationships in a DWC, based on the mole balance at the interface between sections. Specification of L, V, S, r_L, and r_V is sufficient to determine all flow rates in the DWC

Sections	Liquid flow	Vapor flow
Reboiler		$V_6 = V$
From 6 to 2		$V_2 = r_V \cdot V_6$
From 6 to 5		$V_5 = (1 - r_V) V_6$
From 2 to 1		$V_1 = V_2 + (1 - q)F$
From 5 to 4		$V_4 = V_5$
From (1 + 4) to 3		$V_3 = V_1 + V_4$
Condenser	$D = V_3 - L$	
	$L_3 = L$	
From 3 to 1	$L_1 = r_L \cdot L_3$	
From 1 to 2	$L_2 = L_1 + qF$	
From 3 to 4	$L_4 = (1 - r_L)L_3$	
From 4 to 5	$L_5 = L_4 + S$	
From (2 + 5) to 6	$L_6 = L_2 + L_5$	
Reboiler	$B = L_6 - V$	

D and B can be exchanged with that of the flow rates L and V, respectively. Ideally, the purities of all three product streams should be controlled: the amount of intermediate component 2 in distillate and bottom streams (x_{D2} and x_{B2}, respectively) and the amounts of light and heavy components in the side stream (x_{S1} and x_{S3}). However, the last set of specifications is usually replaced by the purity of the side stream (x_{S2}), which means that one degree of freedom can be used to achieve some other objective, such as, for example, to minimize implicitly the energy requirements.

4.3 OPTIMAL OPERATION AND V_{MIN} DIAGRAM

The additional degree of freedom given by the liquid split ratio (r_L) is typically used for optimal operation of a DWC, in an energy efficient manner. As illustrated by a case study described later, plotting the reboiler duty versus the liquid split ratio reveals a minimum point (minimum reboiler duty) corresponding to the optimal liquid split ratio. Note, however, that the minimum energy required during operation may shift due to disturbances in the feed composition or changes in the set point (i.e., different purity targets). Consequently, the liquid split ratio should be used in an additional control loop to achieve minimization of the energy used.

To quantify the energy requirements, the concept of minimum energy or minimum vapor flow rate (V_{min}) can be used—assuming sections with an infinite number of stages in order to be independent of the detailed stage design. In practice, the real energy input will be slightly higher than the minimum requirements and the actual number of stages is selected based on the required separation purity and balancing the total investment and operating costs. However, a V_{min} diagram is a straightforward and common measure used to compare the energy requirements in different distillation arrangements (Halvorsen and Skogestad, 2003a–c, 2011).

Pioneered by Halvorsen and Skogestad (1997, 2003a–c) the V_{min} diagram can be used to estimate the potential energy savings and to establish the stage and reflux requirements for a three-product DWC (Dejanović, Matijašević, and Olujić, 2011). Here we briefly describe the V_{min} diagram—also known as the minimum energy requirements (MER) diagram—and its usability for evaluating multicomponent separations. Let us consider a single two-product column with a multicomponent feed (F). The minimum energy requirements for a given product specification is indirectly determined by the minimum vapor flow through the feed stage in a column with an infinite number of stages. The equivalent heat required to generate this vapor flow can be obtained by multiplying the vapor flow rate with the heat of vaporization. Note that at constant pressure and steady state there are two degrees of freedom in operation. The V_{min} diagram plots the vapor flow rate above the feed (V/F) versus the net flow of product to the top (D/F) per unit of feed. For each given pair (D/F, V/F) all the other properties are completely determined, such as all component recoveries and product compositions. The feed enthalpy condition is given by the liquid fraction (q) in the feed (Halvorsen and Skogestad, 2011).

Figure 4.2 plots the V_{min} diagram for an equimolar ternary system (benzene–toluene–xylene or ABC)—the figure is in fact more generic when the numerical values on the X, Y axes are ignored. The V_{min} diagram shows how the feed components for a ternary feed (ABC) are distributed to the top and bottom products in a simple two-product 'infinite stage' distillation column as a function of the operating point (D/F, V/F). For values of V/F above the upper boundary following the three peaks in the diagram (0,0-P_{AB}-P_{AC}-P_{BC}-1,0), the column is overfractionated meaning that valuable energy is wasted. Note that the point located at coordinates (1,0) in Figure 4.2 is more generically defined as $1,(1-q)$ but in this specific case we simply considered the feed as saturated liquid ($q = 1$).

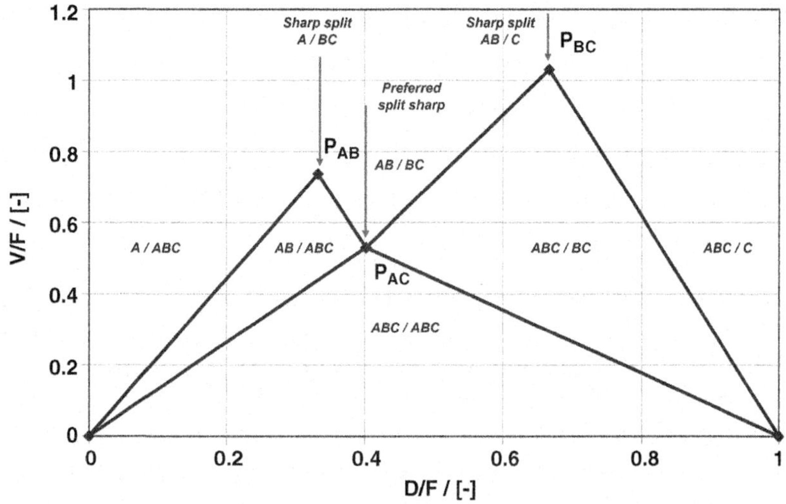

Figure 4.2 V_{min} diagram of a BTX ternary system

The values at the peaks (P_{AB}, P_{AC}, P_{BC}) give the vapor flow for the corresponding sharp neighbor component splits. As the vapor flow (V) is reduced below the line for a given D, one more component becomes distributed as the boundary lines are crossed. The knots are V_{min} for the so-called *preferred splits* where a sharp split between two key-components is specified, while allowing intermediate components to be distributed. Only the sharp split between each possible pair of key components must be solved to find the diagram for a multicomponent feed. Basically, for n components ($n > 1$), one can find the complete diagram by calculating $1+2+3\ldots n-1 = n(n-1)/2$ points. For the three-component example shown in Figure 4.2, only three points are needed: P_{AB}: sharp A/B split, P_{BC}: sharp B/C split, and P_{AC}: sharp A/C split. P_{AC} is the *preferred split* that is the minimum energy operating point for a sharp separation between the heavy and light keys while the intermediate distributes to both column ends. At any operating point at or above the V-shaped P_{AB}-P_{AC}-P_{BC}, a sharp A/C split is obtained but with higher energy than that required at the exact point P_{AC} (Halvorsen and Skogestad, 2011). Note that the minimum energy required for the ternary separation in a DWC corresponds to the highest peak (i.e., P_{BC} in Figure 4.2).

The diagram for real mixtures can be obtained quite simply by simulating the given multicomponent feed in a two-product column with a very large number of stages. The simulation is carried out for different values of the product split (D/F), and computing the corresponding V_{min}/F to obtain the specified or desired purity of the product. The term V_{min} means that it is the minimum vapor flow corresponding to an infinite number of stages, but in practice one can get very close to V_{min} by using a finite number of stages, typically $N = 4N_{min}$. Note that for high-purity separations by distillation the value of V_{min} depends only weakly on the purity. Thus, in practice one may specify small impurities of heavy key in the top and light in the bottom for each pair of keys. In the case of a binary feed, there will be only one peak (Halvorsen and Skogestad, 2011). For ideal mixtures with constant relative volatility and constant molar flows one can obtain the V_{min} diagram without the need to carry out time consuming simulations. In this case, for an infinite number of stages, one may use the classical Underwood equations and the V_{min} diagram can be calculated directly from the feed properties (Halvorsen and Skogestad, 2011). However, due to its nature, the V_{min} diagram is limited to zeotropic mixtures only—hence it cannot be applied to azeotropic mixtures.

4.4 OVERVIEW OF DWC CONTROL STRUCTURES

All the control structures presented hereafter could use or make use of an additional optimization loop that manipulates the liquid split in order to control the heavy component composition in the top of the fractionator, and implicitly minimize the energy requirements. Ling and Luyben (2009) and Halvorsen and Skogestad (2011) have already shown that implicit optimization of energy usage is achieved by controlling the heavy impurity at the top of the prefractionator.

Many authors analyzed the effect of minimizing the energy requirements on the controllability properties of the system—as an optimal design might give the highest energy savings but lack good controllability (Segovia-Hernandez, Hernandez, and Jimenez, 2005; Robles-Zapiain, Segovia-Hernandez, and Bonilla-Petriciolet, 2008; Tamayo-Galvan *et al.*, 2008). Rong and Turunen (2006) reported a reliable synthesis method while Gomez-Castro *et al.* (2008) proposed a robust method for the design of distillation sequences with dividing walls and recommended an analysis of both thermodynamic and controllability properties. Accordingly, it was concluded that the distillation arrangements (e.g., DWC) other than conventional Petlyuk presented the best values for condition number and minimum singular value. As a result, good dynamic closed-loop performance could be expected for these types of thermally coupled distillation systems (Gomez-Castro *et al.*, 2008). Moreover, it was found that thermally coupled distillation systems (TCDSs) are not only well controllable but sometimes they exhibit dynamic responses that are easier to manage than in the case of conventional distillation sequences (Robles-Zapiain, Segovia-Hernández, and Bonilla-Petriciolet, 2008; Alcántara-Ávila *et al.*, 2008; Gomez-Castro *et al.*, 2008).

Table 4.2 gives an overview of the main DWC control structures reported so far in the literature, while the next sections provide a more detailed analysis of each of them (Kiss and Bildea, 2011).

4.4.1 Three-Point Control Structure

The simplest control structure that can be imagined is simply an extension of the control of a regular distillation column with a side stream. This is known also as the *three-point control structure* (Figure 4.3a) (Kiss and Bildea, 2011). The distillate purity (x_{D1}) is controlled by manipulating the reflux rate (L), the side stream purity (x_{S2}) is controlled by manipulating the side stream flow rate (S), and the bottom purity (x_{B3}) is controlled by manipulating the vapor boil-up (V). The three-point structure *LSV* was suggested by Wolff and Skogestad (1995). They studied the separation of an ethanol/propanol/butanol mixture in a system with a 20-tray prefractionator and 40-tray main column. The authors analyzed the controllability of the system by using linear tools, concluding that the system is easy to control. Moreover, they also performed dynamic simulations of the nonlinear model, showing that the column handles well disturbances and some set-point changes. However, sometimes small changes in the purity set-points (e.g., x_{S2}, from 0.994 to 0.996) could not be handled—changes of 100% in L and V being required. The most likely reason for this is that the column does not have enough stages to achieve the degree of separation, at least not without adjusting the liquid (r_L) and vapor (r_V) split ratios. Note that the mole fraction x_{S2} does not define the composition of the side stream, as it can correspond to several levels of the concentration of impurities x_{S1} and x_{S3}. Although this does not appear to be a problem in a practical implementation, the authors tried to find a solution by using the remaining available degree of freedom (liquid split ratio, r_L), hence adding another point to the control structure.

Table 4.2 Overview of various DWC control structures

DWC control structure	Ternary system	References
Three-point control structure	Ethanol/propanol/butanol	Wolff and Skogestad (1995)
Four-point control structure	Benzene/toluene/xylene	Kiss and Rewagad, (2011))
Three-point control structure with nested loops	Methanol/2-propanol/butanol	Abdul Mutalib and Smith (1998); Abdul Mutalib *et al.* (1998)
Three-point control structure with alternative pairing	Hypothetical system with constant relative volatilities: 1 : 2.15 : 4.65	Serra *et al.* (Serra, Espuña, and Puigjaner 1999, 2000; Serra *et al.* 2001)
Performance control of prefractionator sub-system using the liquid split	Hypothetical system with constant-relative volatility. Relative volatilities: 1 : 2 : 4, boiling points 100, 50, 0 °C	Halvorsen and Skogestad (1997, 1999)
Control of the temperature in the top of the prefractionator. Model predictive control (MPC)	Butanol/pentanol/hexanol	Adrian, Schoenmakers, and Boll (2004)
Linear controllability analysis, nonlinear dynamic simulations	Pentane/hexane/heptane	Bildea (2004)
Control of the temperature in the bottom of the prefractionator. Temperature–composition cascade	Ethanol/1-propanol/1-butanol	Wang and Wong (2007)
Null space method. Self-optimizing control	Hypothetical system with constant-relative volatility. Relative volatilities: 1 : 3 : 9	Alstad and Skogestad (2007)
Feedforward control to reject frequent measurable disturbances	Benzene/toluene/xylene	Ling and Luyben (2009)
Feedforward and differential temperature control loops.	Benzene/toluene/xylene	Ling and Luyben (2010)
LQG/LQR, higher order controllers based on H_∞ norm μ-synthesis.	Benzene/toluene/xylene	van Diggelen, Kiss, and Heemink (2010)

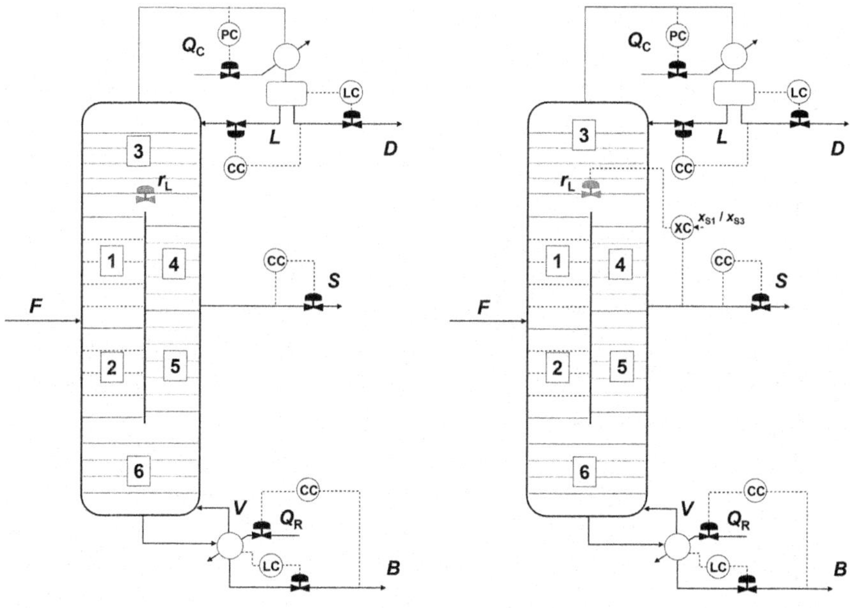

Figure 4.3 Three-point (a) and four-point (b) control structures

4.4.2 Three-Point Control Structure with Alternative Pairing

Serra *et al.* (Serra, Espuña, and Puigjaner, 1999, 2000; Serra *et al.*, 2001; Serra, Espuña, and Puigjaner, 2003) performed a comprehensive study concerning the operation and controllability of dividing-wall distillation columns. They studied a model system with constant relative volatility ($\alpha = 1:2.15:4.65$). Several control structures where investigated, with each structure using a different set of manipulated variables to control the products purities. However, the liquid split ratio (r_L) was not considered. Different controllability indices were used to assess the performance of the pairing in a three-point control structure. The results showed that the best structure was the one previously introduced by Skogestad and Wolff—as illustrated in Figure 4.3a (Kiss and Bildea, 2011). Moreover, Serra *et al.* (Serra, Espuña, and Puigjaner 1999, 2000; Serra *et al.* 2001) compared the performance of the PI control with dynamic matrix control (DMC), revealing the better performance of the PI control and the limitations of the DMC (Lundström *et al.*, 1995; Serra *et al.*, 2001).

The authors also pointed out that the performance is dependent on the design of the column, but alas they did not elaborate further on this observation.

4.4.3 Four-Point Control Structure

In addition to the reflux, side stream, and boil-up that control the concentration of the main components in the product streams, the liquid split ratio (r_L) can be added to the set of manipulated variables with the goal of controlling the levels of both impurities in the side stream. The resulting four-point control structure is shown in Figure 4.3b (Kiss and Bildea, 2011). Wolff and Skogestad (1995) tested this structure on the separation of the same system previously described. Both linear and nonlinear tools predicted difficult control. Besides a small gain from the manipulated variables toward the controlled variables, the steady state feasibility space shows regions where no steady states exists.

4.4.4 Three-Point Control Structure with Nested Loops

As an alternative, Wolff and Skogestad (1995) switched the V–x_{B3} and S–x_{S2} control loops to the pairing V–x_{S2} and S–x_{B3} (Figure 4.4a) (Kiss and Bildea, 2011). Although the linear controllability tool did not predict control difficulties, the structure proved unworkable under mild disturbances. The authors attributed this to the strong nonlinearity of the pairing from V to x_{S2}. However, it should be observed that the loops V–x_{S2} and S–x_{B3} are nested: a change of the side stream flow rate is expected to affect first the side stream purity x_{S2} and only later the bottom purity x_{B3}. Therefore, strong interactions between the loops are expected.

The same control structure was also considered by Abdul Mutalib, Zeglam, and Smith (1998). However, they presented a simulation study of the methanol/2-propanol/butanol separation system. Unlike Wolff and Skogestad (1995), it was found that the pairing $V - x_{S2}$ and $S - x_{B3}$ worked well in the three-composition control structure. Moreover, they switched the reflux and distillate in the level and distillate-concentration loops, as shown in Figure 4.4b (Kiss and Bildea, 2011). Comparison of the performance between the two control structures, for set-point changes and feed disturbances, showed that the system was indeed controllable.

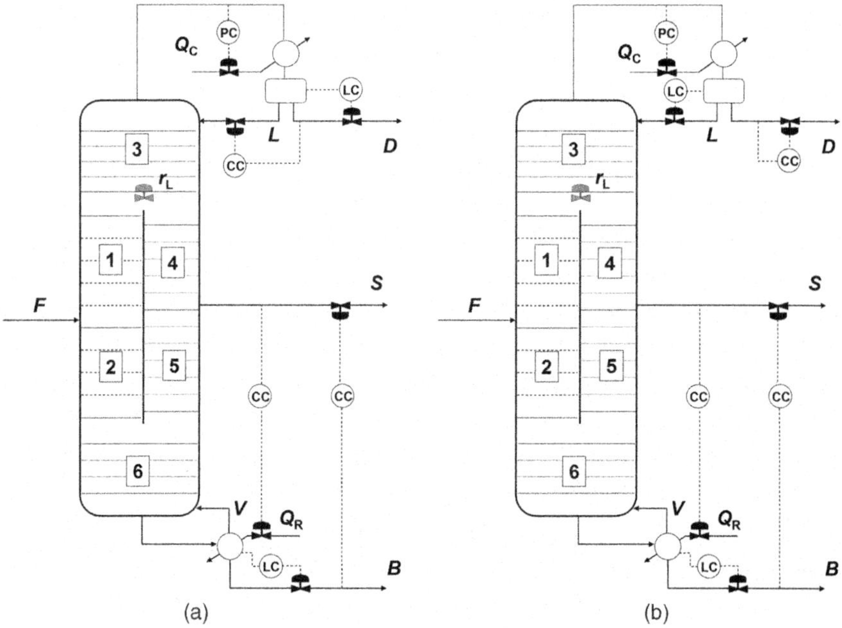

Figure 4.4 Three-point control structure with reversed loops (a) or with switched loops pairing (b)

4.4.5 Performance Control of Prefractionator Sub-system using the Liquid Split

Halvorsen and Skogestad (1997, 1999, 2011) pointed out two key tasks that should be achieved by the prefractionator when the system is facing disturbances or set-point changes are required:

1. Keep the heaviest component from going out the top of the prefractionator section;
2. keep the lightest component from going out the bottom of the prefractionator section.

Any amount of heavy component going out the top of the wall will end up also in the liquid flowing down in the main column and thus strongly affect the purity of the side stream (S). Similarly, any portion of the lightest component that goes out the bottom of the prefractionator section will flow up through the side stream section, mostly in the vapor phase, with a lesser effect on the composition of the side stream. Since the side stream is collected as a liquid product, small amounts of light impurity in the vapor

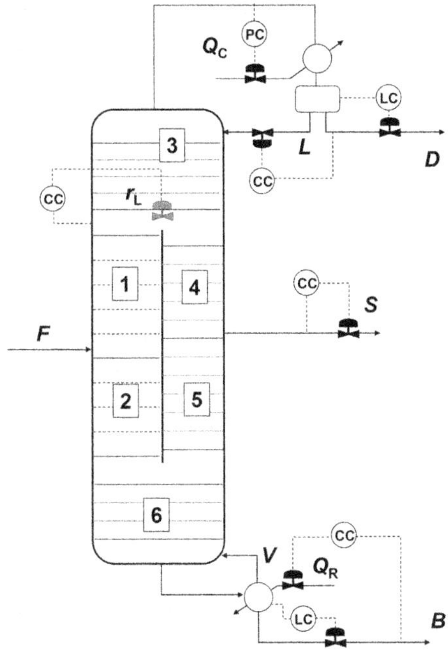

Figure 4.5 Control structure using the liquid split to control the heavy impurity at the top of the prefractionator

phase will not significantly affect the side stream composition. However, even tiny amounts of heavy impurity in the liquid phase will greatly affect the side stream composition (Ling and Luyben, 2009).

Based on these considerations, the authors added a fourth control loop where the liquid split ratio (r_L) is used to control the level of the heavy impurity in the top of the prefractionator (Figure 4.5) (Kiss and Bildea, 2011). This approach was also successfully used later by other authors (Ling and Luyben, 2009, 2010). Note that the mixture considered for separation was a hypothetical constant-relative volatility system. The relative volatilities were 1 : 2 : 4, and the boiling points 100, 50, 0 °C (Halvorsen and Skogestad, 1997, 1999).

4.4.6 Control Structures Based on Inferential Temperature Measurements

4.4.6.1 *Controlling the Temperature in the Top of the Prefractionator*

Adrian, Schoenmakers, and Boll (2004) reported interesting experimental results concerning the control of a butanol/pentanol/hexanol system.

Compared to previous studies, temperature control was used instead of composition control. The location of the controlled temperatures included the top of the prefractionator, a point above the side draw, and the lower part of the column (Figure 4.6a) (Kiss and Bildea, 2011). It should be also remarked that the heat input (reboiler duty, Q_r) was fixed. An alternative configuration was proposed later by Wang and Wong (2007), in which one of the controlled temperatures is located in the bottom of the prefractionator side (Figure 4.6b) (Kiss and Bildea, 2011).

Model predictive control (MPC) was employed by a second control structure, shown in Figure 4.7 (Kiss and Bildea, 2011)—where the heat input was added for a total of four manipulated variables used to control three temperatures. The MPC controller showed better performance for disturbances not only in the feed flow-rate but also in the feed composition. Nevertheless, the additional manipulated variable casts some doubt on the conclusion of Adrian, Schoenmakers, and Boll (2004) that MPC alone is generally better than a multi-SISO control (e.g., conventional multi-loops PID).

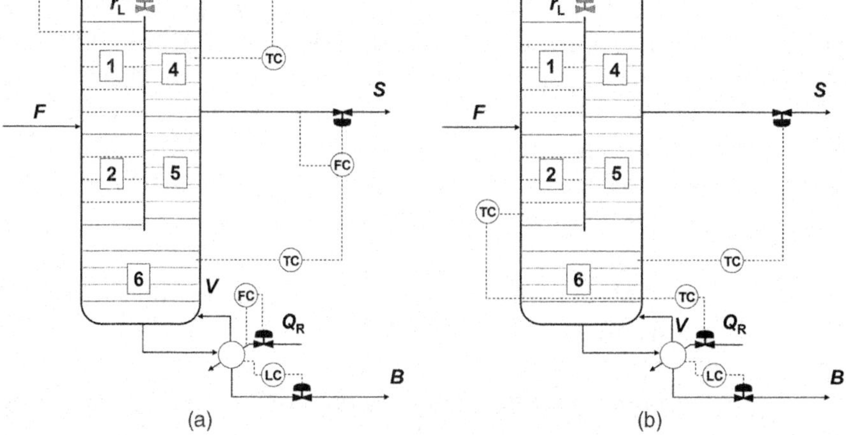

Figure 4.6 DWC control structure using temperature measurements for the top (a) or bottom (b) of the prefractionator

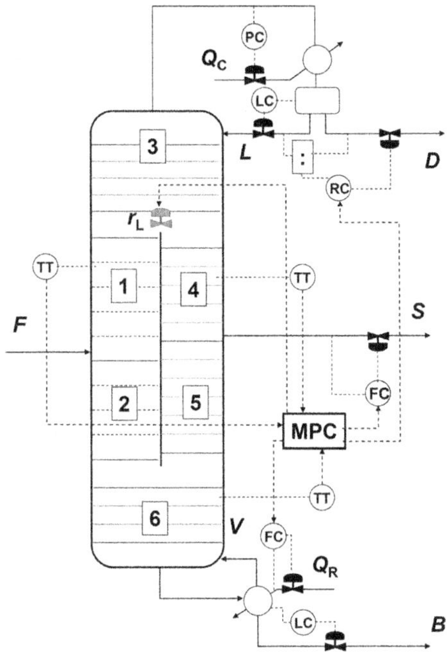

Figure 4.7 Model predictive control (MPC) of a DWC, using temperature measurements

4.4.6.2 Controlling the Temperature in the Bottom of the Prefractionator

Wang and Wong (2007) studied a high-purity ethanol/1-propanol/1-butanol column. PID algorithms and temperature control were used, instead of composition control. A temperature in the prefractionator and two temperatures in the main column were selected. The pairing was completely different than that previously proposed by Adrian, Schoenmakers, and Boll (2004). Figure 4.7 shows that a temperature in the bottom section of the prefractionator was controlled by manipulating reboiler heat duty while a temperature in the rectifying section was controlled by manipulating the reflux flow rate (Kiss and Bildea, 2011). Moreover, a temperature near the base of the column was controlled by manipulating the side stream flow rate. Again, the liquid split ratio (r_L) was not used for control. Stable control was achieved and products returned to their desired purity levels for feed flow rate changes. However, large product purity deviations were reported for feed composition disturbances (Wang and Wong, 2007). The authors recommended a temperature/composition cascade control structure (Figure 4.8a) to solve this problem (Kiss and Bildea, 2011).

4.4.7 Feedforward Control to Reject Frequent Measurable Disturbances

Ling and Luyben (2009) proposed to control the impurity levels in the three product streams and one composition in the prefractionator. This implicitly also achieves minimization of the energy requirements. The four manipulated variables are liquid split, reflux flow rate, side stream flow rate, and vapor boil-up. Figure 4.8b shows the four-point control structure using combined feedback—feed-forward to account for feed flow disturbances (Kiss and Bildea, 2011). In addition, the authors suggest combining feedback and feed-forward to improve the per-

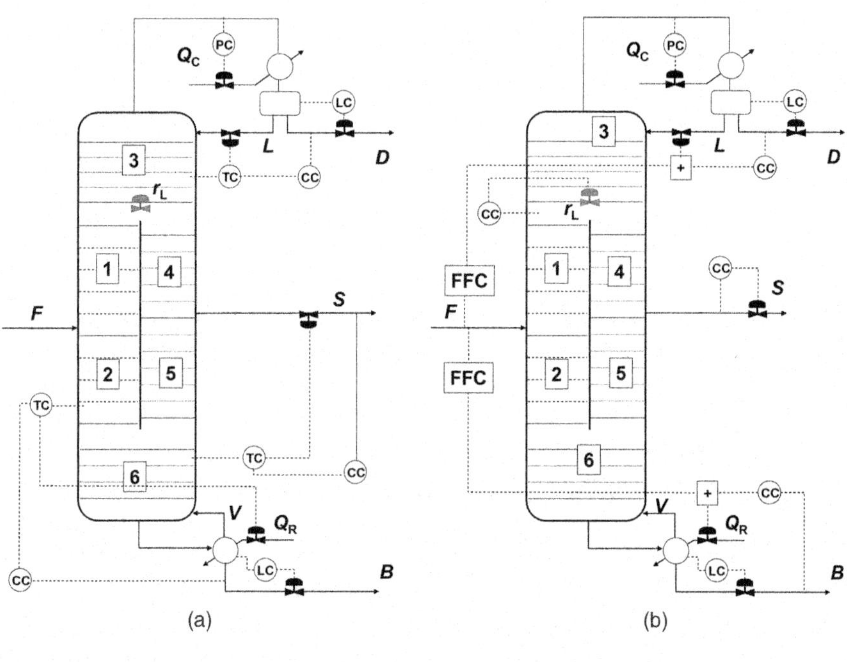

Figure 4.8 DWC control structure using a concentration—temperature cascade (a); four-point control structure using combined feedback—feed-forward to account for feed flow disturbances (b)

formance with respect to feed flow rate disturbances. The improved performance was demonstrated by the results of the dynamic simulation (Ling and Luyben, 2009).

In a follow-up study (Ling and Luyben, 2010), the authors also explored the use of temperature measurements to avoid expensive and high-maintenance composition analyzers. Two types of temperature control structures were studied. In the first, three temperatures located in the main column and one temperature on the prefractionator side of the wall were used to adjust the four manipulated variables. Feed flow rate disturbances are well handled with this structure, but product purities start to deviate significantly from their desired values for feed composition changes greater than 10%. In a second proposed control structure, four differential temperature control loops were used. The performance is improved and disturbances of 20% in feed composition are also well handled with only small deviations in product purities. This second structure also handles large changes in the column operating pressure (Ling and Luyben, 2010).

4.4.8 Advanced Control Techniques

Alstad and Skogestad (2007) described the null space method as a self-optimizing control method that selects the control variables as combinations of measurements. For the case of a Petlyuk distillation setup this resulted in the following candidate measurements: temperature at all stages and all flow rates. Using the null space method a subset of six measurements was obtained, resulting in a practical implementation.

Woinaroschy and Isopescu (2010) showed the ability of iterative dynamic programming to solve time optimal control of a DWC system, and their study focused on the startup control of a DWC.

These different studies investigate different separation systems and types of disturbances and hence a common conclusion to identify the best controller cannot be drawn. In a comparative study, van Diggelen, Kiss, and Heemink (2010) applied more advanced controllers such as LQG/LQR, GMC, and higher order controllers based on H_∞ norm μ-synthesis. The performance was compared to other structures involving PID controllers in a multi-loop framework. The controllers were applied to a DWC used in the ternary separation of benzene–toluene–xylene (BTX). More details are provided in Chapter 5, on advanced control of dividing-wall columns.

4.5 CONTROL GUIDELINES AND RULES

Some broad guidelines and rules for the design of a DWC as well as standards for selecting control structures for a DWC were recommended (Il Kim, Lee, and Park, 2007). These were used in several studies described previously. Hereby, the following guidelines can be recommended in order to select the appropriate control strategy for a DWC. Note that they correspond to the history of developing control structures for the dividing-wall distillation systems:

- Use the reflux and boil-up streams to control the distillate and bottom compositions;
- alternatively, use the distillate and bottoms flow rates for composition control;
- use the side stream flow rate to control the side stream composition;
- alternatively, use the distillate and bottoms flow rates for composition control;
- use the liquid split to control the amount of heavy component leaving in the top of the prefractionator—this preserves the optimality characteristics of a given design;
- use (inferential) temperature measurements to avoid composition measurements, combined when required with feed-forward ratio schemes;
- use concentration–temperature cascade loops to improve disturbance rejection;
- combine conventional feedback control with feed-forward control loops;
- apply model predictive control (MPC) and/or other MIMO (multi-input multi-output) control strategies, when SISO loops are not sufficient.

The overview presented previously indicates that dividing-wall distillation columns have good controllability properties, provided that an appropriate control structure is implemented. Among the multi-loop PID control strategies, LSV and DSV were the best, being able to handle persistent disturbances in reasonably short times. However, significantly shorter settling times and better control performance were achieved using advanced MIMO controllers such as MPC. Notably, information about control structures for industrial applications described in the open literature is very scarce, and so it is very desirable that column manufactures make such information available (Yildirim, Kiss, and Kenig, 2011).

4.6 CASE STUDY: PENTANE–HEXANE–HEPTANE SEPARATION

This section presents the results of design, dynamic simulation, and control of a DWC for the separation of 100 kmol h^{-1} of a pentane–hexane–heptane mixture 20 : 60 : 20 (molar). For the same mixture, the controllability analysis of a Petlyuk arrangement (Bildea, 2004) predicted easy rejection of disturbances, which was confirmed later by rigorous dynamic simulations. However, the system could not cope with a large and fast change of the hexane feed flow rate. Remarkably, this control problem was apparent during pressure driven simulation, but could not be identified by linear analysis or by flow-driven simulations in Aspen Dynamics.

The preliminary design of the DWC—obtained by applying the short-cut method recommended by Triantafyllou and Smith (1992)—was the starting point of rigorous simulations in Aspen Plus and Aspen Dynamics. The distillate, reflux, and side stream flow rates were adjusted to meet the purity specifications. Weir heights of 5 cm and residence times of 5 min in the reflux drum and 10 min in column sump were assumed. Condenser and tray pressure drops of 0.05 and 0.01 bar, respectively, were specified. Pumps were chosen for a 0.3 bar pressure drop over the corresponding valve. The complete flow-sheet is presented in Figure 4.9 —although not all pumps and valves are shown—while Table 4.3 summarizes the results of sizing the columns (Kiss and Bildea, 2011). The energy requirement of the design was 1.164×10^6 kcal h^{-1}, which could be further reduced to

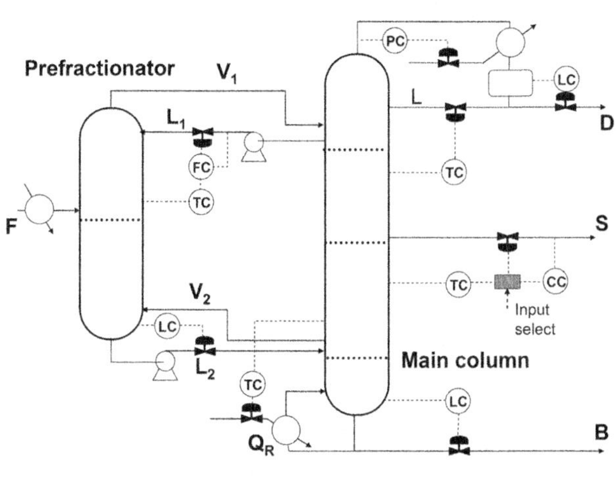

Figure 4.9 Two-column model of a dividing-wall distillation column

Table 4.3 Design of the equivalent Petlyuk distillation system

Design parameter	Value
Prefractionator	
Number of trays	12
Feed tray	5
Diameter (m)	0.8
Top stage pressure (bar)	1.17
Main column	
Number of trays	42
Top feed/withdraw tray	6
Side stream product	15
Bottom feed/withdraw tray	32
Diameter (m)	1.2
Top stage pressure (bar)	1
Distillate rate (kmol h^{-1})	19.77
Side stream rate (kmol h^{-1})	60.20
Reflux rate (kmol h^{-1})	160.22
Liquid to prefractionator, L_1 (kmol h^{-1})	31
Vapor to prefractionator, V_2 (kmol h^{-1})	62
Reboiler duty (10^6 kcal h^{-1})	1.164
Condenser duty (10^6 kcal h^{-1})	1.123

1.074×10^6 kcal h^{-1} by adjusting the liquid and vapor flow rates from the main column to the prefractionator, $L_1 = 27.55$ kcal h^{-1} and $V_2 = 68.51$ kcal h^{-1} (Figure 4.10) (Kiss and Bildea, 2011). However, in this section we restrict ourselves to the sub-optimal design presented in Table 4.3 (Kiss and Bildea, 2011).

A full pressure-driven dynamic simulation was built in Aspen Dynamics. Four control strategies denoted by CS1, CS2, CS3, and CS4 were

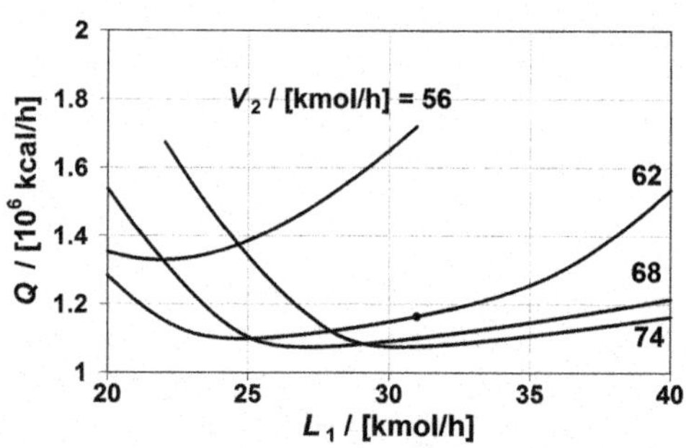

Figure 4.10 Pentane–hexane–heptane separation in a DWC: energy requirements versus liquid flow rate for different values of the vapor flow rate (the dot marker shows the operating point)

investigated. In addition to the standard pressure and level control loops, temperatures at the top (tray 4) and bottom (tray 36) of the main column were controlled by the reflux and reboiler duty, respectively. The side-stream flow rate controlled a temperature below the side-stream (tray 27) in CS1 and CS2 and the side-stream purity in CS3 and CS4. The flow rate of liquid going from the main column to the prefractionator (L_1) was kept constant in CS1 and CS3 or used to control the temperature on the prefractionator top (tray 3) in CS2 and CS4 (Kiss and Bildea, 2011).

Figure 4.11 (Kiss and Bildea, 2011) presents the dynamic simulation results for the following scenario. The simulation starts from the steady state. After 0.5 h, the feed rate is increased by 10%, from 8617 to 9400 kg h^{-1}. At time $t = 3$ h the feed rate is reduced to 90% of the initial value (7600 kg h^{-1}). For all control structures, the composition of top and bottom streams is almost constant, even if only temperature measurements are used. However, controlling a temperature in the middle section of the main column (CS1 and CS2) does not ensure the purity of the side stream product, which drops from 0.99 to 0.9 for a 10% increase of the feed flow rate. Moreover, when the feed rate is brought back to the initial value, or even reduced to 90% of it, the purity does not recover. This behavior suggests the existence of multiple steady states.

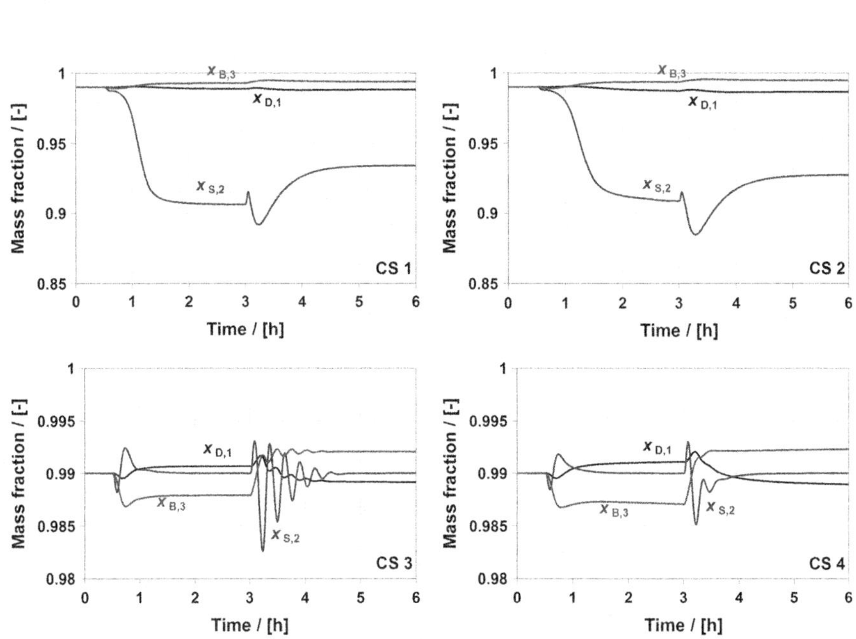

Figure 4.11 Results of dynamic simulations for the separation of a pentane–hexane–heptane mixture in a DWC (control structures CS 1–4)

The performance of the system dramatically improves when composition measurement is available and the side-stream flow rate is used to control its purity (CS3 and CS4). In this case, the purities of top and bottom products deviate by less than 0.3% from the design values. The performance could be further improved by adding composition–temperature cascade loops. Moreover, excellent control is achieved for the side-stream composition. Control of the prefractionator top temperature (CS4) improves the dynamics of the side-stream composition control loop. Note that, for all control structures investigated, the system becomes unstable if instead of the liquid rate (L_1) the liquid split ratio (r_L) is kept constant.

The controllability of the design with minimum energy requirements was also investigated. This design appeared to be more sensitive to disturbances even if the purity of the side stream was controlled. For example, a 10% increase of the flow rate destabilized the system if performed stepwise. This change could be implemented only by ramping it over 6 h (Kiss and Bildea, 2011).

4.7 CASE STUDY: ENERGY EFFICIENT CONTROL OF A BTX DWC

Despite being a technology already implemented at the industrial scale, the dynamic control and optimization of DWC has been explored in only a few papers (van Diggelen *et al.*, 2010; Ling and Luyben, 2009, 2010; Woinaroschy and Isopescu, 2010). Compared to a conventional separation sequences, the control of a DWC is more difficult due to the increased interaction among the controlled and manipulated variables. This chapter proposes several multi-loop PID control structures (DB/LSV, DV/LSB, LB/DSV, LV/DSB) that keep under control the product purities while at the same time implicitly minimizing the energy requirements. This is achieved by manipulating the liquid split (r_L) to control the composition of the heaviest component (C) in the top of the prefractionator. The performances of the control strategies and the dynamic response of the DWC are investigated in terms of products composition time profiles for various persistent disturbances in the feed flow rate and composition. To allow a fair comparison of the results with previously published references, this section considers as case study the industrially relevant ternary separation of the mixture benzene–toluene–xylene (BTX) in a DWC. Table 4.4 lists the physical properties of the BTX mixture (Kiss and Rewagad, 2011).

Table 4.4 Physical properties of the investigated ternary system: benzene–toluene–xylene

Physical property	Benzene	Toluene	Xylene
Molecular formula	C_6H_6	C_7H_8	C_8H_{10}
Molecular weight	78.11	92.14	106.17
Density (kg m^{-3})	878.6	866.9	860.0
Viscosity (cP at 20 °C)	0.652	0.590	0.620
Critical pressure (bar)	48.95	41.08	35.11
Critical temperature (°C)	288.9	318.6	343.05
Melting temperature (°C)	5.53	−94.97	13.26
Boiling temperature (°C)	80.09	110.63	138.36

The integration of two columns into one shell also leads to more interactions among the controlled and manipulated variables, and ultimately in the controllability of the system. Although much of the literature focuses on the control of binary distillation columns, there are a few studies on the controllability and dynamic optimization of DWCs (Halvorsen and Skogestad, 1997; Serra, Espuña, and Puigjaner, 1999, 2000; Serra *et al.*, 2001; Adrian, Schoenmakers, and Boll, 2004; Ling and Luyben, 2009, 2010; Diggelen *et al.*, 2010; Kiss and Bildea, 2011). The problem is that different DWC separation systems were used and, hence, no fair comparison of optimal controllers was possible. To solve this problem, we explore here the DWC control issues on one system (BTX) and compare various multi-loop PID control strategies enhanced with implicit dynamic optimization aimed at minimizing the energy requirements.

Thermodynamically equivalent structures for Petlyuk/DWC arrangements can be generated using a synthesis method proposed by Rong and Turunen (2006). Here we limit ourselves to the study of only the conventional DWC structure. Aspen Plus and Aspen Dynamics were used as powerful computer aided process engineering (CAPE) tools to build the rigorous steady-state and dynamic simulations. Non-random two liquids (NRTL) was used as property model in simulations based on RadFrac as an accurate distillation unit. Figure 4.12 illustrates the schematics of the modeled DWC, consisting of six column sections of eight stages each, as well as the composition profile along the column (Kiss and Rewagad, 2011). The feed stream consisting of an equimolar mixture of benzene–toluene–xylene (noted as ABC for convenience) is fed into the prefractionator side, between sections 1 and 2. Benzene is obtained as top distillate, xylene as bottom product, and toluene is withdrawn as side stream of the main column (between column sections 4 and 5). No azeotropes are present in the ternary system BTX.

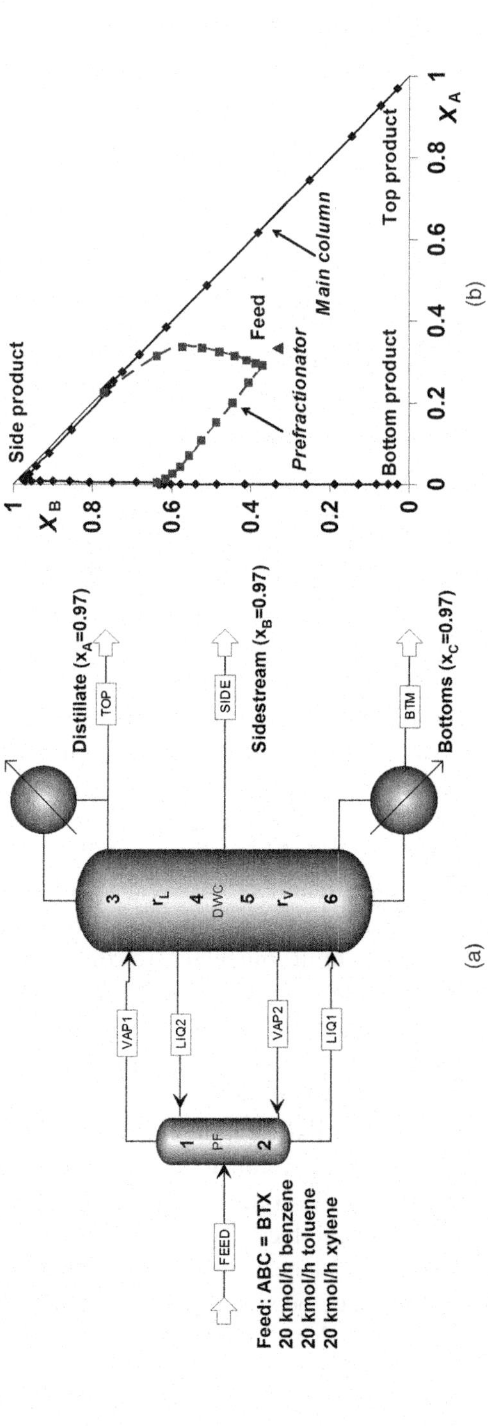

Figure 4.12 Schematics of the simulated BTX dividing-wall column: six sections of eight stages each (a), and the composition profile inside the DWC as a ternary diagram (b)

The steady-state purity of all product streams is considered to be 97% in this work, to allow a fair comparison with previously reported results (Halvorsen and Skogestad, 1997; Diggelen *et al.*, 2010). The converged Aspen Plus simulation was exported to Aspen Dynamics (Ling and Luyben, 2009, 2010), where several PID loops within a multi-loop framework were applied. Remarkably, the PID controllers remain the most used controllers in the chemical industry, for several practical reasons: simplicity of control structure, robustness with respect to model uncertainties and disturbances, and easy manual stabilization of the process when an actuator or sensor fails.

4.7.1 Energy Efficient Control Strategies

For a DWC, two multi-loops are needed to stabilize the column and another three to maintain the set points specifying the product purities. From a practical viewpoint, there are only a few configurations that make sense. The level of the reflux drum and the reboiler can be controlled by the variables L (liquid reflux), D (distillate), V (vapor boil-up), or B (bottoms). Consequently, there are four *inventory control* options to stabilize the column and to control the level in the reflux tank and the level in the reboiler, namely, the combinations: D/B, L/V, L/B, and V/D (Diggelen *et al.*, 2010; Kiss and Bildea, 2011).

Figure 4.13 shows the multi-loop PID control structures considered in this section: DB/LSV, DV/LSB, LB/DSV, and LV/DSB (Kiss and Rewagad, 2011). The part for the control of product purities is often called *regulatory control.* One actuator is left (r_L) that can be used for optimization purposes such as minimizing the energy requirements. All these control structures are based on PID loops within a multi-loop framework, with an additional optimization loop that manipulates the liquid split in order to control the heavy component composition in the top of fractionator, and implicitly achieving minimization of the energy requirements. Ling and Luyben (2009), as well as Halvorsen and Skogestad (2011), have demonstrated in their studies that implicit optimization of energy usage is achieved by controlling the heavy impurity at the top of the prefractionator. Note that any heavy component (C) going out the top of the wall will appear also in the liquid flowing down in the main column and thus strongly affect the purity of the side-stream (S). Since the side-stream is collected as a liquid product, small amounts of light impurity in the vapor phase will not significantly affect its composition. However, even tiny amounts of

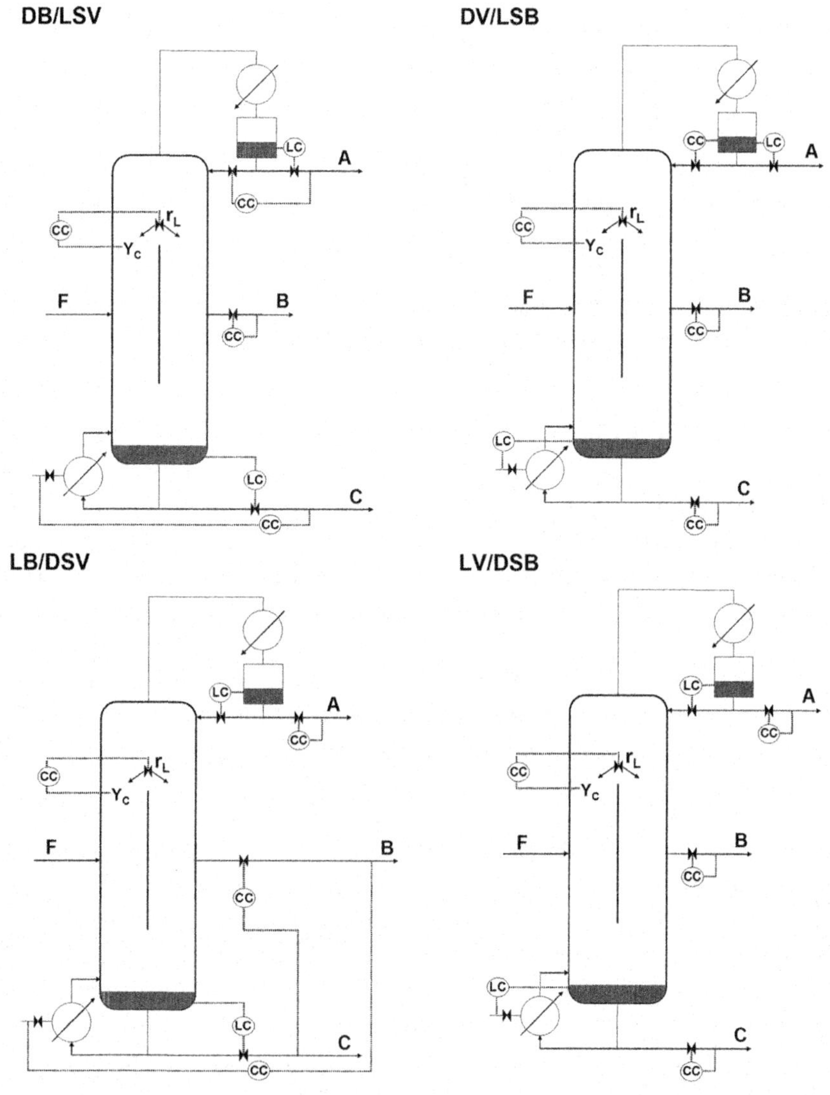

Figure 4.13 Energy efficient control structures based on PID loops within a multi-loop framework

heavy impurity in the liquid phase will greatly affect the composition of the side stream.

At the bottom of the dividing-wall section, the vapor flow is split proportionally to the cross-sectional area of each side and the hydrodynamic conditions (e.g., flow resistance). The cross-sectional area of each side is fixed by the physical location of the wall, which is set at the design stage hence it cannot be changed later on during operation (Dejanović, Matijašević, and Olujić, 2010). Because the location of the wall fixes how the vapor flow splits between the two sides of the column, the vapor split (r_V) variable is not adjustable during operation for control purposes (Kiss and Bildea, 2011). Therefore, the vapor split ratio is not used as a manipulated variable in the dynamic simulations presented hereafter.

In addition, sensitivity analysis was used as a powerful tool to determine the optimal parameters corresponding to the minimum energy requirements. Plotting the V_{min} diagrams pioneered by Halvorsen and Skogestad (1997) contributes to understanding the optimal operating point with the minimum energy requirements. Figure 4.14 (Kiss and Rewagad, 2011) illustrates the optimal liquid split ratio (r_L)—as well as the heavy component mole fraction in the top of fractionator ($Y_{C\text{-}PF1}$)—corresponding to the minimum reboiler duty (Q_{reb}).

To choose the best control structure, controllability indices can be applied to give information about the behavior of the system (Segovia-Hernandez, Hernandez-Vargas, and Marquez-Munoz, 2007). To obtain these indices, a methodology proposed by Gabor and Mizsey (2008) for a multiple input and multiple output (MIMO) system was used. This methodology is based on using a linearized state space model of the process that can be calculated with the Control Design Interface module of Aspen Dynamics. The results were then processed in MathWorks MATLAB®. In this study, steady state and dynamic controllability measures are employed in terms of relative gain array (RGA) as previously described (Gross *et al.*, 1998; McAvoy *et al.*, 2003; Skogestad and Postlethwaite, 2005).

The RGA gives information about the interactions among the controlled and manipulated variables. The RGA of a nonsingular matrix G, is a square complex matrix defined as: $\text{RGA}(G) = G \times (G^{-1})^T$, where G is the multivariable state space process model and T is the transpose of the corresponding matrix. The RGA element is defined as the ratio of open loop gain to the closed loop gain for a pair of variables. For a selected pair of variables, values of RGA element close to 1 are preferred and any other deviation suggests a weak relationship. The application of RGA can be extended to the frequency domain, which gives a deeper insight into process dynamics.

To compare the different control structures, we make use of RGA number versus frequency plots. The RGA number is defined as following: $\text{RGA number} = |\text{RGA} - I|_{sum}$, where I is the unity matrix representing the control structure. Note that pairings with low values of RGA number at a given frequency are preferred due to weak interactions (i.e., RGA is close

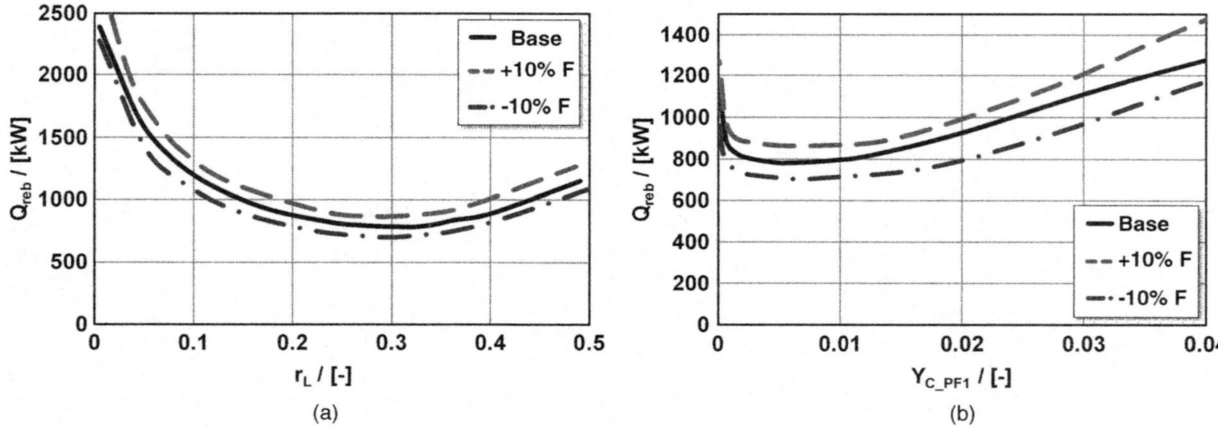

Figure 4.14 Control structures based on PID loops within a multi-loop framework: DB/LSV, DV/LSB, LB/DSV, and LV/DSB

to the unity matrix). For all the structures, the RGA number values are almost constant up to 0.5 rad min^{-1}, and start to change only after that. For all values of frequency, the DB/LSV configuration has the lowest RGA number, indicating that it as the most stable control structure. LV/DSV and LB/DSV control structures have the same RGA numbers for all frequencies, suggesting similar dynamic behavior.

4.7.2 Dynamic Simulations

For the dynamic simulations performed in this study, the purity set points (SPs) are 97% for all product specifications, while persistent disturbances of +10% in the feed flow rate (F) and +10% in the feed composition (x_A) were used for the dynamic scenarios. Even when the reported disturbances are exerted at the same time, no serious problems—such as instability or lack of capability in reaching the set points—were observed, as illustrated by the following figures.

Table 4.5 shows the tuning parameters of the PID controllers (Kiss and Rewagad, 2011). The control loops were tuned by a simple version of the direct synthesis method (Luyben and Luyben, 1997). According to this method, the desired closed-loop response for a given input is specified. Then, with the model of the process known, the required form and the tuning of the feedback controller are back-calculated. For all controllers, the acceptable control error ($\Delta\varepsilon_{max}$) and the maximum available control action (Δu_{max}) were specified. Then the controller gain, expressed in engineering units, was calculated as $K_c = \Delta u_{max}/\Delta\varepsilon_{max}$ and translated into percentage units. First order open-loop models were assumed to calculate the integral time of the pressure and concentration control loops. As fairly accurate evaluations of the process time constants τ, 20, 40, and 60 min were used, respectively. It can be shown (Luyben and Luyben, 1997) that the direct synthesis method requires that the reset time of a PI controller is equal to the time constant of the process (i.e., $\tau_i = \tau$). For the level controllers, a large reset time $\tau_i = 100$ min was chosen as no tight control is required for level set points.

The mole fractions of components A in the top distillate (x_A), B in the side stream (x_B), and C in the bottom product (x_C) return to their SP within reasonably short settling times. The dynamic response of the DB/LSV control structure, shown in Figure 4.15 (Kiss and Rewagad, 2011), is characterized by low overshooting and short settling times. However, the control structure DV/LSB exhibits a longer settling time as well as oscillations (Figure 4.16) (Kiss and Rewagad, 2011).

Table 4.5 Tuning parameters of the PID controllers

DB/LSV	P (%/%)	I (min)	D (min)	Control direction
$xA \rightarrow L$	3	40	0	−
$xB \rightarrow S$	3	20	0	+
$xC \rightarrow V$	3	40	0	−
$y_C \rightarrow r_L$	1	20	0	+
Tank level $\rightarrow D$	1	100	0	+
Reboiler level $\rightarrow B$	1	100	0	+
		DV/LSB		
$xA \rightarrow L$	1	60	0	−
$xB \rightarrow S$	1	40	0	+
$xC \rightarrow B$	1	60	0	+
$y_C \rightarrow r_L$	1	20	0	+
Tank level $\rightarrow D$	1	100	0	+
Reboiler level $\rightarrow V$	1	100	0	+
		LB/DSV		
$xA \rightarrow D$	3	60	0	+
$xB \rightarrow V$	3	40	0	−
$xC \rightarrow S$	3	40	0	−
$y_C \rightarrow r_L$	1	20	0	+
Tank level $\rightarrow L$	1	100	0	+
Reboiler level $\rightarrow B$	1	100	0	+
		LV/DSB		
$xA \rightarrow D$	1	40	0	+
$xB \rightarrow S$	1	20	0	+
$xC \rightarrow B$	1	40	0	+
$y_C \rightarrow r_L$	1	20	0	+
Tank level $\rightarrow L$	1	100	0	+
Reboiler level $\rightarrow V$	1	100	0	+

Figure 4.17 illustrates the case of the LB/DSV control structure, which shows settling times similar to the DB/LSV control structure (Kiss and Rewagad, 2011). However, the LV/DSB control structure shows a similar response to that of DB/LSV, being characterized by a longer settling time and oscillations (Figure 4.18) (Kiss and Rewagad, 2011).

The overall results of the dynamic simulations demonstrate that these control structures cope well with persistent disturbances in the feed flowrate and in the feed composition. Moreover, the DV/LSB control structure has a dynamic response similar to DB/LSV, while the LV/DSB control structure is similar to LB/DSV. However, the LV/DSB control structure leads to oscillations and longer settling times—which are in fact in line with the previous reports (Diggelen *et al.*, 2010). Controlling the liquid level in the reboiler by using the reboiler duty instead of the bottom flow rate, combined with controlling the level in the reflux drum by manipulating the reflux stream (when $L \gg D$), leads to oscillation in

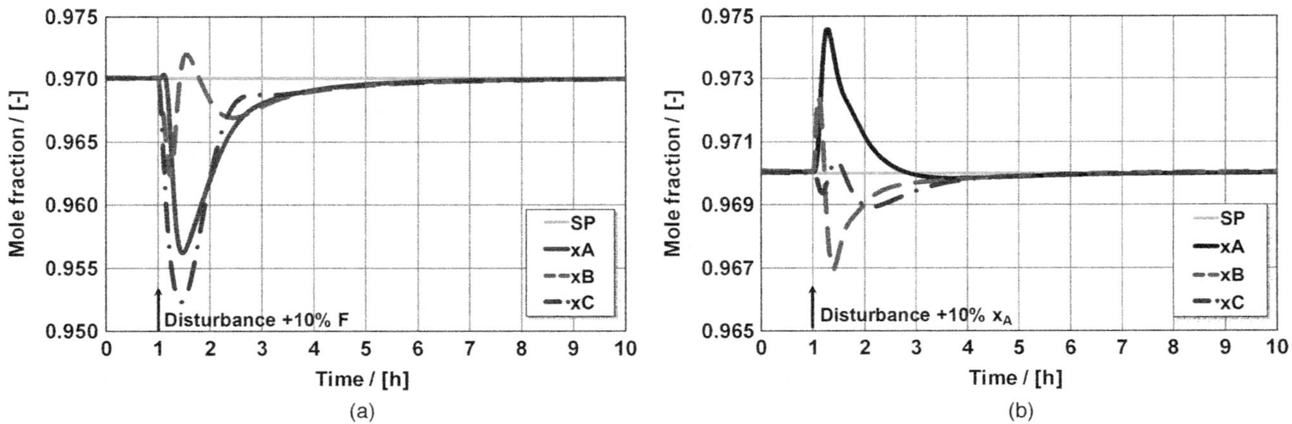

Figure 4.15 Dynamic response of the DB/LSV control structure, at a persistent disturbance of +10% in the feed flow rate (a) and +10% x_A in the feed composition (b)

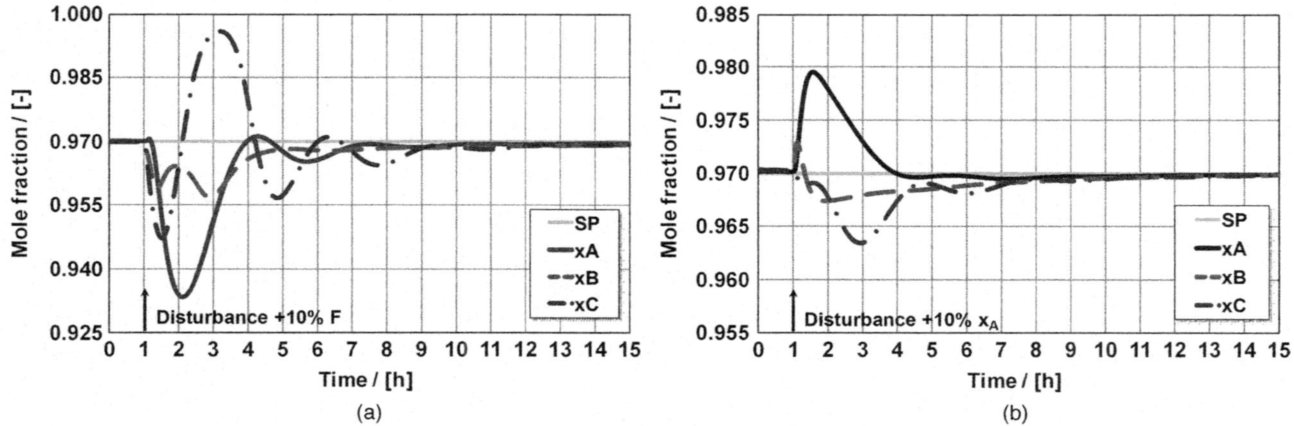

Figure 4.16 Dynamic response of the DV/LSB control structure, at a persistent disturbance of +10% in the feed flow rate (a) and +10% x_A in the feed composition (b)

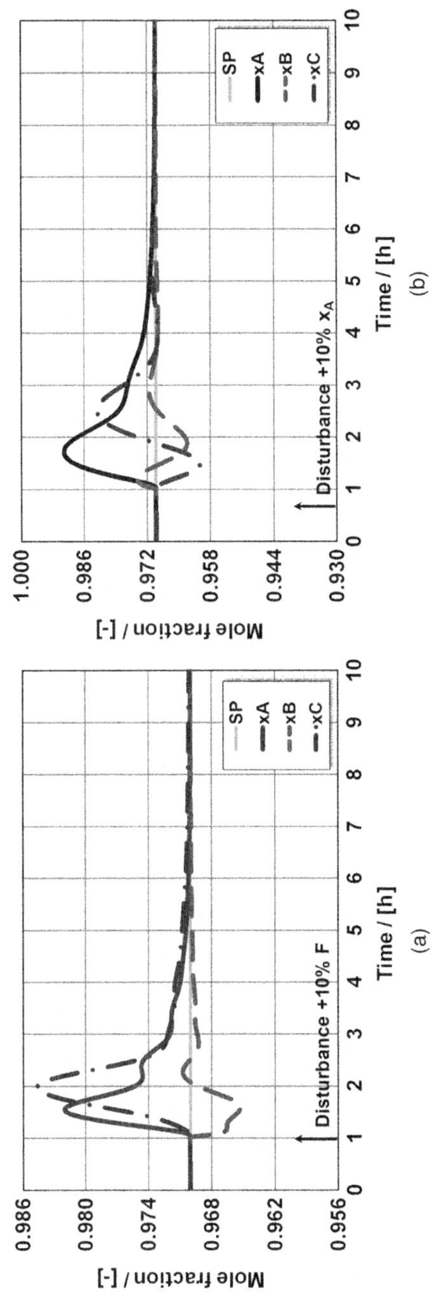

Figure 4.17 Dynamic response of the LB/DSV control structure, at a persistent disturbance of +10% in the feed flow rate (a) and +10% x_A in the feed composition (b)

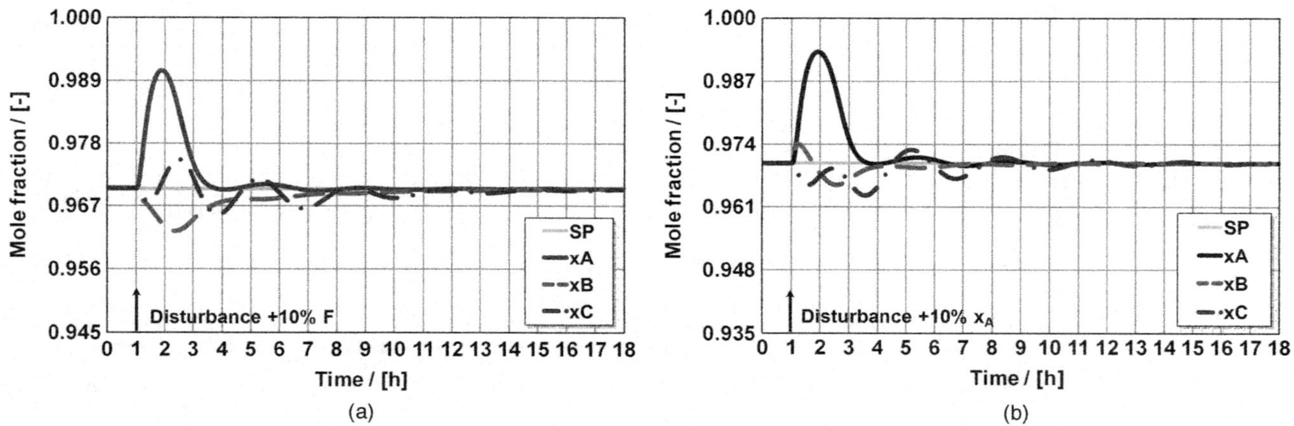

Figure 4.18 Dynamic response of the LV/DSB control structure, at a persistent disturbance of +10% in the feed flow rate (a) and +10% x_A in the feed composition (b)

the dynamic behavior. Even in case of simultaneous disturbances (Figure 4.19a), no serious problems are observed (Kiss and Rewagad, 2011). The same also holds true for changes in the set-point (Figure 4.19b) (Kiss and Rewagad, 2011).

Figure 4.20 compares the performance of all control structures (Kiss and Rewagad, 2011), in terms of the integral absolute error (IEA). The IAE value conveniently takes into account both the overshooting and the settling time of various dynamic responses. Basically, IAE is defined as:

$$\mathrm{IAE} = \int_0^t |e(t)|\mathrm{d}t \tag{4.1}$$

where e is the error as compared to the set point, over the time period t.

According to Figure 4.20, DB/LSV is the most stable control structure with a settling time of less than 7 h for all components, whereas LV/DSB is the most unstable with a settling time of over 14 h for all components (Kiss and Rewagad, 2011). LB/DSV and DV/LSB have settling times of less than 7 and 12 h, respectively—but with oscillations. Note that all control structures cope well with the exerted disturbances, exhibiting relatively low IAE values.

Looking at DB/LSV and DV/LSB structures shown in Figure 4.13 one can observe that there is a change in the control structure configuration to control the reboiler level (Kiss and Rewagad, 2011). In DV/LSB, changing the level using a steam (heat) supply leads to time delays due to the heat transfer and column operation that eventually serves as a cause for the oscillations. The composition profiles of both these structures are similar, except for the oscillations in p-xylene composition.

In the case of LB/DSV, the level in the reflux drum is controlled by manipulating the reflux, which is a potential cause of the minor oscillations and larger settling times. Note that controlling the slightly smaller flow-rate stream (in this case the distillate) is a better control strategy as any changes in smaller flow rate will be larger on a relative basis, hence providing better sensitivity for the controller. However, controlling the larger stream (in this case the reflux stream) leads to sensitivity and oscillations mainly in the toluene and xylene compositions. This is specific reasoning for the columns with larger reflux ratio and, hence, it should not be generalized.

Controlling the inventories in the column by manipulating the reboiler duty (for the general case) instead of the bottom flow rate, and liquid reflux (when $L \gg D$), generates oscillation and makes the column unstable. As LV/DSB uses both it is the most unstable configuration with largest settling

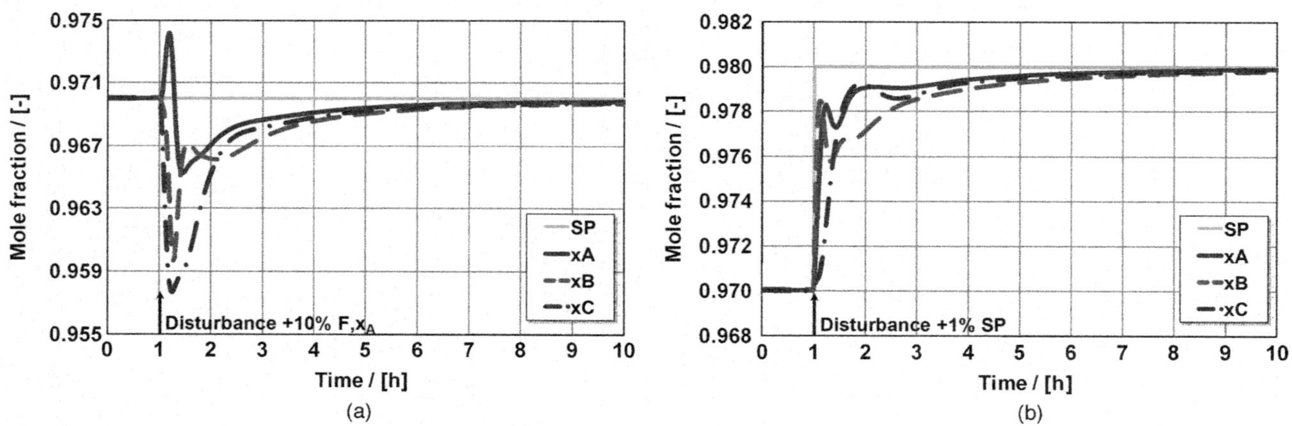

Figure 4.19 Dynamic response of the DB/LSV control structure, at a persistent combined disturbance of +10% in feed flow rate and +10% x_A in the feed composition (a) and +1% change in the set point (SP) (b)

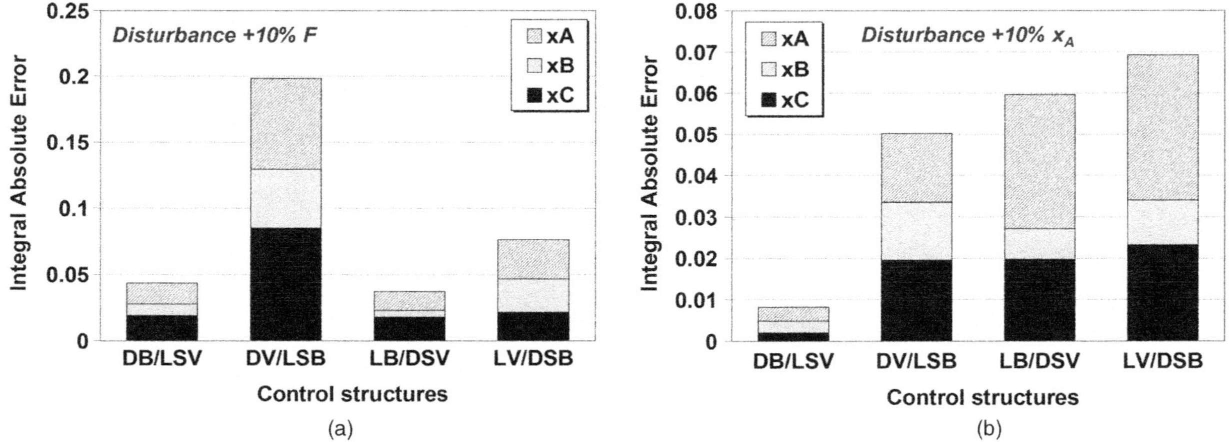

Figure 4.20 Comparison of performance of controllers in terms of integral absolute error (IEA) for a persistent disturbance of +10% in the feed flow rate (a) and +10% x_A in the feed composition (b)

Table 4.6 Absolute values of the relative gain array (RGA) for the multivariable process model

	Manipulated variables					
Controlled variables	r_L	L	V	S	D	B
x_A	0.0064	0.8474	0.2146	0.0700	0.0016	0
x_B	0.2817	1.5936	3.4802	3.1600	0	0
x_C	0.2483	1.4526	4.3175	2.1131	0	0
$y_{C,PF1}$	1.027	0.0112	0.0532	0.0150	0	0
H_T	0	0.0004	0.0012	0	0.9985	0
H_R	0	0	0	0	0	1

time and oscillations. This is similar to previously reported findings (Diggelen *et al.*, 2010). The dynamic response of LV/DSB is similar to that of LB/DSV but with larger oscillations. Oscillations in the xylene composition are comparatively larger than oscillations in benzene and toluene, which are also exhibited by the LB/DSV control structure.

Table 4.6 presents the absolute values of RGA for the system studied in this work (Kiss and Rewagad, 2011). These values suggest that manipulating DB for inventory control and LSV for regulatory control is the most favorable pairing. Among the proposed inventory control loops, the pairing of $h_t - L$ (liquid hold up in reflux tank and liquid reflux) and $h_r - V$ (liquid hold up in reboiler and vapor flow rate) has the lowest relative gain values and hence suggest weak coupling. This serves as a proof of the physical explanation provided above for the oscillations observed in case of DV/LSB and LV/DSB control structures.

4.8 CONCLUDING REMARKS

This literature overview indicates that dividing-wall columns have very good controllability properties, provided that an appropriate control structure is implemented. Among the multi-loop PID control strategies, DB/LSV and LB/DSV perform best, being able to handle persistent disturbances in reasonably short times. However, significantly shorter settling times and better control performance were achieved using advanced multi-input multi-output (MIMO) controllers such as model predictive control (MPC).

The DWC control structures presented here for the BTX separation—based on PID controllers in a multi-loop framework—are able to simultaneously control the product compositions and minimize the energy requirements in a very practical way. The energy optimization is based on a simple strategy, namely, to control the heavy component composition at the top of the prefractionator side of the DWC by manipulating the liquid split ratio. Remarkably, this optimal control condition is implicitly sufficient. The steady-state relationships show that maintaining or minimizing this composition leads to energy requirements that are near or at the minimum possible values as feed composition changes (Kiss and Rewagad, 2011).

The results of the dynamic simulations illustrate the feasibility of the proposed control structures. The DB/LSV and LB/DSV control structures give best performance in terms of low overshooting and short settling times. Moreover, based on the successful application to other relevant mixtures, it can be considered that these PID control structures are very well applicable to many other separations of ternary systems in a dividing-wall column.

REFERENCES

Abdul Mutalib, M.I. and Smith, R. (1998) Operation and control of dividing wall distillation columns, Part 1: Degrees of freedom and dynamic simulation. *Transactions of the Institution of Chemical Engineers, Part A*, *76*, 308–318.

Abdul Mutalib, M.I., Zeglam, A.O., and Smith, R. (1998) Operation and control of dividing wall distillation columns, Part 2: Simulation and pilot plant studies using temperature control. *Transactions of the Institution of Chemical Engineers, Part A*, *76*, 319–334.

Adrian, R., Schoenmakers, H., and Boll, M. (2004) MPC of integrated unit operations: Control of a DWC. *Chemical Engineering & Processing*, *43*, 347–355.

Alcántara-Ávila, J.R., Cabrera-Ruiz, J., Segovia-Hernández, J.G. *et al.* (2008) Controllability analysis of thermodynamically equivalent thermally coupled arrangements for quaternary distillations. *Chemical Engineering Research & Design*, *86*, 23–37.

Alstad, V. and Skogestad, S. (2007) Null space method for selecting optimal measurement combinations as controlled variables. *Industrial & Engineering Chemistry Research*, *46*, 846–853.

Bildea, C.S. (2004) On the controllability of Petlyuk distillation systems. *NPT Process Technologie*, *11*, 21–25.

Dejanović, I., Matijašević, Lj., and Olujić, Z. (2010) Dividing wall column - A breakthrough towards sustainable distilling. *Chemical Engineering and Processing: Process Intensification*, *49*, 559–580.

Dejanović, I., Matijašević, Lj., and Olujić, Z. (2011) An effective method for establishing the stage and reflux requirement of three-product dividing wall columns. *Chemical and Biochemical Engineering Quarterly*, *25*, 147–157.

van Diggelen, R.C., Kiss, A.A., and Heemink, A.W. (2010) Comparison of control strategies for dividing-wall columns. *Industrial & Engineering Chemistry Research*, *49*, 288–307.

Gabor, M. and Mizsey, P. (2008) A methodology to determine controllability indices in the frequency domain. *Industrial & Engineering Chemistry Research*, *47*, 4807–4816.

Gomez-Castro, F.I., Segovia-Hernandez, J.G., Hernandez, S. *et al.* (2008) Dividing wall distillation columns: Optimization and control properties. *Chemical Engineering & Technology*, *31*, 1246–1260.

Gross, F., Baumann, E., Geser, A. *et al.* (1998) Modelling, simulation and controllability analysis of an industrial heat-integrated distillation process. *Computers & Chemical Engineering*, *22*, 223–237.

Halvorsen, I.J. and Skogestad, S. (1999) Optimal operation of Petlyuk distillation: Steady-state behavior. *Journal of Process Control*, *9*, 407–424.

Halvorsen, I.J. and Skogestad, S. (1997) Optimizing control of Petlyuk distillation: Understanding the steady-state behaviour. *Computers & Chemical Engineering*, *21*, 249–254.

Halvorsen, I.J. and Skogestad, S. (2003a) Minimum energy consumption in multicomponent distillation. 1. Vmin diagram for a two-product column. *Industrial and Engineering Chemistry Research*, *42*, 596–604.

Halvorsen, I.J. and Skogestad, S. (2003b) Minimum energy consumption in multicomponent distillation. 2. Three-product Petlyuk arrangements. *Industrial and Engineering Chemistry Research*, *42*, 605–615.

Halvorsen, I.J. and Skogestad, S. (2003c) Minimum energy consumption in multicomponent distillation. 3. More than three products and generalized Petlyuk arrangements. *Industrial and Engineering Chemistry Research*, *42*, 616–629.

Halvorsen, I.J. and Skogestad, S. (2011) Energy efficient distillation. *Journal of Natural Gas Science and Engineering*, *3*, 571–580.

Harmsen, J. (2010) Process intensification in the petrochemicals industry: Drivers and hurdles for commercial implementation. *Chemical Engineering and Processing*, *49*, 70–73.

Il Kim, K., Lee, M., and Park, S. (2007) Dynamic simulation for the structural design of the divided wall column for different feed composition and various separation features, in *International Conference on Control, Automation and Systems, 2007, ICCAS '07*, IEEE, Piscataway, 235–239.

Kiss, A.A. and Bildea, C.S. (2011) A control perspective on process intensification in dividing-wall columns. *Chemical Engineering and Processing: Process Intensification*, *50*, 281–292.

Kiss, A.A. and Rewagad, R.R. (2011) Energy efficient control of a BTX dividing-wall column. *Computers & Chemical Engineering*, *35*, 2896–2904.

Ling, H. and Luyben, W.L. (2009) New control structure for divided-wall columns. *Industrial & Engineering Chemistry Research*, *48*, 6034–6049.

Ling, H. and Luyben, W.L. (2010) Temperature control of the BTX divided-wall column. *Industrial & Engineering Chemistry Research*, *49*, 189–203.

Lundström, P., Lee, J.H., Morari, M., and Skogestad, S. (1995) Limitations of dynamic matrix control. *Computers & Chemical Engineering*, *19*, 409–421.

Luyben, W.L. and Luyben, M.L. (1997) *Essentials of Process Control*, McGraw-Hill, New York.

McAvoy, T., Arkun, Y., Chen, R. *et al.* (2003) A new approach to defining a dynamic relative gain. *Control Engineering Practice*, *11*, 907–914.

Robles-Zapiain, S., Segovia-Hernandez, J.G., Bonilla-Petriciolet, A., and Maya-Yescas, R. (2008) Energy-efficient complex distillation sequences: Control properties. *Canadian Journal of Chemical Engineering*, *86*, 249–259.

Rong, B.G. and Turunen, I. (2006) A new method for synthesis of thermodynamically equivalent structures for Petlyuk arrangements. *Chemical Engineering Research & Design*, *84*, 1095–1116.

Segovia-Hernandez, J.G., Hernandez, S., and Jimenez, A. (2005) Analysis of dynamic properties of alternative sequences to the Petlyuk column. *Computers & Chemical Engineering*, *29*, 1389–1399.

Segovia-Hernandez, J.G., Hernandez-Vargas, E.A., and Marquez-Munoz, J.A. (2007) Control properties of thermally coupled distillation sequences for different operating conditions. *Computers & Chemical Engineering*, *31*, 867–874.

Serra, M., Espuña, A., and Puigjaner, L. (1999) Control and optimization of the divided wall column. *Chemical Engineering and Processing*, *38*, 549–562.

Serra, M., Espuña, A., and Puigjaner, L. (2003) Controllability of different multicomponent distillation arrangements. *Industrial and Engineering Chemistry Research*, *42*, 1773–1782.

Serra, M., Espuña, A., and Puigjaner, L. (2000) Study of the divided wall column controllability: influence of design and operation. *Computers & Chemical Engineering*, *24*, 901–907.

Serra, M., Perrier, M., Espuña, A., and Puigjaner, L. (2001) Analysis of different control possibilities for the divided wall column: feedback diagonal and dynamic matrix control. *Computers & Chemical Engineering*, *25*, 859–866.

Skogestad, S. and Postlethwaite, I. (2005) *Multivariable Feedback Control. Analysis and Design*, 2nd edn, John Wiley & Sons, Ltd., Chichester.

Tamayo-Galvan, V.E., Segovia-Hernandez, J.G., Hernandez, S. *et al.* (2008) Controllability analysis of alternate schemes to complex column arrangements with thermal coupling for the separation of ternary mixtures. *Computers & Chemical Engineering*, *32*, 3057–3066.

Triantafyllou, C. and Smith, R. (1992) The design and optimisation of fully thermally coupled distillation columns. *Transactions IChemE, Part A*, *70*, 118–132.

Wang, S.J. and Wong, D.S.H. (2007) Controllability and energy efficiency of a high-purity divided wall column. *Chemical Engineering Science*, *62*, 1010–1025.

Woinaroschy, A. and Isopescu, R. (2010) Time-optimal control of dividing-wall distillation columns. *Industrial & Engineering Chemistry Research*, *49*, 9195–9208.

Wolff, E.A. and Skogestad, S. (1995) Operation of integrated three-product (Petlyuk) distillation columns. *Industrial & Engineering Chemistry Research*, *34*, 2094–2103.

Yildirim, O., Kiss, A.A., and Kenig, E.Y. (2011) Dividing wall columns in chemical process industry: A review on current activities. *Separation and Purification Technology*, *80*, 403–417.

5

Advanced Control Strategies for DWC

5.1 INTRODUCTION

Today's manufacturing processes present many challenging control problems. Among these are nonlinear dynamic behavior, uncertain and time varying parameters, and unmeasured disturbances. During the recent decades, the control of these systems received considerable attention in both academia and industry. While advanced control strategies made the nonlinear process control much more practical, there is still a considerable gap between the control theory and the industrial practice. It is frustrating for the control theory community that elegant and comprehensive frameworks for system analysis and design are rarely applied in the chemical industry, which still applies the PID controllers (90% of cases), and relies on manual control in difficult situations (Rewagad and Kiss, 2012).

Much of the literature focuses on the control of binary distillation columns and there are only a limited number of studies on the (advanced) control of dividing-wall columns (DWCs) (Kiss and Bildea, 2011). The only advanced control strategy that has made a significant impact on the industrial scale is model predictive control (MPC) (Maciejowski, 2002; Agachi *et al.*, 2006; Mathur *et al.*, 2008). The success of MPC in the process industry can be attributed to several factors, such as (i) MPC handles multivariable control problems well, while taking into account the actuator limitations; (ii) MPC allows operation with inputs and outputs constraints while providing a robust optimization routine;

Advanced Distillation Technologies: Design, Control and Applications, First Edition.
Anton Alexandru Kiss.

(iii) control update rates are relatively low in chemical process industry applications and hence there is plenty of online computation time available; (iv) unlike PID controllers, MPC takes into account simultaneously the effects of all manipulated variables to all controlled variables. The application of MPC to DWC can amplify the advantages of both technologies in terms of process stability and optimal and improved performance (Morari and Lee, 1999).

In this chapter, the performances of control strategies and the dynamic response of the DWC is investigated in terms of product compositions for various persistent disturbances in the feed flow rate, feed composition, and set point changes in the product specifications with and without noise in the measurements. An industrially relevant case study—namely the ternary separation of benzene–toluene–xylene (BTX)—is used along this chapter, to illustrate the approach and make a fair comparison against the best conventional control alternatives.

5.2 OVERVIEW OF PREVIOUS WORK

Distillation is a classic example of a process that can be quite nonlinear (Skogestad and Postlethwaite, 2005). While various controllers are used for binary distillation columns, only several control structures were studied for DWC (Kiss and Bildea, 2011). In most cases, PID loops within a multiloop framework controllers were used to steer the system to the desired steady state and reach the goal of dynamic optimization (Halvorsen and Skogestad, 1997, 1999; Serra *et al.*, 1999, 2000; Hernández and Jiménez, 1999; Kim, 2002; Segovia-Hernandez *et al.*, 2007; Gabor and Mizsey, 2008; Cho *et al.*, 2009; Ling and Luyben, 2009, 2010). Despite the complex design and controllability issues, the use of advanced controllers for DWCs is even more limited. Serra *et al.* (2001) found the dynamic matrix control (DMC) to be deficient in DWC control. Compared to PID controllers, the application of DMC showed longer settling time for set point tracking and disturbance rejection. DMCs heavily depend on the linear step response model of the plant. Based on this model, optimal inputs are computed as solutions to a quadratic program. Depending on the size and sign of the step changes and nonlinearity of the system during identification of the response model, the control may converge or diverge (Lundström *et al.*, 1995; Dayal and MacGregor, 1996).

Woinaroschy and Isopescu (2010) showed the ability of iterative dynamic programming to solve time optimal startup control of a DWC, while Niggemann *et al.* (2011) focused on the modeling and in-depth analysis of the start-up of DWCs. All these different studies investigate different separation systems and types of disturbances and, hence, a common conclusion to identify the best controller could not be drawn (Kiss and Bildea, 2011).

Van Diggelen, Kiss, and Heemink (2010) published a study comparing various control structures based on PID loops versus more advanced controllers, including LQG/LQR, GMC, and high order controllers obtained by H_∞ and μ synthesis—but no optimal energy control was used. The LQG with integral action and reference inputs was found to deliver the best control performance.

In a study reported by Adrian *et al.* (2004), MPC outperformed the PID controller while simultaneously taking into account a larger number of manipulated variables. A black box approach using commercial software was applied in the identification of prediction model and development of the controller, which restricts the understanding to a large extent. In addition, the tuning parameters of the MPC controller were not provided. More recently, Buck *et al.* (2011) also reported the experimental implementation of MPC using the temperature profile of the column. Although these experimental studies prove the real-life practicability of MPC, they do not provide an analysis of the transient behavior of DWC under persistent disturbances.

Kvernland *et al.* (2010) applied MPC only to a particular case of DWC, namely, the Kaibel column that separates a feed stream into four product streams using only a single column shell. The objective for optimal operation of the column was chosen to be the minimization of the total impurity flow. When the Kaibel column was exposed to disturbances, the MPC obtained typically less total impurity flow than conventional decentralized control and it was also able to counteract process interactions better than decentralized control.

Following a literature review, it is clear that efforts are being made to develop the control strategies for DWC. As the distillation process is a multivariable process, this leads to a multivariable control problem. Based on the benefits mentioned earlier, MPC seems to be a worthwhile option to optimally control a multivariate, nonlinear, and constrained process such as DWC. However, the control of a DWC using MPC has been successfully studied (but only to a certain degree) to date only by Adrian *et al.* (2004), Kvernland *et al.* (2010), and Buck *et al.* (2011). Moreover, their results do now allow a direct and fair comparison with other separation systems previously reported Van Diggelen, Kiss, and Heemink (2010).

Therefore, there is a strong need for further investigation of the application of MPC to DWC. Rewagad and Kiss (2012) made use of one ternary system (BTX) and compared MPC (alone or not) with the best multi-loop PID control strategy. The internal prediction model used by MPC was derived from linearization of the nonlinear distillation model, and not from step–response experiments. Such a method is more accurate as the resulting first-principle linear model represents all the states, the same as the nonlinear model and it is not limited to the range of identification experiments (Zhu, 1998; Maciejowski, 2002; Gabor and Mizsey, 2008).

5.3 DYNAMIC MODEL OF A DWC

To allow a fair comparison of the results with previously published references, we consider as case-study the industrially relevant ternary separation in a DWC of a BTX (benzene–toluene–xylene) mixture. Table 5.1 lists the physical properties of these components Van Diggelen, Kiss, and Heemink (2010).

The modeled DWC consists of six sections of eight stages each (Figure 5.1) (Rewagad and Kiss, 2012). The feed stream is an equimolar mixture of benzene, toluene, and xylene (denoted ABC for convenience) that is fed into the prefractionator side, between section 1 and 2. Benzene is obtained as top distillate, xylene as bottom product, while toluene is withdrawn as side stream of the main column (between sections 4 and 5).

The dynamic model proposed here is used to develop control strategies—hence it is recommended to use linearized liquid dynamics instead of neglecting the liquid dynamics. When the liquid dynamics are not neglected but simplified by a linearization, the initial response is more realistic. Hence, linearized liquid dynamics are incorporated in the model. Note that it is rather impractical to control the vapor split

Table 5.1 Physical properties of the investigated ternary system: benzene–toluene–xylene (BTX)

Physical property (units)	Benzene	Toluene	Xylene
Molecular formula	C_6H_6	C_7H_8	C_8H_{10}
Molecular weight	78.11	92.14	106.17
Density (kg m^{-3})	878.6	866.9	860.0
Viscosity (cP at 20 °C)	0.652	0.590	0.620
Critical pressure (bar)	48.95	41.08	35.11
Critical temperature (°C)	288.9	318.6	343.05
Melting temperature (°C)	5.53	−94.97	13.26
Boiling temperature (°C)	80.09	110.63	138.36

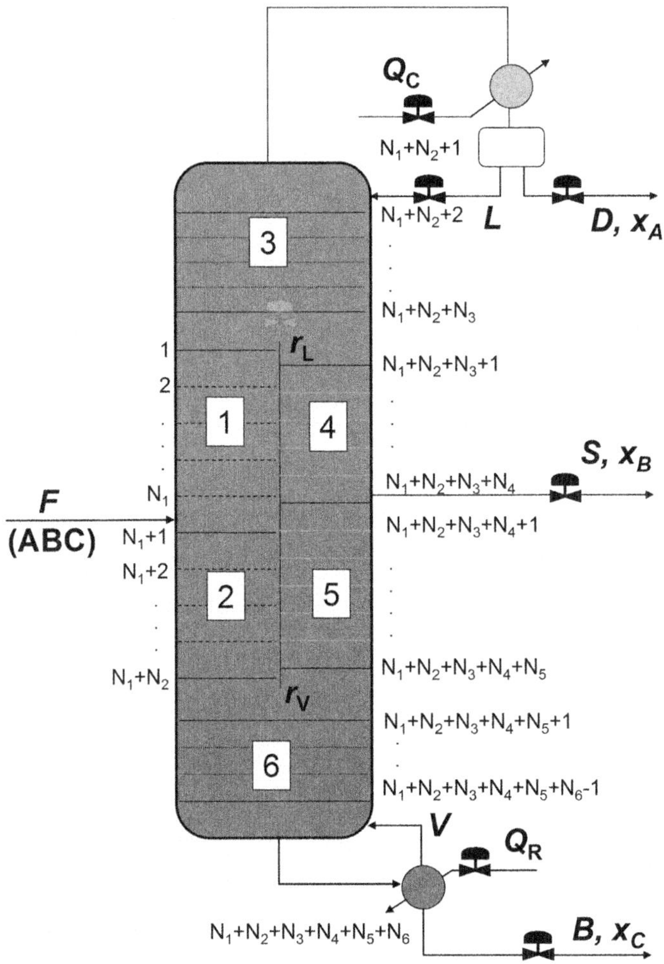

Figure 5.1 Schematics of the simulated dividing-wall column (DWC): six sections of eight stages each

and, hence, it is considered as a design variable—although with a variable vapor split the energy loss in the presence of feed disturbances is significantly lower than with a fixed vapor split (Dwivedi *et al.*, 2012).

For the dynamic model several reasonable simplifying assumptions were made Van Diggelen, Kiss, and Heemink (2010): (i) constant pressure, (ii) no vapor flow dynamics, (iii) linearized liquid dynamics, and (iv) the energy balances and changes in enthalpy were neglected. Although the model is relatively simple, it does capture all the essential elements required to describe the system. The full dynamic model was implemented in MathWorks MATLAB® and Simulink® and is based on the Petlyuk model previously reported in the literature by Halvorsen and Skogestad (1997). Note that the DWC is thermodynamically equivalent to the Petlyuk system that is modeled using the following equations:

$$\begin{aligned} \dot{x} &= \mathrm{f}(\boldsymbol{x}, \boldsymbol{u}, \boldsymbol{d}, \boldsymbol{t}) \\ y &= \mathrm{g}(\boldsymbol{x}) \end{aligned} \tag{5.1}$$

where $\boldsymbol{u} = [L_0 S\ V_0 D\ B\ R_L R_V]$ is the input vector, $\boldsymbol{d} = [F\ z_1 z_2 q]$ is the disturbance vector, $\boldsymbol{x}$ is the state vector consisting of 104 compositions for the first two components and 52 liquid hold-ups, and $\boldsymbol{y} = [x_A x_B x_C H_T H_R]$ is the output vector.

For the *state equations*, numbering is from top to bottom starting in the prefractionator and then continuing in the main column. Overall, the DWC has six sections that are conveniently illustrated in Figure 5.1:

$$H_i \frac{\mathrm{d}x_{i,j}}{\mathrm{d}t} = L_{i,j}(x_{\text{liquid\_split},j} - x_{i,j}) + V_{i-1,j}(y_{i-1,j} - y_i);\ \text{for}\ i = 1 \tag{5.2}$$

$$H_i \frac{\mathrm{d}x_{i,j}}{\mathrm{d}t} = L_{i,j}(x_{i+1,j} - x_{i,j}) + V_{i-1,j}(y_{i-1,j} - y_i);\ \text{for}\ i = 2 \ldots 8 \tag{5.3}$$

$$H_i \frac{\mathrm{d}x_{i,j}}{\mathrm{d}t} = L_{i,j}(x_{\text{in\_2},j} - x_{i,j}) + V_{i-1,j}(y_{i-1,j} - y_i);\ \text{for}\ i = 9 \tag{5.4}$$

$$H_i \frac{\mathrm{d}x_{i,j}}{\mathrm{d}t} = L_{i,j}(x_{i+1,j} - x_{i,j}) + V_{i-1,j}(y_{i-1,j} - y_i);\ \text{for}\ i = 10 \ldots 16 \tag{5.5}$$

$$H_i \frac{\mathrm{d}x_{i,j}}{\mathrm{d}t} = L_{i,j}(x_{\text{in\_3},1} - x_{i,j}) + V_{i-1,j}(y_{i-1,j} - y_\mathrm{i});\ \text{for}\ i = 17 \tag{5.6}$$

$$H_i \frac{\mathrm{d}x_{i,j}}{\mathrm{d}t} = L_{i,j}(x_{i+1,j} - x_{i,j}) + V_{i-1,j}(y_{i-1,j} - y_i);\ \text{for}\ i = 18 \ldots 23 \tag{5.7}$$

$$H_i \frac{\mathrm{d}x_{i,j}}{\mathrm{d}t} = L_{i,j}(x_{i+1,j} - x_{i,j}) + V_{i-1,j}(y_{\text{in\_3},j} - y_i);\ \text{for}\ i = 24 \tag{5.8}$$

$$H_i \frac{\mathrm{d}x_{i,j}}{\mathrm{d}t} = L_{i,j}(x_{\text{in\_4},j} - x_{i,j}) + V_{i-1,j}(y_{i-1,j} - y_i);\ \text{for}\ i = 25 \tag{5.9}$$

$$H_i \frac{\mathrm{d}x_{i,j}}{\mathrm{d}t} = L_{i,j}(x_{i+1,j} - x_{i,j}) + V_{i-1,j}(y_{i-1,j} - y_i);\ \text{for}\ i = 26 \ldots 32 \tag{5.10}$$

$$H_i \frac{\mathrm{d}x_{i,j}}{\mathrm{d}t} = L_{i,j}(x_{\text{in\_25},j} - x_{i,j}) + V_{i-1,j}(y_{i-1,j} - y_i);\ \text{for } i = 33 \tag{5.11}$$

$$H_i \frac{\mathrm{d}x_{i,j}}{\mathrm{d}t} = L_{i,j}(x_{i+1,j} - x_{i,j}) + V_{i-1,j}(y_{i-1,j} - y_i);\ \text{for } i = 34\ldots40 \tag{5.12}$$

$$H_i \frac{\mathrm{d}x_{i,j}}{\mathrm{d}t} = L_{i,j}(x_{\text{in\_6},j} - x_{i,j}) + V_{i-1,j}(y_{i-1,j} - y_i);\ \text{for } i = 41 \tag{5.13}$$

$$H_i \frac{\mathrm{d}x_{i,j}}{\mathrm{d}t} = L_{i,j}(x_{i+1,j} - x_{i,j}) + V_{i-1,j}(y_{i-1,j} - y_i);\ \text{for } i = 42\ldots48 \tag{5.14}$$

where the liquid holdup is given by:

$$\frac{\mathrm{d}H_i}{\mathrm{d}t} = L_{i+1} - L_i \tag{5.15}$$

and the vapor composition is calculated by:

$$y_{i,j} = \frac{\alpha_j}{\sum_j \alpha_j x_{i,j}} \tag{5.16}$$

for the components $j \in \{1, 2, 3\}$ and relative volatilities α.

Furthermore, some special concentrations have to be specified:

$$x_{\text{in\_2},j} = (L_8 x_{8,j} + F_0 z_j - (1-q)F_0 y_{9,j})/L_9;\quad \text{for input section 2, tray no. 9} \tag{5.17}$$

$$x_{\text{in\_1},j} = x_{\text{liquid\_split}};\ \text{for input section 1, tray no. 1} \tag{5.18}$$

$$x_{\text{in\_4},j} = x_{\text{liquid\_split}};\ \text{for input section 4, tray no. 25} \tag{5.19}$$

$$x_{\text{in\_6},j} = (L_{16}x_{16} + L_{40}x_{40})/x_{41};\ \text{liquid input section 6, tray no. 41} \tag{5.20}$$

$$y_{\text{in\_3},j} = (V_1 y_1 + V_4 y_{25})/V_{24};\ \text{vapor input section 3, tray no. 24} \tag{5.21}$$

$$x_{\text{in\_5},j} = x_{\text{side\_splitter}};\ \text{liquid input section 5, after liquid splitter} \tag{5.22}$$

The *dynamics of the liquid splitter*, located between section 3 and sections 1/4, are:

$$\begin{aligned} \frac{d}{dt}H_{\text{liquid\_split}} &= L_{24} - (L_1 + L_{25}) \\ H_{\text{liquid\_split}}\frac{d}{dt}x_{\text{liquid\_split}} &= L_{24}(x_{24} - x_{\text{liquid\_split}}) \end{aligned} \tag{5.23}$$

The *dynamics of the side splitter*, of the liquid side splitter located between section 3 and sections 4/5, are given by:

$$\begin{aligned} \frac{d}{dt}H_{\text{side\_split}} &= L_{32} - L_{33} - S \\ H_{\text{side\_split}}\frac{d}{dt}x_{\text{side\_split}} &= L_{32}(x_{32} - x_{\text{side\_split}}) \end{aligned} \tag{5.24}$$

The *dynamics of the reboiler*, situated between section 3 and sections 1/4, are given by:

$$\begin{aligned} \frac{d}{dt}H_{\text{reboiler}} &= L_6 - V_0 - B \\ H_{\text{reboiler}}\frac{d}{dt}x_{\text{reboiler}} &= L_{\text{in}}(x_{\text{in}} - x_{\text{reboiler}}) - V_0(y_{\text{reboiler}} - x_{\text{reboiler}}) \end{aligned} \tag{5.25}$$

The *dynamics of the reflux tank*, situated between section 3 and sections 4/5, are given by:

$$\begin{aligned} \frac{d}{dt}H_{\text{reflux\_tank}} &= V_0 - L_0 - D \\ H_{\text{reflux\_tank}}\frac{d}{dt}x_{\text{reflux\_tank}} &= V_0(x_{\text{in}} - x_{\text{reflux\_tank}}) \end{aligned} \tag{5.26}$$

Liquid flow rates:

$$L_{i,j} = L_0 R_L;\ \text{for}\, i = 1\ldots 8 \tag{5.27}$$

$$L_{i,j} = L_0 R_L + qF_0;\ \text{for}\, i = 9\ldots 16 \tag{5.28}$$

$$L_{i,j} = L_0;\ \text{for}\, i = 17\ldots 24 \tag{5.29}$$

$$L_{i,j} = L_0(1 - R_L);\ \text{for}\, i = 25\ldots 32 \tag{5.30}$$

$$L_{i,j} = L_0(1 - R_L) - S;\ \text{for}\ i = 33 \ldots 40 \tag{5.31}$$

$$L_{i,j} = L_0 + qF_0 - S;\ \text{for}\ i = 41 \ldots 48 \tag{5.32}$$

Vapor flow rates:

$$V_{i,j} = V_0 R_v + (1 - q)F_0;\ \text{for}\ i = 1 \ldots 8 \tag{5.33}$$

$$V_{i,j} = V_0 R_v;\ \text{for}\ i = 9 \ldots 16 \tag{5.34}$$

$$V_{i,j} = V_0 R_v + (1 - q)F_0 + V_0(1 - R_v);\ \text{for}\ i = 17 \ldots 24 \tag{5.35}$$

$$V_{i,j} = V_0(1 - R_v);\ \text{for}\ i = 25 \ldots 32 \tag{5.36}$$

$$V_{i,j} = V_0(1 - R_v);\ \text{for}\ i = 33 \ldots 40 \tag{5.37}$$

$$V_{i,j} = V_0 R_v;\ \text{for}\ i = 9 \ldots 16 \tag{5.38}$$

Note that the DWC model considered here makes use of theoretical stages, hence there is no difference made in terms of column internals (trays or packing). Nevertheless, from a practical viewpoint, the HETP is assumed to be the same on both sides of the column if packing is used as internals. Moreover, the potential HETP differences between the two sides of the column can be avoided by proper design as well as control measures Van Diggelen, Kiss, and Heemink (2010).

Figure 5.2 provides the residue curve map (RCM) and the composition profile inside the DWC by means of a ternary diagram (Rewagad and Kiss, 2012). The bottom, side, and top product are close to the left-land, top, and right-hand corners, respectively. In this work, the steady state purity of all the product streams is considered to be 97% to allow a direct and fair comparison with previous reports.

A local *linearized model* around the steady state $(\boldsymbol{x}^*, \boldsymbol{u}^*)$ is obtained by numerical differentiation using the formula:

$$\frac{\partial f(x)}{\partial x} = \frac{f(x + h) - f(x - h)}{2h} \tag{5.39}$$

Hence a state space model is obtained by computing the derivative of the functions f and g with respect to alternating x or u. The model has 156 states (96 compositions for A and B, eight compositions for A and B in the two splitters, reflux tank and reboiler; 48 tray hold-ups plus four

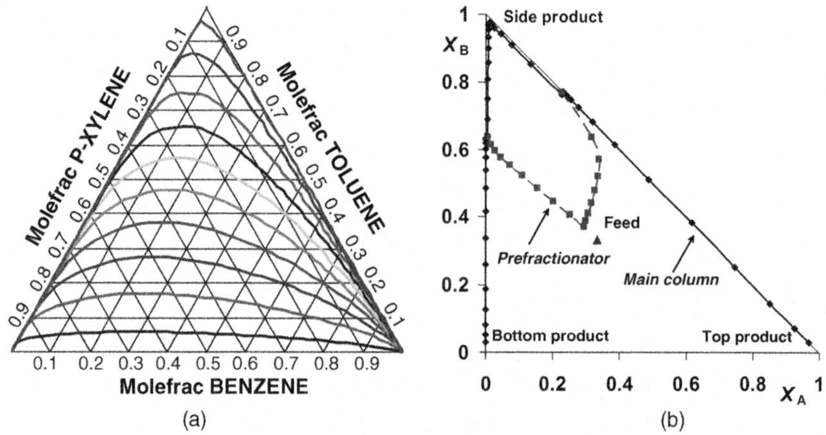

Figure 5.2 Residue curve map (RCM) of the BTX ternary mixture (a) and composition profile inside the dividing-wall column as a ternary diagram (b)

additional hold-ups for the two splitters, reflux tank and reboiler) and 11 inputs (five control, two design, and four disturbance variables), just as the nonlinear model. The approximation of the linear model of the process must be checked to avoid any mismatch with the nonlinear process model and consequent failure of the control system. Notably, the linear models are generally good around the nominal operating point where the linearization is performed. The full nonlinear process model was linearized to obtain the continuous state space model. The resulting state space model has 156 states (96 compositions for A and B; eight compositions of A and B in the vapor, liquid splitters, reflux tank, and reboiler; 48 tray hold ups; four hold ups for the vapor, liquid splitters, reflux tank, and reboiler), six inputs and six outputs representing the controlled and manipulated variables chosen.

The quality of the linearization is evaluated by performing open loop simulations and exerting disturbances. The feed flow-rate (F) was varied up to $\pm 20\%$ of their respective nominal value and the deviations in the bottom xylene composition (x_C) were analyzed. This disturbance and test variable were selected due to its dominant first-order time constant. Therefore, it serves as a worst case scenario. Figure 5.3 shows the open loop response of the linearized and nonlinear model (Rewagad and Kiss, 2012). For $\pm 10\%$ disturbances, there is a perfect match between the responses of the linearized and nonlinear model. Minor differences can be observed for the larger disturbances only, and these differences increase when moving away from the nominal operating point.

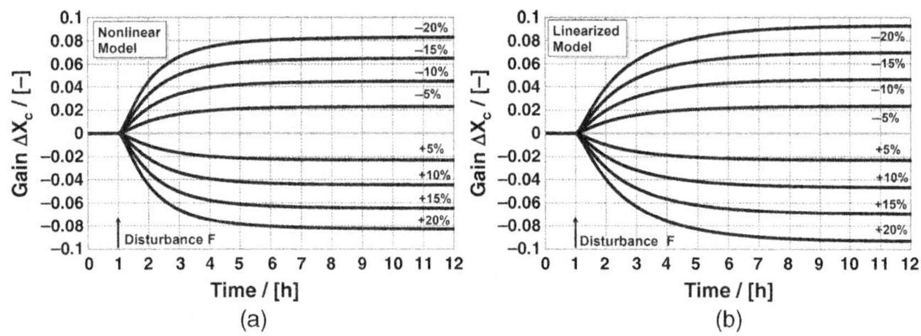

Figure 5.3 Open loop response of the nonlinear (a) and linearized model (b), at various persistent disturbances in the feed flow rate

5.4 CONVENTIONAL VERSUS ADVANCED CONTROL STRATEGIES

This section compares conventional three-point control strategies based on PID loops, within a multi-loop framework, and more advanced controllers such as LQG/LQR, GMC, and high order controllers obtained by H_∞ controller synthesis and μ-synthesis.

5.4.1 PID Loops within a Multi-loop Framework

The controllers used most in industry are the PID controllers (Johnson and Moradi, 2005). For a DWC, two multi-loops are needed to stabilize the column and another three loops to maintain the set points specifying the product purities. While there are six actuators (D S B $L_0 V_0$ R_L) using PID loops within a multi-loop framework, many combinations are possible. However, there are only a few configurations that make sense from a practical viewpoint. The level of the reflux tank and the reboiler can be controlled by the variables L_0, D, V_0, and B. Hence, there are four so-called *inventory control* options to stabilize the column, the combinations: D/B, L/V, L/B, and V/D to control the level in the reflux tank and the level in the reboiler (Figure 5.4) Van Diggelen, Kiss, and Heemink (2010). The part for the control of product purities is often called the *regulatory control*. One actuator is left (R_L), which can be used for optimization purposes such as minimizing the energy requirements, by controlling the heavy impurity in the top of the prefractionator section (Halvorsen and Skogestad, 1997; Ling and Luyben, 2009; Kiss and Rewagad, 2011).

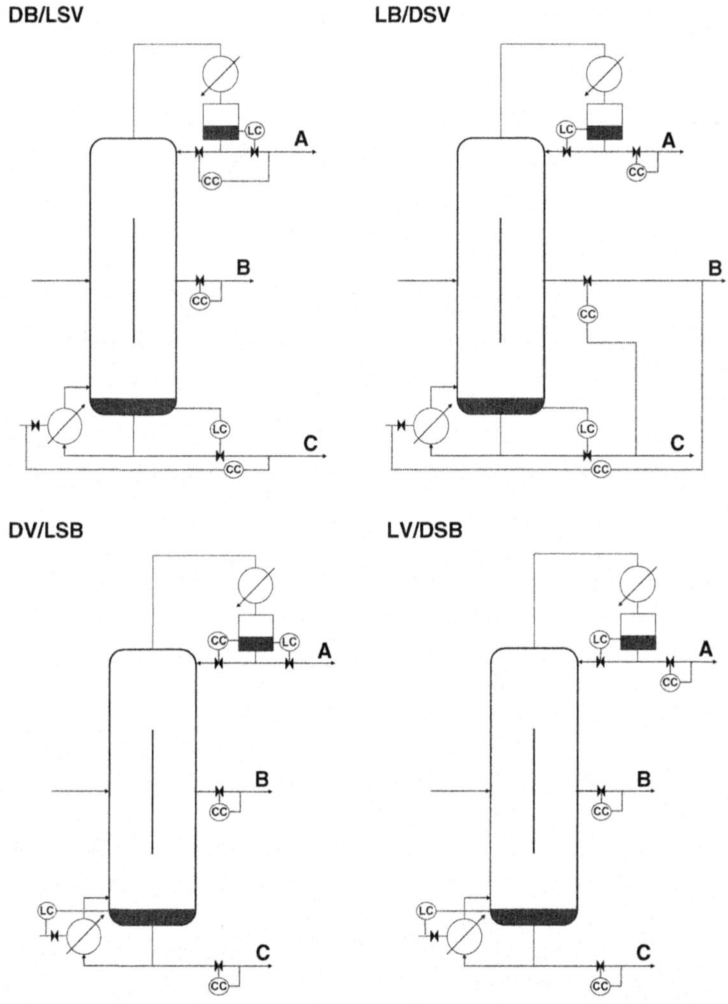

Figure 5.4 Control structures based on PID loops within a multi-loop framework: DB/LSV, DV/LSB, LB/DSV, and LV/DSB

Figure 5.5 shows the RGA number versus frequency plot Van Diggelen, Kiss, and Heemink (2010). This clearly distinguishes between the LV/DSB and DB/LSV control structures, where the DB/LSV option is preferable to LV/DSB. However, the RGA numbers for the other structures (LB/DSV and DV/LSB) have similar values, located between the RGA number of the LV/DSB and DB/LSV structures.

Note that the closed-loop stability of the decentralized PID controls structure is still an unsolved problem (Johnson and Moradi, 2005).

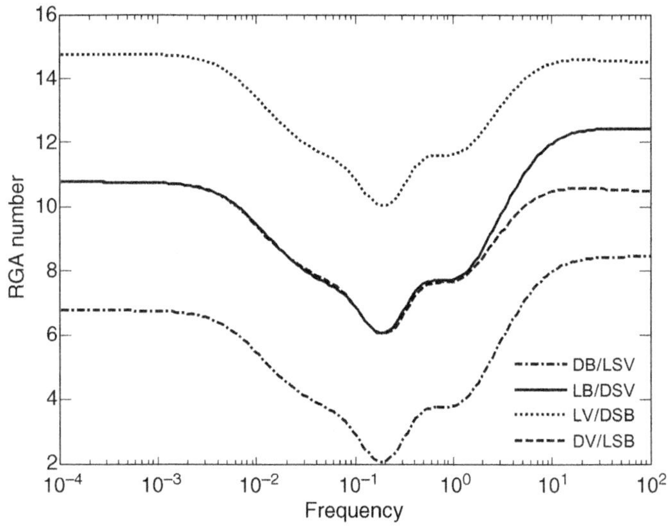

Figure 5.5 RGA number versus frequency, for the PID loops within a multi-loop framework

PID loops within a multi-loop framework imply a tuning problem with many solutions and, hence, they are difficult to solve. Nevertheless, the PID loops within a multi-loop framework are more or less model independent. Hence, if the true plant is quite different from the model, it is likely that the control system will still work. PI controllers in a multi-loop framework control the system via a matrix structure with only one PI controller on each column. The full order multi-input multi-output (MIMO) problem has been successfully solved and it has in addition a useful cost criterion: linear quadratic Gaussian control.

5.4.2 Linear Quadratic Gaussian Control

Linear quadratic Gaussian control (LQG) is a combination of an optimal controller LQR *linear quadratic regulation* and optimal state estimator (Kalman filter) based on a linear state-space model with measurement and process noise that minimizes the cost function:

$$J_{LQ} = \int_0^{\infty} \boldsymbol{x}(t)^{\mathrm{T}} Q \boldsymbol{x}(t) + \boldsymbol{u}(t)^{\mathrm{T}} R \boldsymbol{u}(t) \mathrm{d}t \tag{5.40}$$

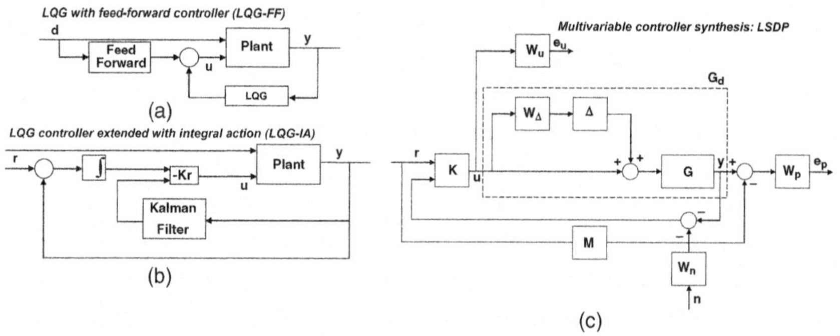

Figure 5.6 LQG with feed-forward controller (a), LQG controller extended with integral action (b); closed-loop interconnection structure of the DWC system with weighted outputs (c)—the dashed box represents a plant G_d from the uncertainty set

LQG is an extension of optimal state feedback that is a solution of the LQR that assumes no process noise and that the full state is available for control. Since in the case of a DWC the full state is a priori unavailable, and measurement noise and disturbances are assumed in the feed, the state should be estimated taking into account the disturbances. While the LQG control deals only with zero-mean stochastic noise it is not suitable for dealing with persistent disturbances in the feed. The resulting offset can be solved using an additional feed-forward controller—structure as shown in Figure 5.6 Van Diggelen, Kiss, and Heemink (2010). For example, if the feed flow rate increases persistently by 10% the product flow rates also increase by 10% in order to reach a steady state. By measuring the feed flow rate the changes can be used directly to adapt the product flow rates with the same percentage. However, for persistent disturbances in feed composition and condition it is more difficult to tune the feed-forward controller.

A working solution is to extend the LQG controller with an integral action (Skogestad and Postlethwaite, 2005). The resulting controller structure is also shown in Figure 5.6 Van Diggelen, Kiss, and Heemink (2010). With an LQR controller there is no tuning problem while the optimal feedback controller is given via an algebraic Ricatti equation (ARE). In addition, the closed-loop system is stable with respect to zero-mean white noise. However, the obtained control structure depends heavily on the linear model used. Hence, a realistic linearized model is needed: with multivariate controller synthesis, robust stability and robust performance can be obtained with respect to model uncertainty, and with nonlinear control the linearization step can be avoided.

Nonlinear Control

For distillation columns, nonlinear control was previously explored by Rouchon (1990). Basically, the PID based controllers and LQG are linear controllers that need a linearization step to control a nonlinear system. Most likely this leads to a loss in control action while the plant behaves nonlinearly. For the control of binary distillation, nonlinear control has been successfully performed (Isidori, 1989; Rouchon, 1990). For example, it is possible to control a binary distillation column using an input–output linearizing (IOL) controller (Biswas *et al.*, 2007). However the controller is based on a reduced model that only considers the bottom purity in the reboiler and the top purity in the reflux drum.

5.4.3 Generic Model Control

Generic model control (GMC) is a process model-based control algorithm using the nonlinear state-space model of the process, and it is a special IOL case if the system has a relative order of one (Lee and Sullivan, 1988; Signal and Lee, 1993; Van Diggelen, Kiss, and Heemink 2010). This can be done directly by solving the nonlinear equation for the input u:

$$\frac{\partial \widehat{g}}{\partial x^{\mathrm{T}}} f(x, u, d, t) = K_1(y^* - y) + K_2 \int_0^{\mathrm{t}} (y^* - y)\mathrm{d}t \tag{5.41}$$

The interpretation is that the derivative of the output y with respect to time follows the predefined PI-control signal at the right-hand side. The full state is needed and in cases where the plant is approximated by a linear model the left-hand side can be replaced with the linear equivalent, where $y*$ is a vector of set points. The closed loop-nominal system is stable when the open-loop model is minimum phase. A linear continuous time system is minimal phase when all poles and zeros are in the left-hand plane. From the pole zero map it can be concluded that this is not the case Van Diggelen, Kiss, and Heemink (2010). As a result, nonlinear control techniques like IOL and GMC are not used hereafter.

5.4.4 Multivariable Controller Synthesis

After selecting a pairing, the design of a diagonal PI structure leads to a suboptimal design. In addition, the LQG/LQR controller has no guaranteed stability margin, which possibly leads to problems in the case of model uncertainties. The following two advanced controller synthesis methods were used to obtain a robust controller: the loop shaping design procedure (LSDP) and the μ-synthesis procedure. Both methods were successfully applied for controller synthesis for a binary distillation column (Gu, 2005). However, in contrast to the approach of Gu (2005), the inventory control and regulatory control problems are solved simultaneously here.

5.4.4.1 Loop Shaping Design Procedure (LSDP)

By carrying out the H_∞ loop shaping design procedure—performed in MATLAB using the command *ncfsyn*—the plant is shaped with a pre-compensator ($\boldsymbol{W}_1$) and a post-compensator ($\boldsymbol{W}_2$) that is the identity matrix. $\boldsymbol{W}_1$ is a diagonal matrix with the following transfer function on the diagonal (for $i = 1 \ldots 5$):

$$\boldsymbol{W}_1(i,i) = 2 \cdot \frac{s+1}{10s} \tag{5.42}$$

The value 2 is chosen for the gain of the filter, to ensure a small steady-state error. Larger gains lead to smaller steady-state errors but worse transient response (Gu, 2005). In addition, for larger values the closed loop system is unstable in the presence of the measurements noise and time delay. The structured value μ has a maximum of 0.7686. Hence, the closed loop system is stable with respect to the modeled uncertainty. Figure 5.6c shows the closed loop system and the weightings Van Diggelen, Kiss, and Heemink (2010).

5.4.4.2 Multivariable Controller μ-Synthesis (DK Iteration Procedure)

The linear model plant G is expanded with input multiplicative uncertainty to obtain a disturbed plant G_d. The input disturbance is an uncertain gain combined with an uncertain delay:

$$\boldsymbol{W}_\mathrm{u} = \begin{pmatrix} k_1 e^{-\Theta_1 s} & & \\ & \ddots & \\ & & k_1 e^{-\Theta_1 s} \end{pmatrix} \tag{5.43}$$

where $k_i \in [0.8, 1.2]$ and $\Theta_i \in [0,1]$ for $i = 1, \ldots, 5$. The matrix $\boldsymbol{W}_u$ can be split into two matrices:

$$\boldsymbol{\Delta} = \begin{pmatrix} \Delta_1 & & \\ & \ddots & \\ & & \Delta_5 \end{pmatrix} \qquad \mathbf{W}_{\Delta} = \begin{pmatrix} W_{\Delta_1} & & \\ & \ddots & \\ & & W_{\Delta_5} \end{pmatrix} \tag{5.44}$$

where $|\Delta_i| \leq 1$ for $i = 1, \ldots, 5$. The functions in the matrix $\boldsymbol{W}_{\Delta}$ are obtained via a fitting procedure in the frequency domain (Gu, 2005):

$$W_{\Delta_i} = \frac{2.2138s^3 + 15.9537s^2 + 27.6702s + 4.9050}{1s^3 + 8.3412s^2 + 21.2393s + 22.6705}, \; i = 1, \ldots, 5 \tag{5.45}$$

This function is the upper bound of 200 realizations of the relative uncertainty. Hence, an uncertainty set consisting of plants G_d is obtained, while the parameters of the uncertainty are within certain ranges. The next step is to synthesize a controller K that remains stable for all the plants G_d in the uncertainty set (*robust stability*). Robust performance is guaranteed if the structured singular value μ of the closed-loop transfer function satisfies at each frequency the condition:

$$\mu_{\hat{\Delta}}(F_L(P, K)(j\omega)) < 1, \quad \forall \omega \tag{5.46}$$

The DK-iteration searches for a controller that satisfies the above condition and stabilizes the closed loop system for all plants in the uncertainty set. The performance weighting function is a diagonal matrix with wp on the diagonal:

$$\boldsymbol{W}_p = \begin{pmatrix} wp & 0.03 & \cdots & \cdots & 0.03 \\ 0.03 & wp & 0.03 & & \vdots \\ \vdots & 0.03 & wp & 0.03 & \vdots \\ \vdots & & 0.03 & wp & 0.03 \\ 0.03 & \cdots & \cdots & 0.03 & wp \end{pmatrix} \qquad wp = 0.1\frac{s+3}{s+10^{-4}} \tag{5.47}$$

The off-diagonal elements are 0.03, such that the products and liquid levels will go to their prescribed set points. The function wp has the effect that for a low frequency range the set points are achieved. The functions W_u limit the control action over the frequency range $\omega \geq 150$ and the gains are chosen independently, such that in the case of strong measurement noise over-steering is avoided:

$$W_u = \begin{pmatrix} wu_1 & & & & \\ & wu_2 & & & \\ & & wu_3 & & \\ & & & wu_4 & \\ & & & & wu_5 \end{pmatrix} \quad \begin{aligned} wu_i &= g_i \frac{s+1}{s+1} \; i = 1...3 \\ wu_i &= g_i \frac{s+1}{0.01s+1} \; i = 4...5 \end{aligned} \tag{5.48}$$

The weights g_i are chosen to ensure good performance and robust control: $g_1 = g_2 = g_3 = 10.44$ and $g_4 = g_5 = 0.2175$. The measurement noise on the five measurements is filtered by:

$$W_n = \begin{pmatrix} wn & & & & \\ & wn & & & \\ & & wn & & \\ & & & wn_4 & \\ & & & & wn_5 \end{pmatrix}, \quad wn = 0.01 \frac{s}{s+1}, \tag{5.49}$$

$$wn_4 = wn_5 = 0.2 \frac{s}{s+1}$$

The reference is linked to the output of the plant G_d via a model M. The model is represented by a diagonal matrix with zero off-diagonal elements and the transfer functions (wm), on the diagonal:

$$wm = \frac{1}{1080s^2 + 288s + 1} \tag{5.50}$$

The off-diagonal elements are zero to avoid interaction and the constants are chosen such that the settling time after an impulse is ~1500 min. Inclusion of such a model makes it easier to achieve the desired dynamics. The μ-synthesis (DK iteration procedure)—performed in MATLAB using the command *dksyn*—results in only few iteration steps. Robust stability analysis reveals that the maximum value of μ is 0.3629. Hence the system is stable under perturbations that satisfy the condition $||\Delta|| < 1/0.3629$. Likewise, the maximum value of μ in the case of the robust performance analysis is 0.9847. Hence the system achieves robust performance for all the specified uncertainties.

5.4.4.3 Performance Comparison

The performance of all the controllers previously discussed is described in great details by Van Diggelen, Kiss, and Heemink (2010). Figure 5.7a,b shows the settling times for +10% disturbances in the feed flow-rate and

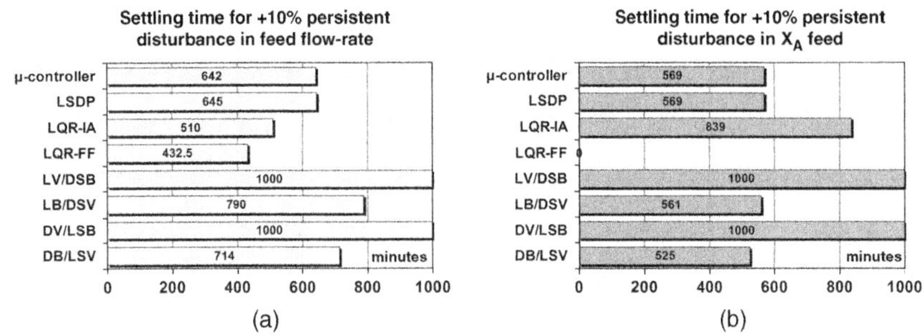

Figure 5.7 Settling time for +10% disturbance in (a) feed flow-rate and (b) feed composition for various controllers

feed composition, respectively (Van Diggelen, Kiss, and Heemink 2010; Yildirim *et al.*, 2011). While PI control structures were also able to control the DWC, significantly shorter settling times and better control were achieved using MIMO controllers. The LQG with integral action and reference inputs was found to deliver the best control performance. However, LQG has been unable so far to make a significant industrial impact, mainly because it cannot address well the constraints on the process inputs, states, and outputs but also due to other limitations in handling the process nonlinearities and re-optimization algorithm required at every step (Qin and Badgwell, 2003). Among the multi-loop PID strategies, DB/LSV and LB/DSV were the best, being able to handle persistent disturbances in reasonably-short times, as conveniently shown in Figure 5.7 (Kiss and Bildea, 2011).

5.5 ENERGY EFFICIENT CONTROL STRATEGIES

Based on the results of our previous studies (Van Diggelen, Kiss, and Heemink 2010); Kiss and Rewagad, 2011) we consider in this section only the best PID configuration (DB/LSV) as reference case. For a DWC subjected to persistent disturbances, the DB/LSV structure was found to perform best as compared to all other PID structures. In this configuration, the liquid levels in the reflux tank and reboiler are maintained by means of D (distillate) and B (bottoms flow rate) whereas the product compositions are maintained by manipulating L (liquid reflux), S (side product flow-rate), and V (vapor boil-up) respectively. An additional optimization loop is added here (4th point control) to manipulate the liquid split (r_L) in order to control the heavy component composition in the top of fractionators (Y_{C_PF1}), and implicitly achieving minimization of the energy requirements.

Table 5.2 Absolute values of the relative gain array (RGA) for a multivariable process model

Controlled variables	Manipulated variables					
	L	S	V	r_L	D	B
x_A	1.0700	0.0672	0.0015	0.0012	0	0
x_B	1.6939	2.0186	0.6747	0.0005	0	0
x_C	1.6254	0.9516	0.3268	0.0006	0	0
$y_{C,PF1}$	0.0015	0.0002	0	1	0	0
H_T	0	0	0	0	1	
H_R	0	0	0	0	0	1

Several studies (Christiansen and Skogestad, 1997; Ling and Luyben, 2009, 2010; Kiss and Rewagad, 2011) have already shown that implicit optimization of the energy usage is achieved by controlling the heavy impurity at the top of the prefractionator. Consequently, the advanced MPC controller was designed to handle a 6×6 system of inputs and outputs. The inputs include the controlled variables—mole fraction of A in distillate (x_A), B in the side stream (x_B), C in the bottom product (x_C), and C in the top of the prefractionator (Y_{C_PF1}), as well as liquid holdups in the reflux tank (H_T) and reboiler (H_R). The outputs include the manipulated variables, namely, D, B, L, S, V, and r_L.

Controllability indices such as relative gain array (RGA) can be useful in understanding the behavior of the system (Segovia-Hernandez *et al.*, 2007; Skogestad and Postlethwaite, 2005). The RGA gives information about the interactions among the controlled and manipulated variables. The RGA element is defined as the ratio of open loop gain to the closed loop gain for a pair of variables. For a selected pair of variables, values of the RGA element close to 1 are preferred and any other deviation suggests a weak relationship. Table 5.2 gives the absolute values of RGA for the system (Rewagad and Kiss, 2012).

These values suggest a high level of interaction of the variables, which makes DWC a good candidate for MPC. The control structure manipulating DB for inventory and LSV for regulatory control has the most favorable pairing of the variables if a multi-loop framework is concerned. Although MPC is typically expected to deliver better performance than PID controllers even under the failure of a manipulated variable (e.g., plugging of a valve), one of its practical drawbacks is that if the MPC controller itself fails the system may become unstable (Ranade and Torres, 2009). This is because MPC—unlike PID—takes into account the effects of all variables on each other and hence its failure can affect the prediction and control actions for all variables being manipulated.

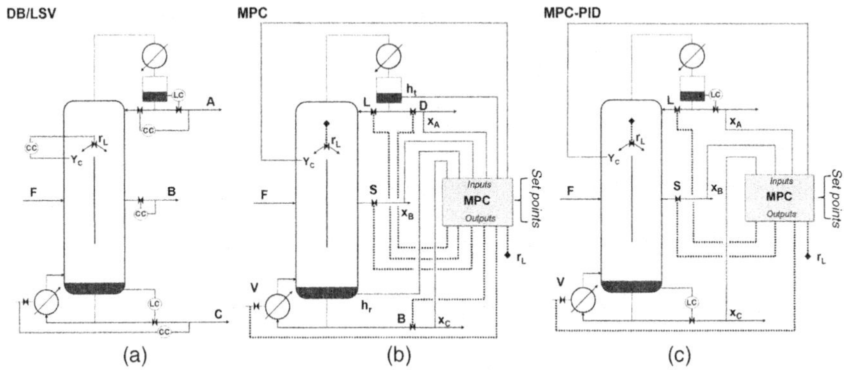

Figure 5.8 Energy efficient control structures based on PID loops (a), MPC (b), and combined MPC and PID (c)

A workaround for this issue is to employ PID to control the inventory and MPC to control the compositions in the column. Figure 5.8 shows the schematics of multi-loop PID, MPC, and combined MPC-PID controllers (Rewagad and Kiss, 2012).

5.5.1 Background of Model Predictive Control

Model predictive control (MPC) is an optimization-based multivariable constrained control technique using a linear or nonlinear process models for the prediction of the process outputs. At each sampling time the model is updated on the basis of new measurements and variables estimated using a Kalman filter taking into account the disturbances and measurement noise. Thus, it inherently assures the feed-forward behavior with an integral action. Then the open-loop optimal manipulated variable moves are calculated over a finite prediction horizon with respect to some cost function, and the manipulated variables for the subsequent prediction horizon are implemented. The prediction horizon is then shifted by, usually, one sampling time into the future and the previous steps are repeated. Figure 5.9 illustrates the generic moving horizon approach of the MPC algorithms (Kiss and Rewagad, 2011).

Solution of the optimization problem for prediction depends on the linear time invariant model used. Nowadays, most of them are in the form of state space models:

$$\begin{aligned} \boldsymbol{x}'(k) &= A\boldsymbol{x}(k) + B\boldsymbol{u}(k) \\ \boldsymbol{y}(k) &= C\boldsymbol{x}(k) + D\boldsymbol{d}(k) \end{aligned} \tag{5.51}$$

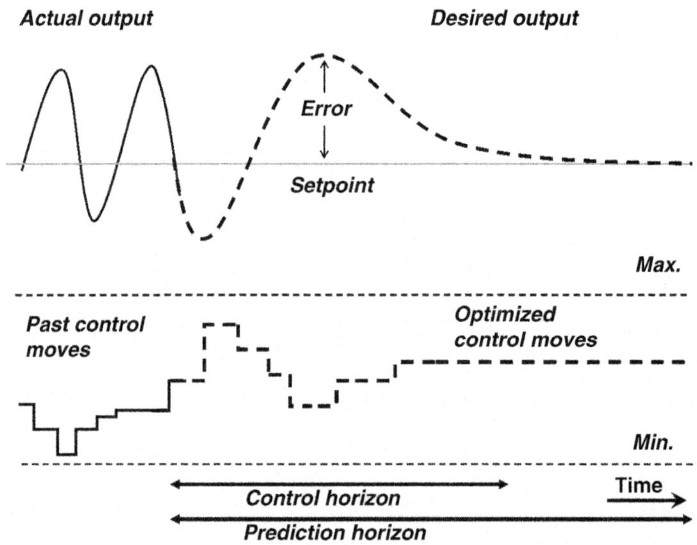

Figure 5.9 Schematic representation of the model predictive control (MPC)

where $\boldsymbol{x}(k)$ is the plant state, $\boldsymbol{u}(k)$ is the vector of manipulated variables, that is, inputs, $\boldsymbol{y}(k)$ is the vector of controlled variables, that is, outputs, and $\boldsymbol{d}(k)$ represents the vector of disturbances. In many practical applications, the plant model matrices $\boldsymbol{A}$, $\boldsymbol{B}$, $\boldsymbol{C}$, and $\boldsymbol{D}$ are obtained by linearizing the nonlinear dynamic models. Based on the space model, the optimal control problem to be solved online—at every sampling time k in the MPC algorithm—can be formulated as follows (Bemporad *et al.*, 2009):

$$\left[\begin{matrix} \min \\ \Delta u(k|k) \\ \vdots \\ \Delta u(m-1+k|k) \\ \varepsilon \end{matrix}\right] \sum_{i=0}^{p-1} \left(\begin{matrix} \sum_{j=1}^{n_y} \left| w_{i+1,j}^{y} \left[y_j(k+i+1|k) - r_j(k+i+1) \right] \right|^2 + \\ \sum_{j=1}^{n_u} \left| w_{i,j}^{u} \left[u_j(k+i|k) - u_{\text{target},j}(k+i) \right] \right|^2 + \\ \sum_{j=1}^{n_u} \left| w_{i,j}^{\Delta u} \Delta u_j(k+i|k) \right|^2 + \rho_\varepsilon \varepsilon^2 \end{matrix} \right) \tag{5.52}$$

where p denotes the length of prediction horizon and m denotes the length of control horizon with respect to the sequence of input increments Δu and slack variable ε. The aim of this optimization function is to minimize the deviation of the predicted controlled variables at time $k+i|k$ from their set point r_j. This is achieved by manipulating the input variables to follow their average value of u_{target} and also considering their deviations. Each input and output is assigned with scalar nonnegative weight values of w and can be penalized on violation of the constraints by factorρ_ε. The MathWorks MATLAB MPC toolbox® was successfully used in this study. For further information on the theory of MPC, the reader is referred to the literature (e.g., Maciejowski, 2002; Camacho and Bordons, 2004).

5.5.2 Controller Tuning Parameters

The PID control loops were tuned by the direct synthesis method described by Luyben and Luyben (1997). As fairly accurate evaluations of the process time constants τ, 20, 40, and 60 min were used. For the level controllers, a large reset time of $\tau_i = 100$ min was chosen as no tight control is required. The tuning parameters of the PID controller for all loops are conveniently listed in Table 5.3 (Rewagad and Kiss, 2012).

As no design rules are available in the literature for tuning MPC controllers, a trial and error method was used. It is contingent upon the number of factors related to the controller as well as process—prediction (p) and control (m) horizon, input (w_u) and output (w_y) weights, sampling time (Δk), operating constraints on inputs and outputs, as well as the rate of change of inputs (Δu). The value of prediction horizon was chosen to be equal to the first-order time constant of the system in a closed loop. The value of the control horizon determines the time period in which the optimization for control is performed. It was found to affect the stability of the system in terms of oscillations and thus an optimum was selected. The computation speed of the MPC solver was found to be inversely dependent on the sampling time. The input and output variables

Table 5.3 Tuning parameters of PID controllers for the energy optimal DB/LSV structure

Controlled variable	Manipulated variable	Gain P (%/%)	Int. time I (min)	Drv. time D (min)	Control direction
x_A	L	3	40	0	−
x_B	S	3	20	0	+
x_C	V	3	40	0	−
y_C	r_L	1	20	0	+
Tank level	D	1	100	0	+
Reboiler level	B	1	100	0	+

Table 5.4 Tuning parameters of MPC controllers

Weights	Manipulated variables						Controlled variables					
	D	B	L	S	V	r_L	x_A	x_B	x_C	y_{C_PF1}	H_T	H_R
w^y	1	0.8	0.3	0.9	0.3	0.1	—	—	—	—	—	—
w^y	—	—	—	—	—	—	1	1	1	1	0.3	0.5
$w^{\Delta u}$	0.2						0.1					
Constraints	(kmol min^{-1})					(–)	(–)				(m)	
(±)	0.2	0.2	0.3	0.2	0.3	0.15	0.02	0.02	0.02	0.005	0.1	0.1

Prediction horizon $p = 20$; control horizon $m = 3$; sampling time $\Delta k = 3$ min.

having a direct effect on the product purities were given the higher values of weights, whereas the variables determining the internal mass balances were given the lower values. The stability of the system was also taken into account by adjusting the weights for input and output rate change. Table 5.4 shows the tuning parameters for the MPC control structure (Rewagad and Kiss, 2012).

5.5.3 Dynamic Simulations

In the dynamic simulations performed in this study, the purity set points (SPs) are 97% for all product specifications. Persistent disturbances of +10% in the feed flow rate (F) and +10% in the feed composition (x_A) were used for the dynamic scenarios. To challenge the stability of the DWC, these disturbances were exerted alone and at the same time. The ability of the controllers to track the set point was also tested by changing all purity SPs from 0.97 to 0.98. The chosen disturbances are unmeasured and hence the controllers are relying only on the feedback action. These unmeasured disturbances and set point changes resemble the most common transitory regimes arising due to planned changes or unexpected disturbances in actual plant operation. The ability of the controllers to cope with inaccurate composition and level measurements was also investigated. Representative disturbances of +10% in the feed flow rate (F) and +1% change of set point were chosen and the dynamic responses with noise were obtained. A filtered white noise with a block signal of mean 0.1 and sample time 1 min was added to the six measurements of controlled variables. The filter equal to gain × signal/(signal + 1) was determined according to the spectrum density of the noise (Gu, 2005). The filter gain was chosen as 0.01 for the composition measurements and 0.2 for the levels in the reboiler and reflux tank.

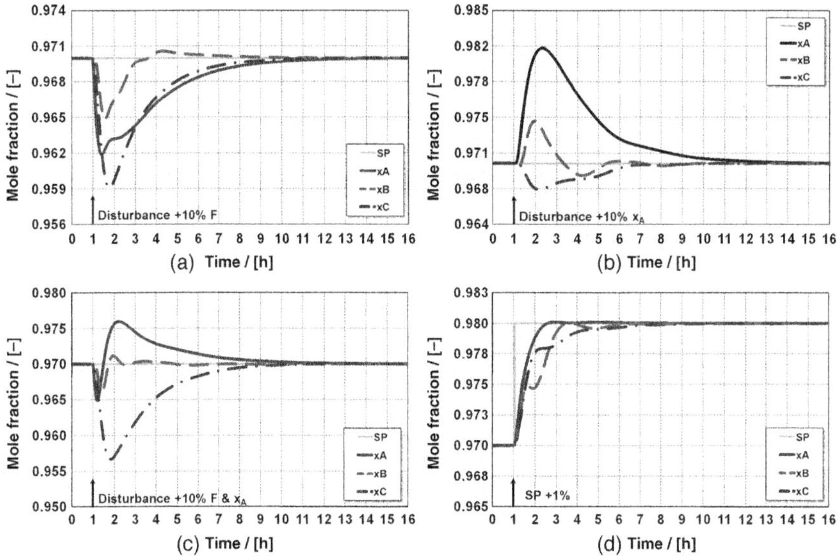

Figure 5.10 Dynamic response of the MPC control structure, at a persistent disturbance of (a) +10% in the feed flow rate, (b) +10% x_{A} in the feed composition, (c) combined +10% in both feed flow rate and composition, and (d) +1% increase of the set point (SP)

As illustrated by the following figures, the mole fractions of components A in the top distillate (x_{A}), B in the side stream (x_{B}), and C in the bottom product (x_{C}) are returning to their SP within reasonably short settling times. The dynamic response of the MPC controller under operating constraints, shown in Figure 5.10, is characterized by low overshooting and short settling times (Rewagad and Kiss, 2012). When compared to the open loop simulation response of the process, the MPC reacts naturally upon the disturbances and steers the system to the given SPs under the specified constrains.

The dynamic response of the combined MPC and PID control structure is practically similar to the MPC response and therefore is not included here. Note that the tuning parameters are the same as the respective constitutional control structures mentioned earlier. This suggests a positive way of overcoming the disadvantages of conventional and advanced controllers and making their combined advantages favorable in practice. The employed PID in such a case can be used for manual stabilization of the process.

Both PID and MPC control structures exhibit a short settling time of <10 h for all components. A common ground to compare the

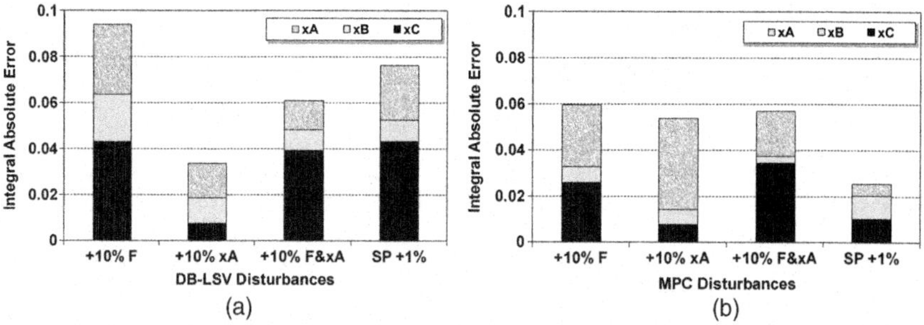

Figure 5.11 Comparison of performance of controllers in terms of integral absolute error (IEA) for various persistent disturbances: DB/LSV PID (a) and MPC controller (b)

performance of these controllers is shown in Figure 5.11, in terms of the integral absolute error (IAE) that conveniently accounts for both overshooting and settling times (Rewagad and Kiss, 2012). The IAE is defined as:

$$\mathrm{IAE} = \int_0^t |e(t)|\mathrm{d}t \tag{5.53}$$

where e is the error as compared to the set point over the time period t.

Accordingly, MPC is the most stable control structure with low values of IAE. The DB/LSV control structure performed better only in the case of $+10\%$ x_A disturbances, whereas the MPC delivered a consistent performance in all cases. The performance of MPC for set-point tracking is excellent, as clearly illustrated by Figure 5.10 (Rewagad and Kiss, 2012). Furthermore, Figure 5.12 proves that the MPC performance is stable and superior to the DB/LSV control structure even when subjected to a measurement noise (Rewagad and Kiss, 2012).

This section proved that MPC based on a linear prediction model is very well able to control the highly nonlinear DWC process. Although the application of nonlinear based controllers is appealing, only a minor improvement in the performance is expected because of the precise linearization of the nonlinear model. It has already been reported that one of the variants of such nonlinear model based controllers—generic model controller (GMC)—is not applicable in practice for the DWC as the open-loop model is non-minimal phase Van Diggelen, Kiss, and Heemink (2010). Any large changes in the operating points and capacity will require re-linearization around the new nominal conditions to ensure

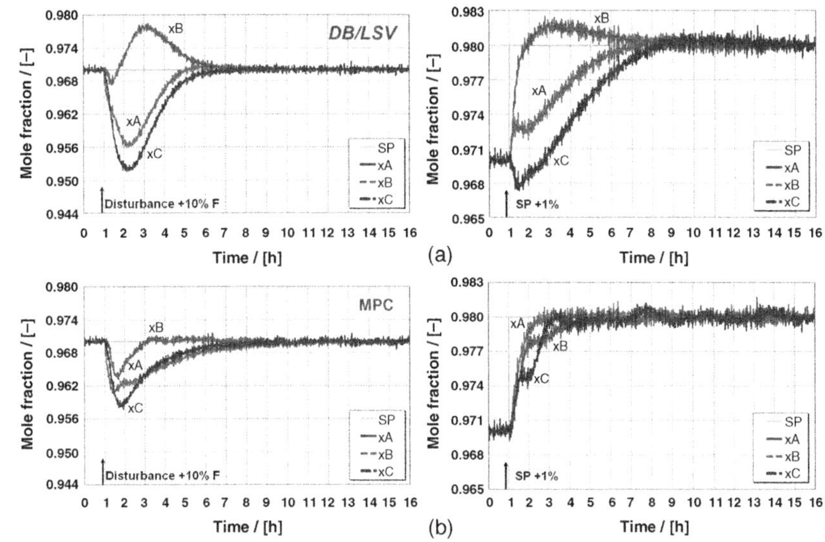

Figure 5.12 Dynamic response of the DB/LSV (a) and MPC (b) control structures, at a persistent disturbance of +10% in the feed flow rate, and +1% increase of the set point, with measurement noise

the robustness of the controller. In such cases, the piecewise linearized (Özkan *et al.*, 2000; Shafiee *et al.*, 2008) and multiple models (Özkan *et al.*, 2003; Porfírio *et al.*, 2003; Chen *et al.*, 2009) predictive control techniques can be applied. By approximating a nonlinear system as a family of affine systems, the analysis of the nonlinear system can be transformed into an analysis of several linear systems.

Another issue is the real-time feasibility of MPC where fast sampling requirements restrict its application. The performance of the MPC can be characterized by the efficiency of the solver in terms of optimization routines, computational capacity of the processor, and complexity of the model used for the predictions. All these introduce a burden in the implementation of the MPC algorithm in industry. Figure 5.3 clearly indicates that the DWC is a process with slow dynamics (Rewagad and Kiss, 2012). Therefore, it should not require fast sampling rates. As a check, the real-time feasibility was tested by comparing the maximum time required by the MATLAB MPC toolbox solver for optimization with the chosen sampling time of 3 min. This time was found to be 2.2 min on an Intel i5 (2.6 GHz processor), thus proving the real-time feasibility. However, in case of a failure to achieve this feasibility, one can use more efficient C-language based optimizers with custom optimization sub-routines.

5.6 CONCLUDING REMARKS

The advanced control strategy based on MPC to perform dynamic optimization has been illustrated in the case of a complex distillation system, namely, a dividing-wall distillation column used for BTX separation. The dynamic model of the DWC that was used here is not a reduced one but a full-size nonlinear model that is representative of industrial applications. The quality of the linearized model used for the predictions inside MPC is derived from and tested against the nonlinear model. Remarkably, although the application of nonlinear based controllers is very appealing—since linearization is not necessary—the generic model controller (GMC) is not applicable in practice for the DWC as the open-loop model is not minimum phase Van Diggelen, Kiss, and Heemink (2010).

The DWC model is not only nonlinear but also a true multi-input multi-output (MIMO) system, hence the applicability of a MIMO control structure starting with a LQG controller was also investigated. With the LQG controller there is an optimal tuning with respect to the corresponding cost function. To cope with persistent disturbances two options were explored: feed-forward control and addition of an integral action. The LQG combined with a feed-forward has good results for a persistent disturbance in the feed flow rate. However, for changes in the feed composition and condition it is difficult to find a good tuning. Moreover, persistent disturbances other than those used for tuning cannot be controlled with LQG. Nevertheless, combining LQG with an integral action and reference input solves the problem Van Diggelen, Kiss, and Heemink (2010).

The loop-shaping design procedure (LSDP) leads to a feasible μ-controller that has some additional benefits, while specific model uncertainties can be incorporated in the control structure. However, reduction of the LSDP controller (order 164) is not possible since the reduced controller is unable to control the column. In contrast, the μ-controller can be reduced from order 218 to only 25 and still have good control performance. In the DWC case described here, the obtained μ-controller is able to minimize the settling time when handling persistent disturbances. While PI control structures are also able to control the DWC, significantly shorter settling times can be achieved using MIMO controllers. Moreover, persistent disturbances are also controlled faster using a MIMO controller Van Diggelen, Kiss, and Heemink (2010).

For energy efficient control, the manipulated variables are selected to achieve the aim of regulatory and inventory control in the column while at the same time minimizing the energy requirements in a very practical way. The optimal energy control is based on a simple strategy to control the heavy component composition at the top of the prefractionator side of the DWC by manipulating the liquid split. The performance of the MPC was evaluated against a conventional PID control structure (DB/LSV) that was previously reported to be the best performing in operating a DWC. A very practical scheme based on the combination of MPC and PID controllers is also depicted, to overcome the disadvantages of individual structures. Overall, MPC delivers an outstanding performance in the case of different industrially relevant disturbances and set point tracking, with and without the measurement noise. The integral absolute error (IEA) measured for MPC performance is the lowest. The major reason for this excellent feature of MPC is its ability to act simultaneously and naturally on all the manipulated variables when the disturbances are exerted. The combined MPC and PID control structure delivered a performance similar to the MPC alone. Thus, in practice, either one of the schemes can be used to serve the control goal (Rewagad and Kiss, 2012).

This chapter proves the ability of linear MPC to control non-minimal phase and nonlinear processes such as DWC. The significant match in the open loop response of linearized and nonlinear model suggests that nonlinear MPC is not expected to deliver a significantly better performance. The functionality and real-time feasibility of the proposed framework of MPC shown in this study provides a platform for its easy transfer to other DWC applications.

NOTATION

B	bottom product flow rate
D	distillate product flow rate
$\mathbf{d}$	disturbance vector
e	error signal
F	feed flow rate
f, g	functions
f_{gain}	filter gain
F_{L}	closed loop transfer function
G_{d}	plant from the uncertainty set
G_{s}	shaped plant
H_i	liquid holdup
H_{R}	liquid holdup in reboiler
H_{T}	liquid holdup in reflux tank

I	identity matrix
IAE	integral absolute error
K	controller
K_1, K_2	control constants
k_i	constant hydraulic factor $i \in \{0, 1, 2\}$
K_r	reference tracking part of controller
K_y	output feedback part of controller
L_0	reflux flow rate
m	control horizon
n	noise signal
N	number of trays
p	prediction horizon
P	proportional gain
q	feed condition
$\boldsymbol{Q}$	weight matrix on state
r	reference signal
$\boldsymbol{R}$	weight matrix on input
R_L	ratio liquid split
R_V	ratio vapor split
S	side product flow rate
T	temperature
t	time
$\boldsymbol{u}$	control vector
V_0	vapor flow rate
$\boldsymbol{W}_1$	pre-compensator matrix
$\boldsymbol{W}_2$	post-compensator matrix
W_n	noise shaping filter
W_p	performance action weighting
W_u	control action weighting
$\boldsymbol{x}$	state vector
x_A	top composition, component A
x_B	side composition, component B
x_C	bottom composition, component C
$x_{i,j}$	liquid composition
$\boldsymbol{y}$	output vector
$\boldsymbol{y}^*$	set-point vector
$y_{i,j}$	vapor composition
z_1	feed composition, component A
z_2	feed composition, component B
α_j	relative volatility
Δ	uncertainty matrix

Δk	sampling time
Δu	change of inputs
Θ	time delay constant
μ_Δ	structured singular value

Subscripts

i tray number $i \in \{1 \ldots N\}$
j component $j \in \{1, 2, 3\}$

REFERENCES

Adrian, R., Schoenmakers, H., and Boll, M. (2004) MPC of integrated unit operations: Control of a DWC. *Chemical Engineering & Processing*, *43*, 347–355.

Agachi, P.S., Nagy, Z.K., Cristea, M.V., and Imre-Lucaci, Á. (2006) *Model Based Control: Case Studies in Process Engineering*, Wiley-VCH Verlag GmbH, Weinheim.

Bemporad, A., Morari, M., and Ricker, N.L. (2009) Model Predictive Control Toolbox 3: User's Guide, The MathWorks Inc.

Biswas, P.P., Ray, S., and Samanta, A.N. (2007) Multi-objective constraint optimizing IOL control of distillation column with nonlinear observer. *Journal of Process Control*, *17*, 73–81.

Buck, C., Hiller, C., and Fieg, G. (2011) Applying model predictive control to dividing wall columns. *Chemical Engineering & Technology*, *34*, 663–672.

Camacho, E.F. and Bordons, C. (2004) *Model Predictive Control*, 2nd edn, Springer Verlag, London.

Chen, Q., Gao, L., Dougal, R.A., and Quan, S. (2009) Multiple model predictive control for a hybrid proton exchange membrane fuel cell system. *Journal of Power Resources*, *191*, 473–482.

Cho, Y., Kim, B., Kim, D. *et al.* (2009) Operation of divided wall column with vapor sidedraw using profile position control. *Journal of Process Control*, *19*, 932–941.

Christiansen, A.C. and Skogestad, S. (1997) Energy savings in complex distillation arrangements: importance of using the preferred separation. Paper presented at AIChE Annual Meeting, Los Angeles, November 1997, paper 199d.

Dayal, B.S. and MacGregor, J.F. (1996) Identification of finite impulse response models: methods and robustness issues. *Industrial & Engineering Chemistry Research*, *35*, 4078–4090.

Diggelen van, R.C., Kiss, A.A., and Heemink, A.W. (2010) Comparison of control strategies for dividing-wall columns. *Industrial & Engineering Chemistry Research*, *49*, 288–307.

Dwivedi, D., Strandberg, J.P., Halvorsen, I.J., Preisig, H.A., and Skogestad, S. (2012). Active vapor split control for dividing-wall columns, *Industrial & Engineering Chemistry Research*, *51*, 15176–15183.

Gabor, M. and Mizsey, P. (2008) A methodology to determine controllability indices in the frequency domain. *Industrial & Engineering Chemistry Research*, *47*, 4807–4816.

Gu, D.W. (2005) *Robust Control Design with MATLAB*, Springer Verlag, London.

Halvorsen, I.J. and Skogestad, S. (1997) Optimizing control of Petlyuk distillation: understanding the steady-state behaviour. *Computers & Chemical Engineering*, *21*, 249–254.

Halvorsen, I.J. and Skogestad, S. (1999) Optimal operation of Petlyuk distillation: understanding the steady-state behaviour. *Journal of Process Control*, *9*, 407–424.

Hernández, S. and Jiménez, A. (1999) Controllability analysis of thermally coupled distillation systems. *Industrial & Engineering Chemistry Research*, *38*, 3957–3963.

Isidori, A. (1989) *Nonlinear Control Systems: An Introduction*, Springer Verlag, Heidelberg.

Johnson, M.A. and Moradi, H. (2005) *PID Control: New Identification and Design Methods*, Springer Verlag, London.

Kim, Y.H. (2002) Structural design and operation of a fully thermally coupled distillation column. *Chemical Engineering Journal*, *85*, 289–301.

Kiss, A.A. and Bildea, C.S. (2011) A control perspective on process intensification in dividing-wall columns. *Chemical Engineering and Processing: Process Intensification*, *50*, 281–292.

Kiss, A.A. and Rewagad, R.R. (2011) Energy efficient control of a BTX dividing-wall column. *Computers & Chemical Engineering*, *35*, 2896–2904.

Kiss, A.A. and Rewagad, R.R. (2011) Model predictive control of a dividing-wall column. *Studia Universitatis Babes-Bolyai Chemia*, *562*, 5–18.

Kvernland, M., Halvorsen, I., and Skogestad, S. (2010) Model Predictive Control of a Kaibel distillation column. Presented at The 9th IFAC Symposium on Dynamics and Control of Process Systems (DYCOPS), July 2010, pp. 539–544.

Lee, P.L. and Sullivan, G.R. (1988) Generic model control (GMC). *Computers and Chemical Engineering*, *12*, 573–580.

Ling, H. and Luyben, W.L. (2009) New control structure for divided-wall columns. *Industrial & Engineering Chemistry Research*, *48*, 6034–6049.

Ling, H. and Luyben, W.L. (2010) Temperature control of the BTX divided-wall column. *Industrial & Engineering Chemistry Research*, *49*, 189–203.

Lundström, P., Lee, J.H., Morari, M., and Skogestad, S. (1995) Limitations of dynamic matrix control. *Computers & Chemical Engineering*, *19*, 409–421.

Luyben, W.L. and Luyben, M.L. (1997) *Essentials of Process Control*, McGraw-Hill.

Maciejowski, J.M. (2002) *Predictive Control with Constraints*, Pearson Education.

Mathur, U., Rounding, R.D., Webb, D.R., and Conroy, R. (2008) Use model predictive control to improve distillation operations. *Chemical Engineering Progress*, *104*, 35–41.

Morari, M. and Lee, J.H. (1999) Model predictive control: past, present and future. *Computers & Chemical Engineering*, *23*, 667–682.

Niggemann, G., Hiller, C., and Fieg, G. (2011) Modeling and in-depth analysis of the start-up of dividing-wall columns. *Chemical Engineering Science*, *66*, 5268–5283.

Özkan, L., Kothare, M.V., and Georgakis, C. (2003) Control of a solution copolymerization reactor using multi-model predictive control. *Chemical Engineering Science*, *58*, 1207–1221.

Özkan, L., Kothare, M.V., and Georgakis, C. (2000) Model predictive control of nonlinear systems using piecewise linear models. *Computers & Chemical Engineering*, *24*, 793–799.

Porfírio, C.R., Almeida Neto, E., and Odloak, D. (2003) Multi-model predictive control of an industrial C3/C4 splitter. *Control Engineering Practice*, *11*, 765–779.

Qin, S.J. and Badgwell, T.A. (2003) A survey of industrial model predictive technology. *Control Engineering Practice*, *11*, 733–764.

Rewagad, R.R. and Kiss, A.A. (2012) Dynamic optimization of a dividing-wall column using model predictive control. *Chemical Engineering Science*, *68*, 132–142.

Rouchon, P. (1990) Dynamic simulation and nonlinear control of distillation columns. Ph.D. thesis, Ecole des Mines de Paris, France.

Ranade, S.M. and Torres, E. (2009) From dynamic mysterious control to dynamic manageable control. *Hydrocarbon Processing*, 77–81.

Segovia-Hernandez, J.G., Hernandez-Vargas, E.A., and Marquez-Munoz, J.A. (2007) Control properties of thermally coupled distillation sequences for different operating conditions. *Computers & Chemical Engineering*, *31*, 867–874.

Serra, M., Espuña, A., and Puigjaner, L. (1999) Control and optimization of the divided wall column. *Chemical Engineering and Processing*, *38*, 549–562.

Serra, M., Espuña, A., and Puigjaner, L. (2000) Study of the divided wall column controllability: influence of design and operation. *Computers & Chemical Engineering*, *24*, 901–907.

Serra, M., Perrier, M., Espuña, A., and Puigjaner, L. (2001) Analysis of different control possibilities for the divided wall column: feedback diagonal and dynamic matrix control. *Computers & Chemical Engineering*, *25*, 859–866.

Shafiee, G., Arefi, M.M., Jahed-Motlagh, M.R., and Jalali, A.A. (2008) Nonlinear predictive control of a polymerization reactor based on piecewise linear Wiener model. *Chemical Engineering Journal*, *143*, 282–292.

Signal, P.D. and Lee, P.L. (1993). Robust stability and performance analysis of Generic Model Control, *Chemical Engineering Communications*, *124*, 55–76.

Skogestad, S. and Postlethwaite, I. (2005) *Multivariable Feedback Control. Analysis and Design*, 2nd edn, John Wiley & Sons, Ltd., Chichester.

Woinaroschy, A. and Isopescu, R. (2010) Time-optimal control of dividing-wall distillation columns. *Industrial & Engineering Chemistry Research*, *49*, 9195–9208.

Yildirim, O., Kiss, A.A., and Kenig, E.Y. (2011) Dividing wall columns in chemical process industry: A review on current activities. *Separation and Purification Technology*, *80*, 403–417.

Zhu, Y.C. (1998) Multivariable process identification for MPC: the asymptotic method and its applications. *Journal of Process Control*, *8*, 101–115.

6

Applications of Dividing-Wall Columns

6.1 INTRODUCTION

Dividing-wall columns (DWCs) are nowadays applied for the separation of mixtures consisting of three or more components. Initially, DWC technology found application mainly for systems with low fractions of light and heavy boiling components. Currently, dividing-wall columns are used for a broad spectrum of systems, such as hydrocarbons, alcohols, aldehydes, ketones, acetals, amines, and others (Dejanović, Matijašević, and Olujić, 2010; Dejanović *et al.*, 2011a; Long and Lee, 2011; Yildirim, Kiss, and Kenig, 2011). Furthermore, DWC can also be used to conduct azeotropic, extractive, and reactive distillation without any major changes to the types of internals used (Olujić *et al.*, 2003; Yildirim, Kiss, and Kenig, 2011). However, despite the fact that DWC technology promises a significant reduction in both operating and capital costs, the number of industrial applications is still limited. Moreover, information on the particular applications of DWC is mostly scarce (Yildirim, Kiss, and Kenig, 2011).

The number of implemented DWC units reported in the literature is listed in Table 6.1 (Yildirim, Kiss, and Kenig, 2011). Most of them—over 116—are in fact applied to the separation of three-component mixtures. By 2010, there were more than 125 applications reported in total. In fact, an exponential increase in the number of applications over recent years was observed (Asprion and Kaibel, 2010; Yildirim, Kiss, and Kenig, 2011). Remarkably, if this increasing trend remains constant for the

Advanced Distillation Technologies: Design, Control and Applications, First Edition.
Anton Alexandru Kiss.

Table 6.1 Overview of industrial applications reported in the literature

Industrial system	Number of applications
Ternary mixtures	>116
Multicomponent mixtures (more than three components)	2
Reactive system	—
Azeotropic system	1
Extractive system	2
DWC revamps	5

coming years, about 350 DWC units can be expected by the end of 2015 (Yildirim, Kiss, and Kenig, 2011).

6.2 SEPARATION OF TERNARY AND MULTICOMPONENT MIXTURES

Table 6.2 summarizes the industrially available DWC applications for three-component mixtures (Yildirim, Kiss, and Kenig, 2011). Accordingly, the DWC technology was applied over 116 times for the separation of ternary mixtures. Most of the industrial columns are exploited by BASF SE (more than 70 units) although other companies—such as Bayer AG, Dow Chemical Co, LG Chem Ltd—have started employing DWC technology as well (Pendergast *et al.*, 2008; Dejanović, Matijašević, and Olujić, 2010; Lee, Shin, and Lee, 2011). Almost all these DWC units are packed columns built by Montz GmbH. Detailed information on the systems or on the specifics of these DWC units are, unfortunately, not available. The reader can find more information in the corresponding patents that are conveniently summarized in a review published by Dejanović, Matijašević, and Olujić (2010).

The world largest column (107 m high) was constructed by Linde AG for Sasol in Johannesburg, South Africa. This trayed column recovers hydrocarbons from the Fischer–Tropsch synthesis (Yildirim, Kiss, and Kenig, 2011). Furthermore, Uhde has built two DWC units in Germany and Saudi Arabia, and also applied an extractive DWC. Sumitomo Heavy Industries and Kyowa Yuka designed six DWCs for undisclosed companies—note though that they refer to DWC as "*column in column.*" Further constructors are Sulzer Chemtech Ltd. and Koch-Glitsch. Sulzer has installed more than 20 DWCs and Koch-Glitsch more than 10 (Parkinson, 2007; Yildirim, Kiss, and Kenig, 2011).

UOP has also built dividing wall columns. One application is realized for optimization of the process shown in Figure 6.1a (Yildirim, Kiss, and

Table 6.2 Industrial applications of DWCs for ternary systems

Company	System	Constructor	Description	References
BASF SE, various sites	Mostly undisclosed	Most columns built by Montz GmbH First commercial DWC in 1985	• More than 70 DWCs • Diameter 0.6–4 m • Operating pressure 2 mbar to 10 bar	Amminudin and Smith (2001); Olujić *et al.* (2009); Kaibel *et al.* (2004)
Sasol, Johannesburg, South Africa	Separation of hydrocarbons from Fischer–Tropsch synthesis unit	Linde AG in 1999	• World's largest DWC • Height 107 m • Diameter 5 m • Tray column	Schultz *et al.* (2002); Parkinson (2005)
Veba Oel Ag, Münchsmünster, Germany	Separation of benzene from pyrolysis gasoline	Uhde in 1999	• 170 000 mt yr^{-1} feed capacity	Schultz *et al.* (2002); Yildirim, Kiss, and Kenig (2011)
Saudi Chevron Petrochemical, Al Jubail, Saudi Arabia	Undisclosed	Uhde in 2000	• 140 000 mt yr^{-1} feed capacity	Yildirim, Kiss, and Kenig (2011)
Lonza, Visp, Switzerland	Multipurpose, for various ternary mixtures	Undisclosed	• Height 10 m • Diameter 0.5 m • Sulzer Mellapak™ 350Y • Hastelloy C-22	Grützner *et al.* (2012)
ExxonMobil, Rotterdam, The Netherlands	Benzene–toluene–xylene fractionation	ExxonMobil; was planned for 2008	• No data available	Parkinson (2007)
Undisclosed	Undisclosed	Sumitomo Heavy Industries and Kyowa Yuka	• Six DWCs • No data available	Premkumar and Rangaiah (2009)

Table 6.2 (*Continued*)

Company	Systemv	Constructor	Description	References
Undisclosed	Separation of C_{7+} aromatics from C_{7+} olefin/paraffin	UOP	• Five DWCs • Trap tray	Schultz *et al.* (2002, 2006)
Undisclosed	Undisclosed reactive system consisting of two reactive components and an inert component	UOP	• Split shell column with two walls	Schultz *et al.* (2006)
Undisclosed	Undisclosed	Sulzer Chemtech Ltd.	• 20 DWCs • No data available	Parkinson (2007)
Undisclosed	Undisclosed	Koch Glitsch	• 10 DWCs • No data available	Premkumar and Rangaiah (2009)

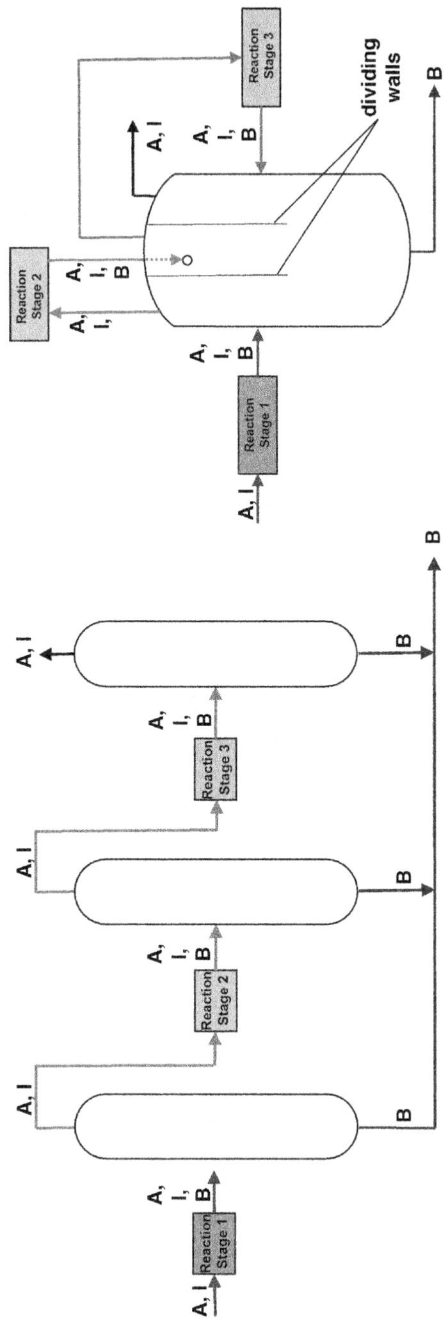

Figure 6.1 Multistage reaction/separation flow sheet with three distillation columns (a) or a single split shell column (b)

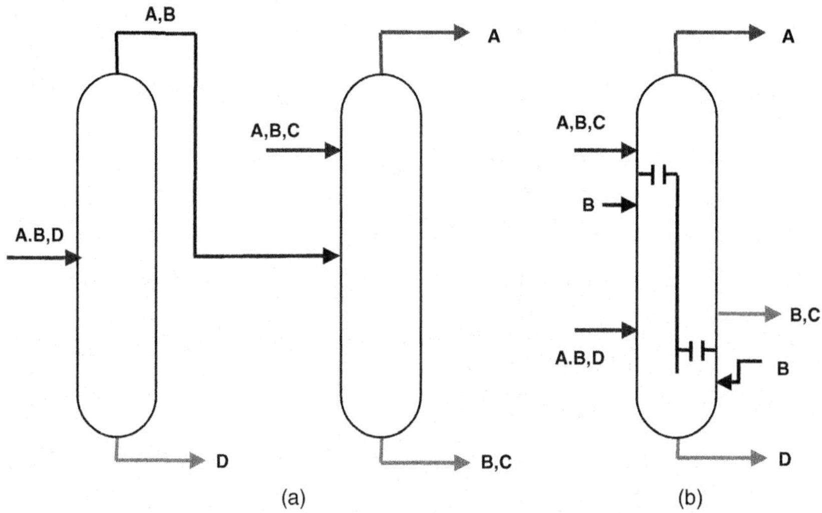

Figure 6.2 PEP fractionation in a two-column design (a) and in a DWC (b)

Kenig, 2011). This consists of a sequence of three reaction and separation units. The reagent A and an inert I are fed to a reactor, where A reacts to form B. Afterwards, the heavy product B is separated from the mixture by distillation. Since the bottom products of all columns are similar, the separation of B can be carried out within only one shell, as illustrated in Figure 6.1b (Yildirim, Kiss, and Kenig, 2011).

The DWC is also applied for the UOP Pacol enhancement process (PEP) shown in Figure 6.2a (Yildirim, Kiss, and Kenig, 2011). Figure 6.2b shows the developed DWC for the same separation problem. In this process, A is pentane, B is benzene, C consists of C_{7+} olefins, and D consists of C_{7+} aromatics (Schultz *et al.*, 2002, 2006). To prevent mixing of C and D, a novel trap tray is applied and excessive B is added. According to Schultz *et al.* (2006), the DWC shown in Figure 6.2b has found over five applications in industry. Energy savings of about 50% and capital savings of around 35% could be achieved as compared to the two-column design (Figure 6.2a).

More recently, Lonza set up a flexible multipurpose DWC unit that fully meets the demands of a steadily changing production, where products change several times a year (Table 6.2). A flexible inlet enables feeding at various locations, and a flexible outlet allows a side draw from different stages. Moreover, simultaneous feeding on both sides of the dividing wall is possible, thus allowing an operation mode comparable to that of traditional distillation columns (Grützner *et al.*, 2012).

Although DWC is now well established in the industrial practice, more quantitative information is needed in the open literature to understand their complex hydraulics. Niggemann, Hiller, and Fieg (2010) reported that they have built a pilot plant to separate a ternary mixture of fatty alcohols (hexanol, octanol, and decanol) into high-purity products (over 99 wt%). The model is able to describe the experiments quantitatively and even account for the self-adjustment of the vapor splits. A case study with the validated model highlighted the strong influence of the heat transfer across the vertical partition wall on the hydrodynamics and vapor distribution. These aspects are of special interest for the design and scale-up of DWC (Niggemann, Hiller, and Fieg, 2010).

Based on the reports so far, it appears probable that most of future implementations of DWC technology will be realized in developing countries with emerging markets—for example, Brazil, Russia, India, and China (the so-called BRIC zone)—rather than in countries with established distillation networks (Yildirim, Kiss, and Kenig, 2011).

For the separation of mixtures with more than three components, there are only two applications of DWC (Table 6.3) (Yildirim, Kiss, and Kenig, 2011). According to Parkinson (2005), BASF SE started in 2002 with the development of DWCs separating quaternary mixtures. The columns are intended to recover fine chemical intermediates. Another DWC for the separation of mixtures with more than three components was developed by UOP. In this process, a split shell column (Figure 6.3) (Yildirim, Kiss, and Kenig, 2011) was introduced for the separation of two streams into five components. The first stream comes from a hydrocracking process that contains different hydrocarbons, such as naphtha, kerosene, diesel,

Table 6.3 Industrial applications of DWC for multicomponent mixtures

Company	System	Constructor	Features	Reference
BASF SE	Recovery of four component mixtures of fine chemical intermediates	BASF SE/Montz GmbH since 2002	• Single wall • Height 34 m • Diameter 3.6 m • Column works under deep vacuum	Dejanović *et al.* (2011a), Olujić *et al.* (2009)
Undisclosed customer in the Far East	Integration of a product separator and an HPNA stripper	Designed by UOP	• Five product streams	Schultz *et al.* (2006), Parkinson (2007)

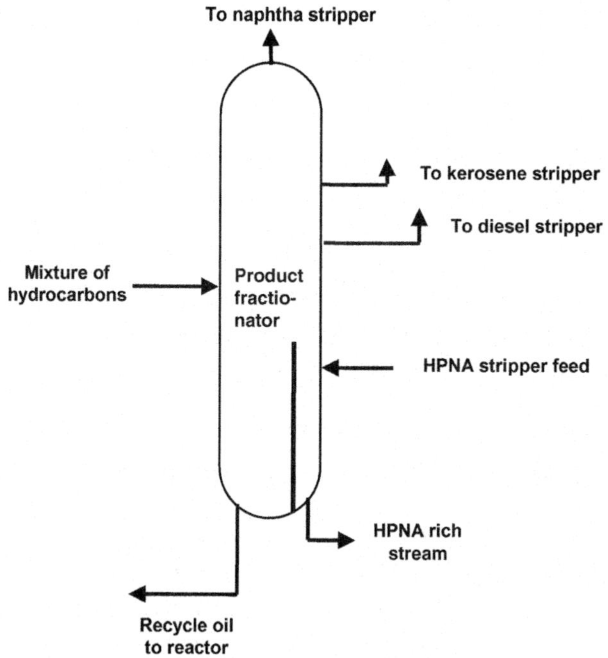

Figure 6.3 Split shell column from UOP

and oil. The second stream is from the reactor section of a plant, and contains heavy polynuclear aromatic (HPNA) and light components. This column was built at an undisclosed company in the Far East (Parkinson, 2007; Schultz *et al.*, 2006).

Recently, Dejanović *et al.* (2011a) have given a theoretical analysis of advanced DWC configurations (Kaibel column and multi-partitioned DWC) and replaced a usual three-column sequence. In their study, the separation of an aromatic mixture of 15 components—from the INA Refinery (Sisak, Croatia)—was used as an industrial case study. The analysis shows that considerable energy and cost savings (43–50%) can be achieved with the DWC technology.

Nevertheless, to date, the application of DWC for the separation of mixtures with more than three components is still very scarce. This is presumably because it is considered to be a more complex technology than for three-component separations. However, due to the clear advantages that hold true also for these systems, further DWC applications can be expected, especially in the BRIC zone and developing countries (Yildirim, Kiss, and Kenig, 2011).

6.3 REACTIVE DIVIDING-WALL COLUMN

A reactive dividing-wall column (R-DWC) represents a combination of a reactor and a separation unit in one DWC—or a combination of reactive distillation (RD) with DWC technology (Figure 6.4) (Yildirim, Kiss, and Kenig, 2011). A R-DWC is a highly integrated operation that is already mentioned in some patents (Kaibel, 1984; Hill *et al.*, 2002). In such columns, reactive systems with more than two products, non-reacting components, or excessive reagents can be separated (Mueller and Kenig, 2007; Kiss, Pragt, and van Strien, 2009; Kiss and Suszwalak, 2012).

So far, only very few industrial applications of R-DWC have been reported (Yildirim, Kiss, and Kenig, 2011; Hernández *et al.*, 2009), but the R-DWC process itself has been analyzed theoretically by several authors—see Table 6.4 for an extensive list (Yildirim, Kiss, and Kenig, 2011). Several comprehensive papers on R-DWC modeling were published by the group of Kenig (Kenig *et al.*, 2006; Mueller and Kenig, 2007, 2010). The modeling was carried out using the rate-based approach and the performance of the R-DWC was only theoretically studied for different systems. The advantages of this integrated process—for example, high conversion, selectivity, and product purity, as well as considerable energy and cost savings—were also demonstrated. Some

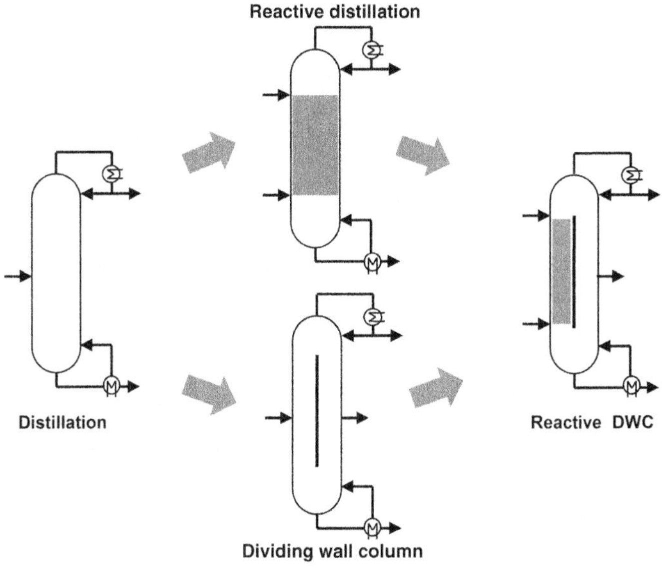

Figure 6.4 Schematic showing the path from a distillation column to a reactive dividing-wall column (where the gray area represents the reactive zone)

Table 6.4 Work published on reactive DWC

Reaction	Simulation	References
Methanol + butanol + acetic acid ↔ methyl acetate + butyl acetate + water Dimethyl carbonate + ethanol ↔ diethyl carbonate + methanol Methyl acetate + water ↔ methanol + acetic acid	Rate-based model/Aspen Custom Modeler	Mueller *et al.* (2007), Mueller and Kenig (2007), Großmann and Kenig (2007)
Methyl acetate + water ↔ methanol + acetic acid	Aspen Plus or PRO/II	Daniel *et al.* (2006)
Isoamylenes + ethanol ↔ tertiary amyl ethyl ether Isoamylenes are:2-methyl-1-butene-1 and 2-methylbut-2-ene	Aspen DISTIL™, Aspen HYSYS™, CSTR	Bumbac, Plesu, and Plesu, 2007
Methyl acetate + water ↔ methanol + acetic acid	Aspen Plus or PRO/II	Sander *et al.* (2007)
Ethanol + acetic acid ↔ ethyl acetate + water	Aspen Plus and Aspen Dynamics	Sandoval-Vergara *et al.* (2008), Hernández *et al.* (2009)
Reactive distillation with ten chemical species	Aspen Plus RADFRAC	Kiss *et al.* (2009)
Isobutane + ethanol ↔ ethyl tertiary butyl ether (ETBE)	Aspen HYSYS	Bumbac *et al.* (2009)
Ethanol + acetic acid ↔ ethyl acetate + water Methanol + isobutylene ↔ methyl tertiary butyl ether Ethanol + ethylene ↔ ethoxyethanol	Aspen Plus	Barroso-Muñoz (2009)
Methanol ↔ dimethyl ether + water	Aspen Plus RADFRAC	Kiss and Suszwalak (2012)

other groups investigated R-DWC systems using standard routines available in commercial software tools, for example, RateSep in Aspen Plus® (AspenTech, 2010a). However, experimental investigations of R-DWC configurations are still very scarce (Sander *et al.*, 2007; Sandoval-Vergara *et al.*, 2008; Yildirim, Kiss, and Kenig, 2011; Hernández *et al.*, 2009).

In a theoretical study, Wang, Huang, and Yu (2011) investigated a novel R-DWC process, designed using different degrees of thermal coupling to achieve energy saving for an ideal quaternary reaction system with the least favorable relative volatility ranking under excess-reactant design. The simulation results showed that R-DWC provides better energy efficiency than reactive distillation without thermal coupling, with the efficiency increasing with the degree of thermal coupling.

Sun *et al.* (2011) reported the design, optimization, and control of a catalytic dividing-wall column (CDWC) for the hydrolysis of methyl acetate (MeAc). The distillate and side rate were used to maintain the desired product purities, while the minimum reboiler duty is obtained by changing the reflux ratio and the vapor split ratio. The results show that energy savings of over 20% are possible.

Kiss *et al.* (2012) proposed a novel biodiesel process based on a reactive DWC that allows the use of only 15% excess of methanol to convert completely the fatty acids feedstock. FAME (fatty acid methyl ester) is produced as pure bottom product and water as side stream, while the methanol excess is recovered as distillate and recycled. The optimal configuration was established by using simulated annealing as optimization method implemented in MATLAB, and coupled with rigorous simulations carried out in AspenTech Aspen Plus. Along with the integrated FAME production, the improved design alternatives allow lower investment costs and high energy savings.

More recently, Kiss and Suszwalak (2012) proposed the enhanced DME production in a single reactive DWC – this novel process being described later in this chapter as a case study for R-DWC. Considering the low number of reports so far, it can be concluded that the application and investigation of DWC for reactive distillation processes is also very limited. This situation can be explained by the clear niche character of the R-DWC technology. However, in the meantime, the DWC concept is rapidly evolving to become a standard distillation tool and consequently further applications for reactive systems can be also expected. Notably, R-DWC studies are somewhat more popular as compared to other integrated DWC applications (e.g., azeotropic or extractive distillation). This can be explained actually by the general interest in reactive distillation itself (Kiss, 2012).

6.4 AZEOTROPIC DIVIDING-WALL COLUMN

In some cases, it is very difficult or even impossible to separate two components (A and B) by distillation, due to the similar boiling points or due to the occurrence of an azeotrope. In such cases different methods can be used, such as pressure swing distillation (PSD) or azeotropic distillation (Huang *et al.*, 2008). PSD relies on the fact that an azeotrope is pressure dependent, so that in order to jump over the azeotrope its composition can be moved by altering the pressure. Azeotropic distillation (AD) uses an additional component—an entrainer (E)—that forms an azeotrope with the components to be separated (Khoury, 2005; Guedes *et al.*, 2007). Entrainers forming heterogeneous azeotropes are preferred, as these azeotropes can be easily separated in a decanter (Sattler and Feindt, 1995). By using a two-column sequence—shown in Figure 6.5 (Yildirim, Kiss, and Kenig, 2011)—an azeotropic mixture A-B can be separated. A detailed description of this process can be found in Stichlmair and Herguijuela (1992).

In principle, it is possible to combine azeotropic distillation with the DWC concept according to the scheme shown in Figure 6.5 (Yildirim,

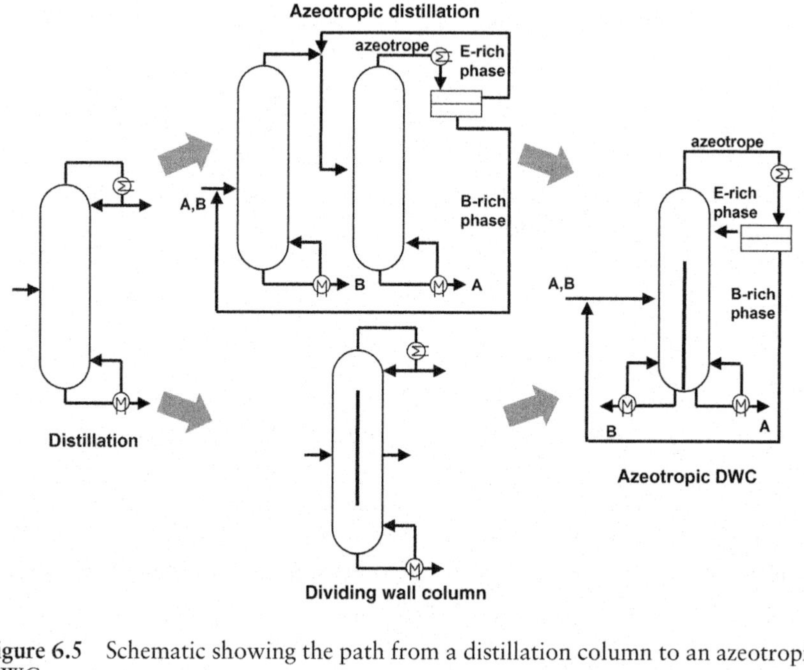

Figure 6.5 Schematic showing the path from a distillation column to an azeotropic DWC

Kiss, and Kenig, 2011). Until now, only a few papers have been published that present a theoretical analysis of azeotropic-DWCs (A-DWC). Midori, Zheng, and Yamada, 2001 simulated an A-DWC with Aspen Plus. The test system was the separation of ethanol and water using cyclohexane as entrainer. Kiss and Suszwalak (2012) also studied the separation in an A-DWC of a near azeotropic ethanol–water mixture from bioethanol production, using *n*-pentane as entrainer. Energy savings of over 20% were reported for the optimal configuration.

Briones-Ramírez *et al.* (2009) also performed theoretical investigations using Aspen Plus. In their study, isopropanol–water–acetone and isopropanol–water–methanol were used as test systems. Furthermore, they applied an optimization procedure using a multi-objective genetic algorithm to find the optimal design. The results of this study show that energy savings of up to 50% can be achieved using A-DWC instead of a conventional two-column sequence.

Nonetheless, industrial application of the A-DWC technology was mentioned only once in the open literature, by Kaibel *et al.* (2006), without any specific information.

6.5 EXTRACTIVE DIVIDING-WALL COLUMN

Azeotropic or narrow boiling mixtures (A, B) can also be separated using extractive distillation. In this process, an additional substance (solvent S), with a boiling point much higher than that of A and B, is added (Huang *et al.*, 2008). By adding this solvent the relative volatility of one of the components, for instance of the heavy boiling component B, decreases. In this case, A can be withdrawn at the top of the first column, whereas S and B can be separated in a second column (Khoury, 2005; Sattler and Feindt, 1995; Schönbucher, 2002; Lei *et al.*, 2004).

The two-column configuration can also be integrated into one column. Figure 6.6 illustrates the integration scheme of an extractive DWC (E-DWC) (Yildirim, Kiss, and Kenig, 2011). The mixture of A and B and the solvent S is fed to the E-DWC (Figure 6.6, right-hand side). The heavy boiling product B interacts with S, which reduces the vapor pressure of B. The light boiling component A is obtained on the feed side of the column, while B and S are separated on the other side of the column, with S being recycled back to the column.

Table 6.5 shows the main industrial applications of E-DWC (Yildirim, Kiss, and Kenig, 2011). Uhde applied the DWC technology for the Morphylane® process, which was initially based on a two-column

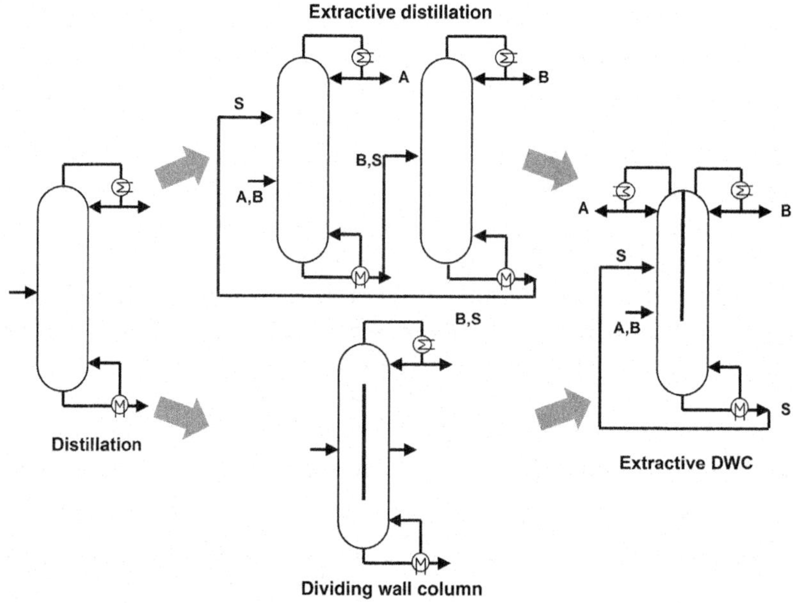

Figure 6.6 Schematic showing the path from a distillation column to an extractive DWC

configuration for the recovery of benzene, toluene, or xylenes from various feed stocks (Diehl, Kolbe, and Gehrke, 2005; Kolbe and Wenzel, 2004). The integrated solution—shown in Figure 6.7 (Yildirim, Kiss, and Kenig, 2011)—was applied in 2004 at Arsol Aromatics GmbH (formerly Aral Aromatics) in Gelsenkirchen (Germany), for a feed capacity of 28 000 mt yr^{-1} (Yildirim, Kiss, and Kenig, 2011). Another E-DWC was applied by BASF SE for the production of butadiene from a C-cut (Heida, Bohner, and Kindler, 2002; Dejanović, Matijašević, and Olujić, 2010; Asprion and Kaibel, 2010). Figure 6.8a shows the extraction/distillation part of the classic butadiene process (Yildirim, Kiss, and Kenig, 2011). At BASF SE, this part of the plant was replaced by the dividing-wall technology shown in Figure 6.8b (Heida, Bohner, and Kindler, 2002; Jobson, 2005).

Although E-DWC has already found industrial application, only a few studies have been published in the open literature. Midori, Zheng, and Yamada (2000) investigated theoretically the application of E-DWC and the results revealed that energy savings of about 36% could be obtained. In their study, acetone–methanol–water was used as a test system and the simulations were carried out using Aspen Plus. Ruffert and Olf (2004) reported on a process development at Bayer Technology Services with

Table 6.5 Industrial applications of extractive DWC

Company	System	Constructor	Features	References
Arsol Aromatics GmbH (formerly Aral Aromatics), Gelsenkirchen, Germany	Separation of toluene and non-aromatics with *N*-formylmorpholine	Constructed by Uhde in 2004	28 000 mt yr^{-1} feed capacity	Parkinson (2007), Diehl, Kolbe, and Gehrke (2005), Yildirim, Kiss, and Kenig (2011)
BASF SE, Ludwigshafen, Germany	Crude butadiene from a crude C_4 using *N*-methylpyrrolidone (NMP) as solvent	BASF SE	Both trays and packings	Jobson (2005), Heida *et al.* (2002)

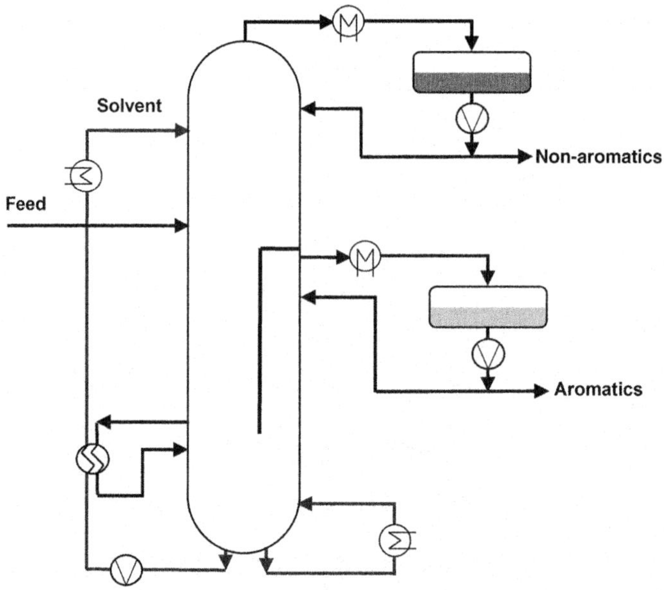

Figure 6.7 Extractive DWC by Uhde

respect to the application of extractive distillation. For the separation of a mixture of isomers, they applied a two-column extractive distillation technology instead of the former crystallization-based process. They also mentioned the possibility of applying the DWC configuration. According to the authors, this technology would have been preferred, if the process

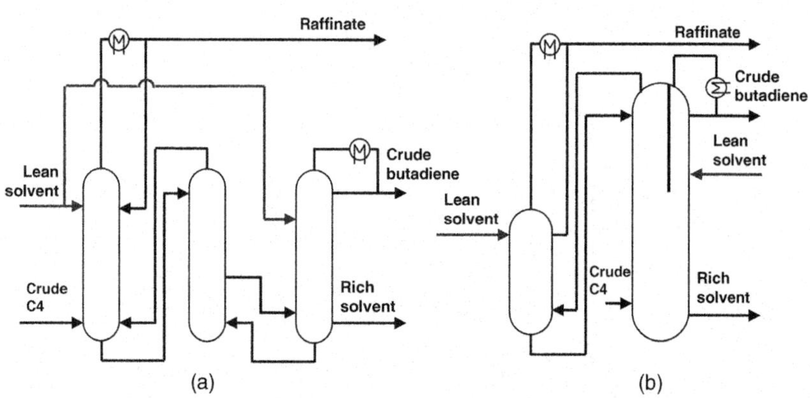

Figure 6.8 Production of butadiene by BASF SE: classic extraction/distillation (a) and dividing-wall technology (b)

could be realized with new equipment. Bravo-Bravo *et al.* (2010) applied a stochastic multi-objective procedure to find the optimal design for three different chemical systems. Kiss and Suszwalak (2012) studied the separation in an E-DWC of a near azeotropic ethanol–water mixture from bioethanol production, using ethylene glycol as solvent—more details about this system are provided in a case study presented later.

6.6 REVAMPING OF CONVENTIONAL COLUMNS TO DWC

The revamping of conventional columns to DWC is a relatively straightforward opportunity to reduce the operating costs. According to Parkinson (2007), the reduction of one column can save up to 30% of the energy costs and the revamping can be paid back within one or two years. Table 6.6 summarizes retrofitted DWCs that are reported in the literature (Yildirim, Kiss, and Kenig, 2011). In academia, the revamp of conventional columns was analyzed by Rangaiah, Ooi, and Premkumar (2009) who studied six industrially important three-component mixtures. For these systems, a two-column distillation configuration and a DWC were designed. According to the results reported, considerable energy and cost savings can be obtained using a revamped DWC. Long, Lee, and Lee (2010) investigated the use of a DWC for the debottlenecking of an existing acetic acid purification process. Several column arrangements were analyzed to show that the DWC requires less investment and energy costs than conventional distillation, the Petlyuk column, or the prefractionator arrangement.

At the industrial scale, Koch-Glitsch and MW Kellogg are so far the only suppliers to have performed retrofits. Koch-Glitsch converted two columns of Exxon-Mobil Refineries into DWC. For instance, the revamping of the Fawley Refinery in England took 30 days, but major energy savings of up to 53% were achieved (Parkinson, 2007). This revamp, including process and mechanical design, is described in detail by Slade, Stober, and Simpson (2006). Another revamp by Koch-Glitsch was made for a CEPSA refinery in Spain (Yildirim, Kiss, and Kenig, 2011). The operating costs dropped by 40% as compared to a conventional two-column system.

Lee, Shin, and Lee (2011) reported a real implementation case in which a conventional column was upgraded to a DWC unit—at LG Chem Ltd (South Korea)—in a 2-ethylhexanol (2-EH) production plant, in which butyraldehyde (BAL) is synthesized from propylene and synthesis gas by the oxo reaction. Normal butyraldehyde (NBAL) and isobutyraldehyde

Table 6.6 Retrofits of conventional columns to DWC as reported in the literature

Company	System	Constructor	Features	References
ExxonMobil Fawley Refinery Southampton, England	Recovers mixed xylenes from reformate motor gasoline	Conversion by Koch-Glitsch in 2005	• 3.8 m/4.3 m diameter • Tray column • 53% energy savings	Parkinson (2007), Slade, Stober, and Simpson (2006)
ExxonMobil Refinery Port Jerome, France	Recovers mixed xylenes from reformate motor gasoline	Conversion by Koch-Glitsch	Reduction of the operating cost up to 40%	Parkinson (2007)
CEPSA Refinery Algeciras, Spain	Separation of paraffins and isoparaffins. Production of isohexane	Conversion by Koch-Glitsch	No data available	Parkinson (2007)
BP (now BP Amoco) Petroleum refinery, Coryton, Great Britain	Undisclosed	Constructed by MW Kellogg	No data available	Amminudin and Smith (2001)
LG Chem Ltd., South Korea	Separation and purification of 2-ethylhexanol (2-EH)	Undisclosed	Dual operation allowing for conventional operation as contingency	Lee, Shin, and Lee (2011)

(IBAL) isomers are then separated through an isomer process, and the NBAL is converted into 2-EH by aldol condensation. Crude 2-EH is then generated by the hydrogenation of 2-ethyl-3-propylacrolein (EPA), and purified to a final 2-EH product in an alcohol purification unit. The conventional alcohol purification unit consists of two sequential simple distillation columns: a heavies-cut and a lights-cut column. By performing a retrofit of the existing heavies-cut column to DWC, considerable energy savings could be achieved at a reasonable installation cost. Moreover, Lee, Shin, and Lee (2011) also discuss how the risks in DWC implementation can be mitigated by establishing contingencies and predicting the performance of the distillation column via modeling.

The design of revamped columns appears generally to be simpler than the design of new DWC, due to the lower degree of freedom. We may expect more columns to be revamped to DWC in the future. However, it remains questionable whether their efficiency is actually superior to a newly designed DWC.

6.7 CASE STUDY: DIMETHYL ETHER SYNTHESIS BY R-DWC

Dimethyl ether (DME) is of great industrial interest due to its use as clean fuel for diesel engines or in combustion cells, as a precursor to other organic compounds, and a green aerosol propellant that can effectively replace chloro-fluoro-carbons. Currently, DME is produced by conversion of various feedstocks such as natural gas, coal, oil residues, and biomass into syngas (CO/H_2), followed by a two-step process: methanol synthesis and then methanol dehydration. Methanol is produced first from syngas over a copper-based catalyst (Cu/Zn, Cu/Zn/Al, Cu/Zn/Co) and then it is dehydrated over a γ-alumina catalyst or zeolites to produce DME (Muller and Hubsch, 2005; Kiss and Suszwalak, 2012). The methanol dehydration step takes place at temperatures of 250–400 °C and pressure of up to 20 bar. The current industrial process involves a fixed-bed reactor, followed by a direct sequence of two distillation columns that deliver high-purity DME (>99.99 wt%) that is virtually odorless (Muller and Hubsch, 2005).

The key problem of the conventional process is the high investment costs for several units that require a large overall plant footprint, as well as the associated energy requirements (Muller and Hubsch, 2005). Consequently, significantly better process alternatives are needed to reduce the capital and operating costs. Recent studies explored the

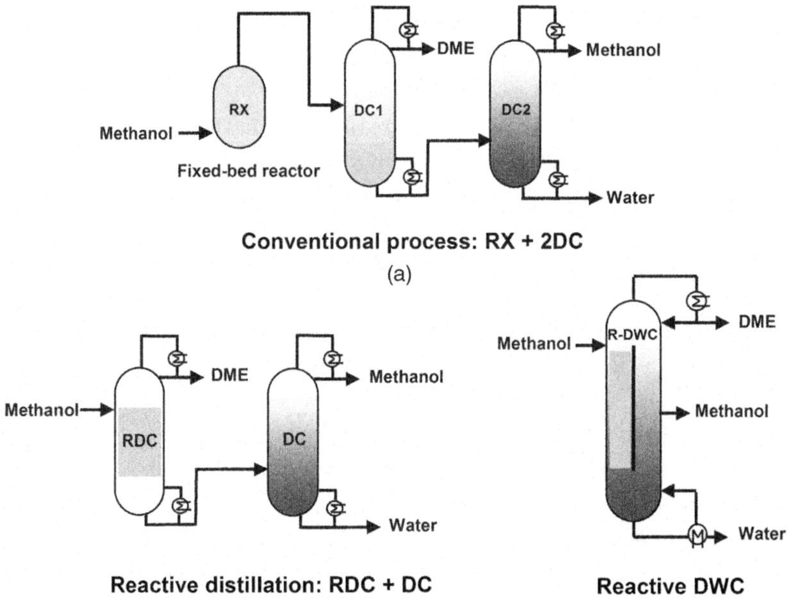

Figure 6.9 Simplified DME production processes alternatives: conventional process (a), reactive distillation (b), and reactive dividing-wall column (c)

possibility of using reactive distillation (RD) for DME production (An, Chuang, and Sanger, 2004; Lei *et al.*, 2011). Although technically feasible, the proposed RD alternatives were not sufficiently attractive economically (Kiss and Suszwalak, 2012).

Figure 6.9a illustrates the simplified conventional flow sheet for methanol dehydration (Kiss and Suszwalak, 2012). The dehydration of pure, vaporized methanol is carried out in a fixed-bed catalytic reactor. The outlet of the reactor consists of DME, water, and methanol. This is cooled and subsequently distilled in the first tower to yield pure DME. The unreacted methanol is separated from water in a second distillation column and recycled back to the reactor (Muller and Hubsch, 2005). The process flow sheet shown in Figure 6.9b involves a RD column followed by an ordinary distillation column (DC) for methanol recovery. The integrated RD column combines in fact the function of the reactor and the DME separation column of the conventional sequence. A further integration step to also include the methanol recovery column leads to a R-DWC system.

This case study presents a novel process for DME production by methanol dehydration, based on a reactive DWC (Figure 6.9c). This integrated system allows the production of high-purity DME in only one unit, with minimal footprint and significantly lower costs. Methanol is fed on top of the reactive zone where the heterogeneous catalyst is located, while DME is produced as top distillate, water as bottom product, and methanol as side stream product that is recycled.

Aspen Plus simulations embedding experimental results were performed using the rigorous RADFRAC distillation unit and explicitly considering three phase balances (AspenTech, 2010a, 2010b). The well-known MESHR equations govern the process. Note that MESHR is an acronym referring to the type of equations: M–mass balance, E–equilibrium relationships, S–summation equations, H–enthalpy balance, and R–reaction rate equations. UNIQUAC–Redlich–Kwong was selected as the most adequate property method in Aspen Plus, and the binary interaction parameters were validated against reported experimental data (Kiss and Suszwalak, 2012). The residue curves map (RCM) and the ternary map of the DME–methanol–water mixture show that no azeotropes are present in this system, but a small liquid phase split envelope is observed, hence the (reactive) distillation columns must be modeled using VLLE data.

The dehydration of methanol is an equilibrium limited reaction leading to DME and water. As verified experimentally, no side-reactions occur at the specified conditions (Lei *et al.*, 2011):

$$CH_3OH \leftrightarrow CH_3OCH_3 + H_2O \tag{6.1}$$

The model of the catalytic distillation column also includes the experimentally determined intrinsic kinetic model parameters previously reported by Lei *et al.* (2011) for methanol dehydration over an ion-exchange resin. The reaction takes place only in the liquid phase. Eley–Rideal and the equivalent power-law models are both suitable for simulation purposes (Lei *et al.*, 2011). The reaction rate, determined for the temperature range 391–423 K, is given by:

$$r = kW_{cat}[\mathrm{MeOH}]^m[\mathrm{H_2O}]^n \tag{6.2}$$

$$k = A\exp(-E_a/RT) \tag{6.3}$$

where W_{cat} is the weight amount of catalyst (e.g., 15 kg of solid catalyst per stage), A is the Arrhenius factor ($A = 5.19 \times 10^9\ \mathrm{m^3 \cdot kg\text{-}cat^{-1} \cdot s^{-1}}$), E_a is the activation energy ($133.8\ \mathrm{kJ \cdot mol^{-1}}$), and m and n are the orders of reaction with respect to methanol and water ($m = 1.51$ and $n = -0.51$).

A comparison is made between the reported conventional DME process and the process alternatives based on a reactive distillation column followed by an ordinary distillation column (RDC + DC) and reactive dividing-wall column (R-DWC) (Kiss and Suszwalak, 2012). For all these alternatives, the same feed stream was used (9 kmol h^{-1} methanol) to allow a fair comparison of the alternative pilot-scale processes with the reported conventional one (Lei *et al.*, 2011).

All processes were optimized in terms of minimal energy requirements, using the sequential quadratic programming (SQP) method implemented in Aspen Plus (Bartholomew-Biggs, 2008; AspenTech, 2010a). This can be linked to the minimization of the total heat duty of the sequence, constraint by the required purities for DME and water, and using several optimization variables such as the total number of stages, number of reactive stages and location of the reactive zone, length of the dividing-wall, location of feed and side-stream, reflux ratio, boil-up rate, and liquid and vapor split. The purity target was selected to be over 99.99 wt% for both DME and water. No hard constraint was set on the purity of the unreacted methanol, as this stream is being recycled in the process. The recent paper of Kiss and Suszwalak (2012) presents more details about all these processes. Here we limit ourselves to presenting the main results for the R-DWC system.

The R-DWC is a highly integrated setup that consists of only one column shell, one reboiler, and one condenser. Owing to the absence of an off-the-shelf DWC unit in Aspen Plus, two coupled RADFRAC units were used as the thermodynamically equivalent of the R-DWC. This method has already proven its applicability in the simulation of DWC systems (Mueller and Kenig, 2007; Kiss, Pragt, and van Strien, 2009; Hernández *et al.*, 2009). The main condition in integrating two distillation columns is that similar operating conditions should be applied. The model of the RDC + DC sequence is used as the starting point for the R-DWC simulation, providing initial estimates for the number of trays, feed tray locations, liquid and vapor split, and size of the reactive zone (Kiss and Suszwalak, 2012). The optimization problem for the minimization of the R-DWC reboiler heat duty is defined as:

$$\begin{aligned} &\mathrm{Min}\,(Q) = f(N_{\mathrm{T}}, N_{\mathrm{F}}, N_{\mathrm{R}}, N_{\mathrm{RZ}}, N_{\mathrm{DWS}}, N_{\mathrm{DWC}}, N_{\mathrm{SS}}, RR, V, F_{\mathrm{SS}}, r_{\mathrm{V}}, r_{\mathrm{L}}) \\ &\text{Subject to } \boldsymbol{y}_m \geq \boldsymbol{x}_m \end{aligned} \tag{6.4}$$

where N_{T} is the total number of stages, N_{F} is the feed stage, N_{R} is the number of reactive stages, N_{RZ} is the location of the reactive zone, N_{DWS}

is the number dividing-wall stages, N_{DWC} is the location of the dividing-wall, N_{SS} is the stage of the side-stream withdrawal, RR is the reflux ratio, V is the boil-up rate, F_{SS} is the flow rate of the side stream product, r_L and r_V are the liquid and vapor split, respectively, and $\mathbf{y}_m$ and $\mathbf{x}_m$ are vectors of the obtained and required purities for the m products. The design problem is a complex optimization problem with both discrete (N_T, N_F, N_R, N_{RZ}, N_{DWS}, N_{DWC}, N_{SS}) and continuous (RR, V, F_{SS}, r_V, r_L) decision variables. Note that in order to determine the optimal ratio between the energy cost and the number of stages an additional objective function was used, Min $N_T\,(RR+1)$, which approximates very well to the minimum of total annualized cost of a conventional distillation column (Dejanović *et al.*, 2011b).

Figure 6.10 plots the temperature and liquid composition profiles in the R-DWC, while the key parameters of the optimal R-DWC design are presented in Table 6.7 (Kiss and Suszwalak, 2012). Remarkably, the temperature difference between the two sides of the wall is very low—less than 15 °C—such conditions being easily achievable in the practical application with little heat transfer expected and negligible effect on the column performance (Dejanović, Matijašević, and Olujić, 2010; Yildirim, Kiss, and Kenig, 2011). The R-DWC unit has 35 stages, with the reactive zone located from stage 8 to 31 on the feed side, and a common stripping section (stage 32 to 35) as well as a common rectifying zone (stage 1 to 7). The methanol stream is fed on stage 8, at the top of the reactive zone—the feed side of the DWC acting as the RD zone where the solid acid catalyst is present. High purity (>99.99 wt%) DME is delivered as top distillate, while similar high-purity water is obtained as bottom product. The unreacted methanol is collected as side product, and then recycled back to the process—mixed with the fresh feed stream of methanol. On the side product part, the methanol concentration remains almost constant over a large range of stages—from stages 10 to 20—thus indicating that the side stream location has only a minor effect on the purities of the products.

The total investment costs (TIC), total operating costs (TOC), and total annual costs (TAC) are calculated for all cases to perform a fair comparison. The equipment costs are estimated using correlations from the Douglas textbook to the price level of 2010, as described by Dejanović *et al.* (2011a). The Marshall and Swift equipment cost index (M&S) considered here has a value of 1468.6. For a carbon steel column, the estimated cost in US$ is given by the relation:

$$C_{\text{shell}} = f_p(\text{M\&S}/280)d_c^{1.066}h_c^{0.802} \qquad (6.5)$$

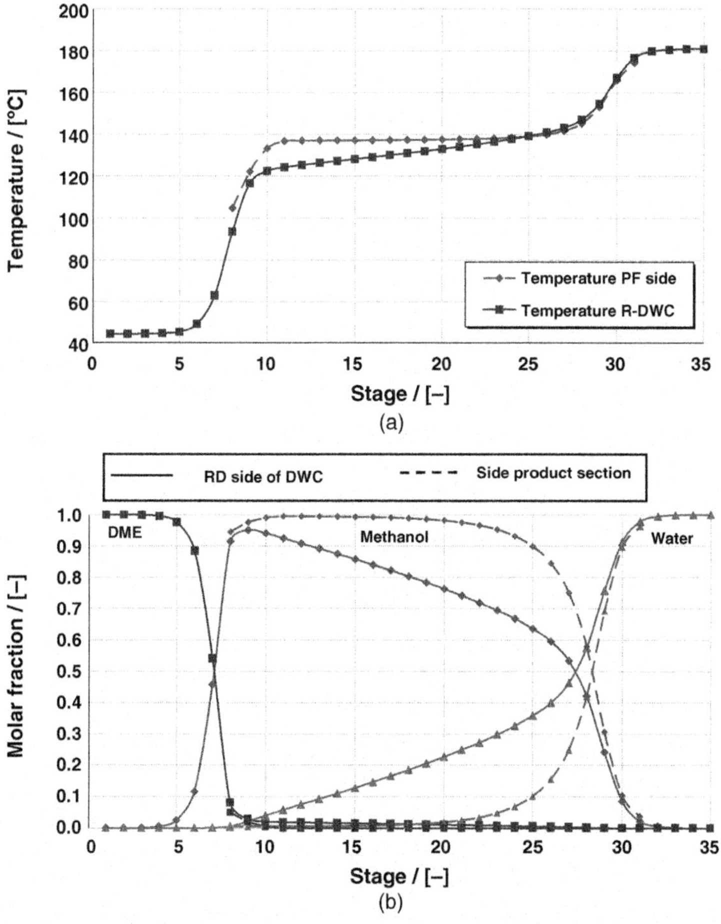

Figure 6.10 Temperature (a) and composition (b) profiles along the reactive DWC (the dashed line is used for the side product section, while the continuous line is used for the main DWC section)

where f_p is the cost factor (equal to 2981.68 in this case), d_c is the column diameter (calculated using the internals-sizing procedure from Aspen Plus), and h_c its height (tangent-to-tangent) considering a tray-spacing of 0.6 m. For heat exchangers (e.g., condensers and reboilers) the following expression was used to calculate the equipment cost (US$):

$$C_{hex} = (M\&S/280)\, c_x\, A^{0.65} \tag{6.6}$$

Table 6.7 Design parameters of an optimized reactive DWC for DME synthesis, and comparison of the key performance indicators for various process alternatives

Design parameters (units)	Value	Design parameters (units)	Value	Key performance indicators	Conventional process	Reactive distillation	Reactive DWC
Flow rate of feed stream (kmol h^{-1})	9	Reflux ratio (kmol $kmol^{-1}$)	2.45	Total investment cost (TIC) ($)	126 075	135 260	96 531
Temperature of feed stream (°C)	25	Operating pressure (bar)	10	Total operating costs (TOC) ($)	31 233	32 186	$13 988
Pressure of feed stream (bar)	10	DME product purity (wt%/mol.%)	>99.99/99.99	Total annual costs (TAC) ($)	43 840	45 712	23 641
Number of stages (–)	35	DME mass flow rate (kg h^{-1})	103.653	Specific energy requirements (kW h per ton DME)	640.5	1367.9	566.4
Stages reactive zone (–)	8–31	Methanol conversion (%)	50.40	CO_2 emissions (kg-CO_2 h^{-1}·per ton DME)	89.57	191.28	79.20
Feed stage (–)	8	Water purity (bottom product) (wt%/mol.%)	>99.99/99.99				
Wall position (from/to stage) (–)	8–31	Reboiler duty (kW)	58.70				
Distillate to feed ratio (kmol $kmol^{-1}$)	0.25	Condenser duty (kW)	–37.13				

where $c_x = 1609.13$ for condensers and 1775.26 for kettle reboilers, while A is the heat transfer area (m^2). In addition, a price of US \$600 m^{-2} was used for the sieve trays cost calculations.

The following utility costs were considered: US \$0.03 per ton of cooling water and US \$13 per ton of steam. For the TAC calculations, a plant lifetime of 10 years was considered. Note that the price of the catalyst in the columns was not accounted for as the ion-exchange resins are rather inexpensive, and the same amount of catalyst was used in all cases described. Moreover, the accuracy of the correlations is in the range of ±30%, which is acceptable and realistic. Clearly, this accuracy is less important when comparing design alternatives since the error is consistent in all cases.

Table 6.7 conveniently illustrates the costs and specific energy requirements of the three processes considered (Kiss and Suszwalak, 2012). Note that the data for the conventional process was calculated based on results reported by Lei *et al.* (2011). Overall, the innovative reactive DWC process has better performance than the conventional or the reactive distillation process: significant energy savings of 12–58%, up to 60% reduced CO_2 emissions, and up to 30% lower total annual costs. Consequently, the novel R-DWC process can be considered as a serious alternative candidate for high-purity DME production in new plants as well as revamped industrial plants.

6.8 CASE STUDY: BIOETHANOL DEHYDRATION BY A-DWC AND E-DWC

Owing to the current shortage of fossil fuels, renewable sources of energy and fuels are intensely investigated. In terms of biofuels, bioethanol is considered as the most promising alternative on short- and medium-terms and its use as a biofuel additive has rapidly increased (Ward and Singh, 2002; Kaminski, Marszalek, and Ciolkowska, 2008). A major advantage of bioethanol over other fuel alternatives, such as hydrogen, is that it can easily be integrated in the existing fuel systems as a 5–85% mixture with gasoline that does not require any modification of current engines. The higher content of oxygen leads to more efficient combustion and thus reduces the carbon footprint. In addition, the raw materials for bioethanol production—such as corn, sugar cane, and wood—capture and convert CO_2 from the atmosphere, thus making bioethanol a carbon neutral source of energy (Kaminski, Marszalek, and Ciolkowska, 2008).

Several processes are used at the industrial scale to produce bioethanol, such as corn-to-ethanol and basic and integrated lignocellulosic biomass-to-ethanol (Balat, Balat, and Oz, 2008). A common feature is the production of diluted bioethanol—around 5–12 wt% ethanol—that needs to be further concentrated (Vane, 2008; Huang *et al.*, 2008; Frolkova and Raeva, 2010). According to current international bioethanol standards, the maximum allowed water content is 0.2 vol.% (EN 15376, Europe), 0.4 vol.% (ABTN/Resolução ANP no. 36/2005, Brazil), or 1.0 vol.% (ASTM D 4806, USA).

Two energy demanding separation steps are required to reach the purity target, mainly due to the presence of the well-known binary azeotrope ethanol–water (95.63 wt% ethanol). The first step is typically an ordinary distillation, also called a *pre-concentration stage,* that concentrates bioethanol up to 92.4–94 wt% (Ward and Singh, 2002; Vane, 2008; Huang *et al.*, 2008; Frolkova and Raeva, 2010). The second step is more complex and of greater interest, as it requires further dehydration of ethanol up to higher concentrations above the azeotropic composition. Several alternatives are available and well described: pervaporation, adsorption, pressure-swing distillation, extractive distillation (ED), and azeotropic distillation (AD), as well as hybrid methods combining these options (Huang *et al.*, 2008; Kaminski, Marszalek, and Ciolkowska, 2008; Vane, 2008; Frolkova and Raeva, 2010).

Pervaporation methods are energy efficient and have a modular design that allows easy maintenance, as well as a smaller surface area than larger equipment such as distillation columns. However, they do reach their limits in the case of large-scale separations (Vane, 2008; Frolkova and Raeva, 2010). Adsorption with molecular sieves has become more popular recently as it requires less energy than distillation. However, the desorption step requires high temperature and/or low pressure, thus leading to very high overall equipment costs (Huang *et al.*, 2008; Vane, 2008; Frolkova and Raeva, 2010). Distillation methods, such as ED and AD, present relatively high energy costs but, despite this major drawback, they are still the option of choice for the large-scale production of bioethanol fuel (Vane, 2008; Frolkova and Raeva, 2010). Usually, ED and AD are performed in a conventional sequence of two columns, the first of them separating ethanol while the other splits water from the recovered mass separating agent (MSA) that is recycled.

This case study presents novel distillation technologies for enhanced bioethanol dehydration, by extending the use of DWCs to energy efficient ED and AD. The novel E-DWC and A-DWC configurations are applied to the enhanced dehydration of bioethanol, from 85 mol.% (93.5 wt% ethanol) to the required standard purity (>99.8 wt%). This particular feed stream is obtained after a pre-concentration step by ordinary distillation that follows the production of bioethanol in a fermentation reactor, which increases the concentration from 5–12 wt% to 93.5 wt% ethanol. Ethylene glycol and *n*-pentane are used here as MSA in an extractive and azeotropic distillation, respectively. Remarkably, the energy requirements are reduced while using fewer equipment units and fewer stages in total, as compared to the conventional ED and AD systems.

Aspen Plus simulations were performed using the rigorous RADFRAC unit, and explicitly considering three phase balances. NRTL and UNIQUAC property methods can be used due to the presence of a non-ideal mixture containing polar components (AspenTech, 2010b). Both methods were successfully used in the past, leading to very similar results (Kiss and Suszwalak, 2012). Note that in the pre-concentration step the diluted ethanol stream (5–12 wt%) obtained by fermentation is rather easily distilled to the near-azeotropic composition of 93.5 wt% ethanol. This is typically performed by an ordinary distillation column that requires significant thermal energy of up to 2.6 kW h kg^{-1} bioethanol, due to the large amount of water that needs to be separated. The feed stream considered here is the one obtained from the pre-concentration stage of bioethanol, and consists of a mixture of 85 kmol h^{-1} (3915.9 kg h^{-1}) ethanol and 15 kmol h^{-1} (270.2 kg h^{-1}) water, thus having a near azeotropic composition (93.5 wt% ethanol). The target purity for the end product was selected as a min. 99.8 wt% ethanol to comply with all bioethanol standards.

All the conventional and novel DWC alternatives were optimized in terms of minimal energy demand using the sequential quadratic programming (SQP) method available in Aspen Plus (Aspen Technology, 2010; Bartholomew-Biggs, 2008). This can be linked to the minimization of the heat duty of the sequence, constraint by the required purity, and recovery of the bioethanol product, using sensitivity analysis and the SQP optimization tool from Aspen Plus. Several optimization variables are used: total number of stages, feed-stage, side-stream and recycle streams location, solvent flow rate, reflux ratio, and liquid and vapor split.

Extractive distillation performs the separation in the presence of a miscible, high boiling, relatively non-volatile component that forms no azeotrope with the other components in the mixture. Ethylene glycol (EG) remains the most common entrainer used in the extractive distillation of ethanol–water, although hyperbranched polymers and ionic liquids were also proposed (Kiss and Suszwalak, 2012). The ternary mixture ethanol–water–glycol presents a single binary azeotrope and no liquid phase splitting. As both distillation columns of the conventional

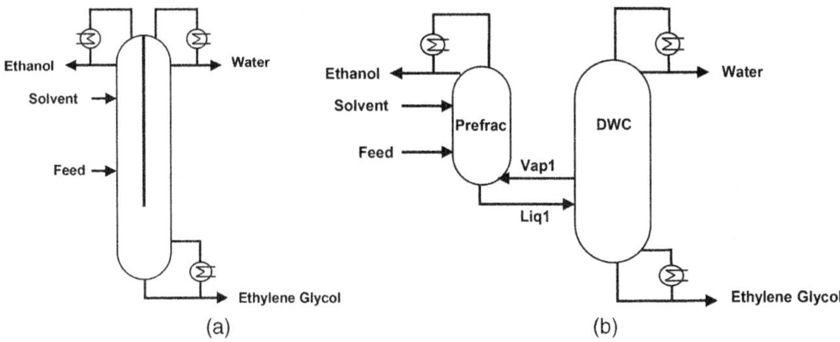

Figure 6.11 Scheme of the proposed E-DWC (a); thermodynamically equivalent, decomposed flow sheet of an E-DWC suitable for simulation (b)

sequence operate at atmospheric pressure, the use of a DWC seems to be an attractive alternative.

Figure 6.11a shows the conceptual design of the proposed E-DWC (Kiss and Suszwalak, 2012). Such a column is usually called a *split shell column with divided overhead section and common bottoms section* (Yildirim, Kiss, and Kenig, 2011). In this column the solvent is separated as a single bottom product, while two distillate products are collected on each side of the wall—ethanol and water, respectively. Since there is no off-the-shelf DWC unit in the currently available process simulators, two coupled RADFRAC units were used in Aspen Plus, as the thermodynamically equivalent of the E-DWC. Figure 6.11b illustrates this decomposed flow sheet, consisting of two column shells and two condensers but only one reboiler (Kiss and Suszwalak, 2012). The Aspen Plus model of the direct sequence was used as the starting point of the E-DWC simulation. The results of the direct sequence simulation provide in fact the initial estimates for the number of trays, feed tray locations, and liquid and vapor split.

Figure 6.12 plots the temperature and liquid composition profiles in the E-DWC, while Table 6.8 presents the key parameters of the optimal design (Kiss and Suszwalak, 2012). The temperature difference between the two sides of the wall is very low, less than 20 °C—such conditions being rather easily realized in the practical application. Moreover, high purity and recovery is obtained for all three products of the extractive dividing-wall column: ethanol and water as top distillates, and EG solvent as recovered bottom product.

Azeotropic distillation is carried out by adding other light chemicals to generate a new, lower-boiling azeotrope that is heterogeneous—thus producing two, immiscible liquid phases. In one sense, adding an

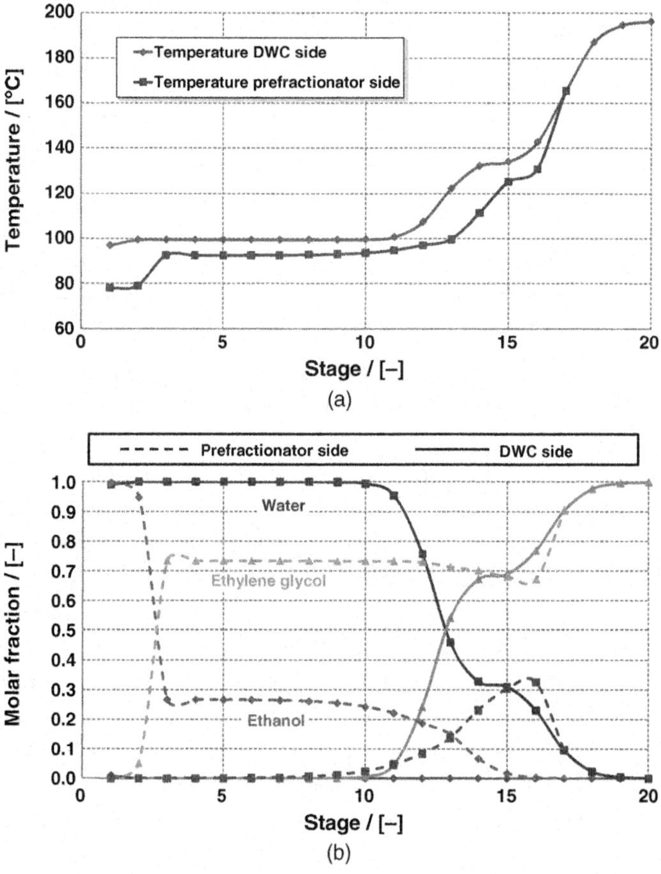

Figure 6.12 Temperature (a) and composition (b) profiles in the E-DWC (the dashed line is used for the prefractionator side, while the continuous line is used for the main DWC side)

entrainer is similar to extractive distillation. However, in the case of extractive distillation, a high-boiling mass separating agent is used, leading to lower energy demands as compared to azeotropic distillation, as the high-boiling solvent does not have to be evaporated (Yildirim, Kiss, and Kenig, 2011). One of the best entrainers for bioethanol dehydration by azeotropic distillation is *n*-pentane, as it forms a low-boiling ternary azeotrope with ethanol and water (Frolkova and Raeva, 2010). Note that cyclohexane was also successfully used as an alternative (Midori *et al.*, 2001; Sun *et al.*, 2011). The mixture ethanol–water–pentane presents three binary azeotropes, one ternary heterogeneous azeotrope, and a significant liquid phase split envelope (Kiss and Suszwalak, 2012).

Table 6.8 Design parameters of an optimized E-DWC for bioethanol dehydration

Design parameters (units)	Value
Reflux ratio prefractionator (kmol kmol^{-1})	0.27
Number of stages prefractionator (–)	16
Feed stage prefractionator (–)	13
Feed stage of extractive agent (–)	3
Reflux ratio DWC (kmol kmol^{-1})	0.2
Number of stages DWC (–)	20
Stage of the interconnection Liq1 (–)	14
Stage of the interconnection Vap1 (–)	14
Interconnection liquid flow (kmol h^{-1})	291
Interconnection vapor flow (kmol h^{-1})	86
Feed flow rate of ethanol (kmol h^{-1})	85
Feed flow rate of water (kmol h^{-1})	15
Feed flow rate of extractive agent (kmol h^{-1})	190
Heat duty prefractionator (kW)	0
Heat duty DWC (kW)	1819.52
Total heat duty (kW)	1819.52
Operating pressure prefractionator (bar)	1
Operating pressure DWC (bar)	1
Ethanol recovery (%)	99.80
Water recovery (%)	99.10
Ethylene glycol recovery (%)	99.90
Purity of ethanol recovered (wt%/mol.%)	99.80/99.84
Purity of water by-product (wt%/mol.%)	97.64/99.07
Purity of ethylene glycol (recycled) (wt%/mol.%)	99.98/99.93

Extension of the conventional sequence to the A-DWC model follows the same simulation and optimization procedure as for the previously described E-DWC case. Figure 6.13a shows a schematic representation of the proposed A-DWC configuration (Kiss and Suszwalak, 2012). This is also known as a *split shell column with common overhead section and divided bottoms section* (Yildirim, Kiss, and Kenig, 2011). This alternative setup consists of a single shell, two reboilers, and only one condenser. Consequently, two bottom products are collected: bioethanol on the feed side and water on the other side. The azeotropic top stream is fed to a decanter from which the organic phase is recycled back to the feed side, while the aqueous phase is returned to the other side of the A-DWC column. A thermodynamically equivalent model using a post-fractionator was implemented in Aspen Plus, using two RADFRAC units thermally coupled. Figure 6.13b illustrates the simulated flow sheet. A-DWC was initialized using the design parameters of the direct sequence, and then it was optimized using the same methodology, as previously described.

Notably, the presence of a high amount of pentane that has a lower boiling point than water and ethanol creates a higher temperature

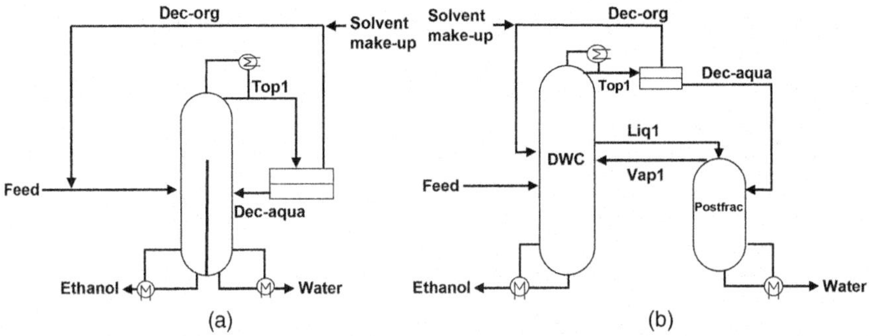

Figure 6.13 Schematics of the proposed A-DWC (a); equivalent A-DWC flow sheet (b)

difference in the DWC design. As compared to E-DWC, the temperature profile—plotted in Figure 6.14 (Kiss and Suszwalak, 2012)—shows a larger difference between the temperatures on the two sides of the dividing-wall. Nevertheless, this difference is still acceptable for practical implementation without the need for an insulating the dividing-wall (Dejanović, Matijašević, and Olujić, 2010). Larger temperature differences would require more attention, as equivalent configurations used for the simulation are thermodynamically equivalent to a DWC only if the heat transfer across the wall can be neglected. Table 6.9 provides the optimal design parameters for the A-DWC system (Kiss and Suszwalak, 2012). The results clearly show that all products can be obtained with the required purities and recoveries.

Table 6.10 compares the energy requirements for the conventional two-column sequence with the proposed E-DWC and A-DWC, respectively (Kiss and Suszwalak, 2012). The specific energy requirements were calculated at 0.51 kW h kg^{-1} for ED and 0.46 kW h kg^{-1} bioethanol for E-DWC. Energy savings of around 10% are possible with the E-DWC, as compared to the conventional ED sequence of two columns. Similarly, the specific energy requirements are 1.78 kW h kg^{-1} for the conventional AD and 1.42 kW h kg^{-1} bioethanol for the A-DWC alternative, respectively. Remarkably, although both cases were optimized, the A-DWC still allows over 20% energy savings as compared to the conventional AD configuration. These savings are actually lower than the usually reported values of 25–40%, because both the conventional ED and AD sequences were also optimized here—existing running plants at industrial scale use in fact more energy than the calculated optimal.

Single-step concentration and dehydration was deployed in a follow-up study by Kiss and Ignat (2012), which extended the scope of the

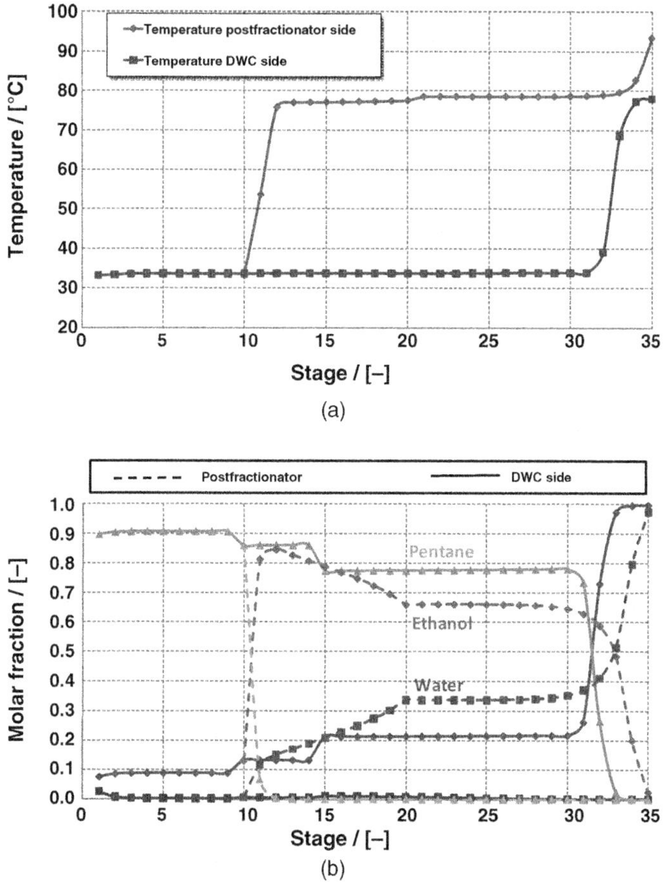

Figure 6.14 Temperature (a) and composition (b) profiles in the A-DWC (the dashed line is used for the post-fractionator side, while the continuous line is used for the main DWC side)

bioethanol dehydration to cover both the dehydration and the pre-concentration steps—the most energy intensive part of the bioethanol production process (Figure 6.15a). A mixture of 10 wt% ethanol (100 ktpy plant) was considered for the concentration and dehydration steps using ethylene glycol as mass separating agent. Similarly to the procedure described earlier, rigorous simulations were carried out in Aspen Plus, and for a direct and fair comparison all alternatives were optimized using the proven SQP method.

The innovative solution proposed by Kiss and Ignat (2012) is based on a novel extractive dividing-wall column (E-DWC) that is able to concentrate and dehydrate bioethanol in a single step, by integrating all units of

Table 6.9 Design parameters of an optimal A-DWC for bioethanol dehydration

Design parameters (units)	Value
Reflux ratio post-fractionator (kmol kmol^{-1})	0.53
Number of stages post-fractionator (–)	25
Feed stage post-fractionator (aqueous phase from decanter) (–)	10
Reflux ratio DWC (kmol kmol^{-1})	0.93
Number of stages DWC (–)	35
Feed stage DWC (recycle organic phase) (–)	15
Feed stage DWC (main feed) (–)	15
Stage of the interconnection Liq1 (–)	10
Stage of the interconnection Vap1 (–)	10
Interconnection liquid flow (kmol h^{-1})	116.1
Interconnection vapor flow (kmol h^{-1})	156.6
Feed flow rate of ethanol (kmol h^{-1})	85
Feed flow rate of water (kmol h^{-1})	15
Feed flow rate of extractive agent (kmol h^{-1})	851.9
Heat duty post-fractionator (kW)	1475.2
Heat duty DWC (kW)	4062.9
Total heat duty (kW)	5538.1
Operating pressure post-fractionator (bar)	1
Operating pressure DWC (bar)	1
Ethanol recovery (%)	99.74
Water recovery (%)	97.40
Purity of bioethanol product (wt%/mol.%)	99.90/99.70
Purity of water by-product (wt%/mol.%)	93.60/97.40

the conventional separation sequence into only one distillation column (PDC: pre-concentration distillation column, EDC: extractive distillation column, and SRC: solvent recovery column). Note that using standard DWC configurations requires about 4–7 times more energy than the conventional distillation sequence—thus rendering them economically unattractive. Technically, the standard DWC configurations (mid and top wall) can produce high purity ethanol. However, the energy required is much higher as compared to the conventional sequence because a huge amount of water (e.g., 90% of the feed stream) must be evaporated and removed as side stream or top distillate. The lesson learned is that water must be removed as bottom product to avoid its complete evaporation. But how to realize this task when the highest boiling component in the system is the heavy solvent and not water?

Figure 6.15b shows a new E-DWC unit that consists of only one column shell, one condenser, and two reboilers—a configuration that is quite counter-intuitive as compared to previous studies on E-DWC (Yildirim, Kiss, and Kenig, 2011; Kiss and Suszwalak, 2012; Kiss and Ignat, 2012). In this column, the feed side (prefractionator) acts as the PDC unit of the conventional sequence. Water is removed as liquid side

Table 6.10 Comparison of heat duties for the extractive (ED) and azeotropic distillation (AD)

	Direct sequence ED			E-DWC		
	DC1	DC2	Total	PreFrac	DWC	Total
Condenser duty (kW)	−1200.62	−367.29	−1567.90	−1175.01	−205.09	−1380.10
Reboiler duty (kW)	1402.43	605.39	2007.82	0.00	1819.52	1819.52
	Direct sequence AD			A-DWC		
	DC1	DC2	Total	DWC	PostFrac	Total
Condenser duty (kW)	−7165.75	−955.88	−8121.63	−12 500.0	0.00	−12 500.0
Reboiler duty (kW)	5948.76	994.02	6942.78	4062.9	1475.20	5538.1

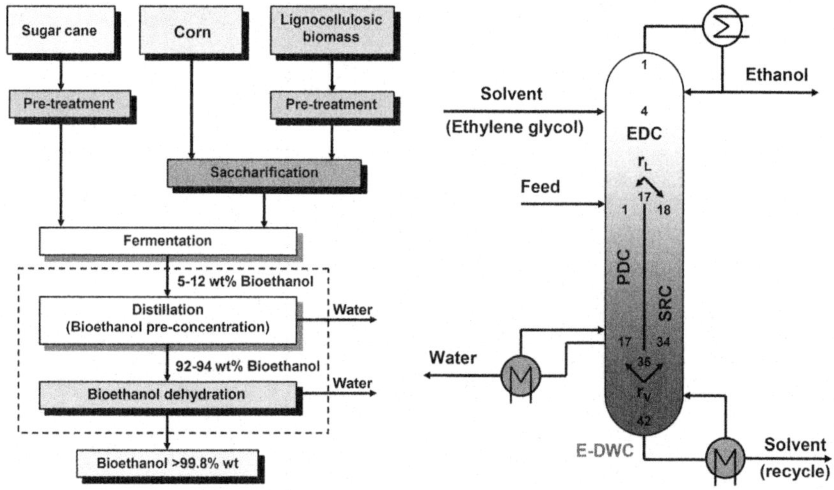

(a) (b)

Figure 6.15 Block flow diagram of the bioethanol production from various feedstock (a). Extractive DWC combining the functions of all conventional columns into one unit (b)

stream, but an additional side reboiler is needed to return the required amount of water vapor to the column. The liquid feed stream is fed on top of the prefractionator side, thus serving as a reflux to the PDC section. The vapor leaving the feed side of the E-DWC has a near azeotropic composition. Solvent is added at the top of the E-DWC; this section acts in fact as the EDC unit of the conventional sequence. Ethanol is separated here as high purity top distillate, and removed as the main product. The liquid flowing down the top section (EDC) is collected and distributed only to the (SRC) side opposite to the feed side (prefractionator) and further down the bottom of the E-DWC. This complete redistribution of the liquid flow is required to avoid the presence and loss of solvent on the feed side (PDC section). In the SRC section, the solvent is separated as bottom product and then recycled in the process. The vapor coming from the bottom of the E-DWC to the lower part of the dividing-wall consists mainly of water. However, this amount is not sufficient for the PDC section and, thus, the requirement for an additional side reboiler.

The results show that energy savings of 17% are possible by using an E-DWC for the single-step bioethanol concentration and dehydration (from 10 to over 99.8 wt%) with specific energy requirements as low as 2.07 kW h kg^{-1} bioethanol, without additional heat-integration. Moreover, a decrease of 18% in the equipment costs is possible for the novel

E-DWC alternative, while using a reduced footprint as compared to the conventional separation process (Kiss and Ignat, 2012).

6.9 CONCLUDING REMARKS

DWC technology has been exploited for the separation of a broad spectrum of chemicals, such as hydrocarbons, alcohols, aldehydes, ketones, acetals, and amines. In principle, DWC units can be built as both trayed and packed columns. However, to date, most DWC units have been constructed by Montz, and used by BASF SE, as packed columns. Details on the specific features of existing DWC are mostly undisclosed. The DWC units are usually larger than common distillation columns, with diameters reaching up to 5 m. For the separation of mixtures with more than three components, only a few industrial applications are reported (Yildirim, Kiss, and Kenig, 2011).

The integration of the DWC concept with azeotropic, extractive, and reactive distillation principles shows a remarkable reduction in terms of investment and operation costs. Currently, industrial applications have been reported only in the field of extractive distillation (BASF SE and UOP). Moreover, the literature on azeotropic and extractive DWC is relatively scarce (Kiss and Suszwalak, 2012; Kiss and Ignat, 2012). For reactive distillation in a DWC, few investigations have been performed, both experimentally and theoretically. Some models have been developed, but no industrial-scale application is available yet (Yildirim, Kiss, and Kenig, 2011).

Compared to conventional distillation towers, DWC units are considerably more energy efficient and require less capital investments, as well as a low plant footprint. The recent rapid expansion of DWC applications allows an educated estimation of about 350 industrial applications that could be expected by 2015.

REFERENCES

Amminudin, K.A. and Smith, R. (2001) Design and optimization of fully thermally coupled distillation columns: Part 2: Application of dividing wall columns in retrofit. *Chemical Engineering Research and Design*, *79*, 716–724.

An, W., Chuang, K., and Sanger, A. (2004) Dehydration of methanol to dimethyl ether by catalytic distillation. *Canadian Journal of Chemical Engineering*, *82*, 948–955.

Aspen Technology (2010a) Aspen Plus: User guide - Volume 1 & 2.

Aspen Technology (2010b) Aspen physical property system - Physical property models.

Asprion, N. and Kaibel, G. (2010) Dividing wall columns: Fundamentals and recent advances. *Chemical Engineering and Processing: Process Intensification*, *49*, 139–146.

Balat, M., Balat, H., and Oz, C. (2008) Progress in bioethanol processing. *Progress in Energy and Combustion Science*, *34*, 551–573.

Barroso-Muñoz, F. (2009) Thermodynamic analysis and hydrodynamic behavior of a reactive dividing wall distillation column. The 9th International Conference on Chemical & Process Engineering Rome, Italy.

Bartholomew-Biggs, M. (2008) *Nonlinear Optimization with Engineering Applications, Springer Optimization and Its Applications*, vol. *19*, Springer Science + Business Media, New York, pp. 1–14.

Bravo-Bravo, C., Segovia-Hernández, J.G., Gutiérrez-Antonio, C. *et al.* (2010) Extractive dividing wall column: Design and optimization. *Industrial & Engineering Chemistry Research*, *49*, 3672–3688.

Briones-Ramírez, A., Gutiérrez-Antonio, C., Jacek, J., and Jan, T. (2009) Dividing wall distillation columns for separation of azeotropic mixtures: Feasibility procedure and rigorous optimization. *Computer Aided Chemical Engineering*, *26*, 555–560.

Bumbac, G., Plesu, A.E., and Plesu, V. (2007) Reactive distillation process analysis in a divided wall column. *Computer Aided Chemical Engineering*, *24*, 443–448.

Bumbac, G., Ene, A., Isopescu, R., and Torma, A. (2009) Process simulation of reactive distillation in dividing wall column for ETBE synthesis process. *Chemical Engineering Transactions*, *18*, 487–492.

Daniel, G., Patil, P., Dragomir, R., and Jobson, M. (2006) Conceptual design of reactive dividing wall columns, in *Distillation & Absorption 2006 (Symposium)*, IChemE, Rugby, pp. 364–372.

Dejanović, I., Matijašević, L., and Olujić, Ž. (2010) Dividing wall column - A breakthrough towards sustainable distilling. *Chemical Engineering and Processing: Process Intensification*, *49*, 559–580.

Dejanović, I., Matijašević, L., Halvorsen, I.J. *et al.* (2011a) Designing four-product dividing wall columns for separation of a multicomponent aromatic mixture. *Chemical Engineering Research and Design*, *89*, 1155–1167.

Dejanović, I., Matijašević, L., and Olujić, Ž. (2011b) An effective method for establishing the stage and reflux requirement of three-product dividing wall columns. *Chemical & Biochemical Engineering Quarterly*, *25*, 147–157.

Diehl, T., Kolbe, B., and Gehrke, H. (2005) Uhde Morphylane extractive distillation - where do we stand? presented at the ERTC Petrochemical Conference, Prague, Czech Republic, 3–5 October 2005.

Frolkova, A.K. and Raeva, V.M. (2010) Bioethanol dehydration: State of the art. *Theoretical Foundations of Chemical Engineering*, *44*, 545–556.

Großmann, C. and Kenig, E.Y. (2007) Trennwandkolonnen - entwicklungsstand und perspektiven. *CIT Plus*, *5*, 38–41.

Grützner, T., Staak, D., Schwegler, B., and Roederer, D. Dividing wall column for industrial multi purpose use. Presented at the 20th International Congress of Chemical and Process Engineering (CHISA), 25-29 August 2012, Prague, Czech Republic, Paper 10.

Guedes, B.P., Feitosa, M.F., Vasconcelos, L.S. *et al.* (2007) Sensitivity and dynamic behavior analysis of an industrial azeotropic distillation column. *Separation and Purification Technology*, *56*, 270–277.

Heida, B., Bohner, G., and Kindler, K. (2002) Consider divided-wall technology for butadiene extraction. *Hydrocarbon Processing*, *81*, 50B–50H.

Hernández, S., Sandoval-Vergara, R., Barroso-Munoz, F.O. *et al.* (2009) Reactive dividing wall distillation columns: Simulation and implementation in a pilot plant. *Chemical Engineering and Processing: Process Intensification*, *48*, 250–258.

Hill, T., Kaibel, G., Meyer, G. *et al.* (2002) Reactive distillation apparatus for production of ethers from hydrocarbon fractions comprises simultaneous hydrogenation, etherification and distillative separation, German Patent DE10033958 A1.

Huang, H.J., Ramaswamy, S., Tschirner, U.W., and Ramarao, B.V. (2008) A review of separation technologies in current and future biorefineries. *Separation and Purification Technology*, *62*, 1–21.

Jobson, M. (2005) Dividing wall distillation comes of age. *The Chemical Engineer*, *766*, 30–31.

Kaibel, B., Jansen, H., Zich, E., and Olujić, Ž. (2006) Unfixed dividing wall technology for packed and tray distillation columns, in *Distillation & Absorption 2006 (Symposium)*, IChemE, Rugby, pp. 252–266.

Kaibel, G. (1984) Method of carrying out chemical reactions and for the simultaneous fractionation of a mixture into several fractions by a distillation column, European Patent EP126288 A2.

Kaibel, G., Miller, C., Stroezel, M. *et al.* (2004) Industrieller einsatz von trennwandkolonnen und thermisch gekoppelten destillationskolonnen. *Chemie Ingineur Technik*, *76*, 258–263.

Kaminski, W., Marszalek, J., and Ciolkowska, A. (2008) Renewable energy source - dehydrated ethanol. *Chemical Engineering Journal*, *135*, 95–102.

Kenig, E.Y., Müller, I., Großmann, C. *et al.* (2006) Trennwandkolonnen: entwicklungsstand und perspektiven. *Chemie Ingenieur Technik*, *78*, 1281–1282.

Khoury, F.M. (2005) *Multistage Separation Processes*, 3rd edn, CRC Press, Boca Raton.

Kiss, A.A., Pragt, H., and van Strien, C. (2009) Reactive dividing-wall columns - how to get more with less resources? *Chemical Engineering Communications*, *196*, 1366–1374.

Kiss, A.A. and Ignat, R.M. (2012) Innovative single step bioethanol dehydration in an extractive dividing-wall column. *Separation & Purification Technology*, *98*, 290–297.

Kiss, A.A. and Suszwalak, D.J.-P.C. (2012) Enhanced bioethanol dehydration by extractive and azeotropic distillation in dividing-wall columns. *Separation & Purification Technology*, *86*, 70–78.

Kiss, A.A. and Suszwalak, D.J-P.C. (2012) Innovative dimethyl ether synthesis in a reactive dividing-wall column. *Computers & Chemical Engineering*, *38*, 74–81.

Kiss, A.A., Segovia-Hernandez, J.G., Bildea, C.S. *et al.* (2012) Reactive DWC leading the way to FAME and fortune. *Fuel*, *95*, 352–359.

Kiss, A.A. (2012) Applying reactive distillation. *NPT Procestechnologie*, *19* (1), 22–24.

Kolbe, B. and Wenzel, S. (2004) Novel distillation concepts using one-shell columns. *Chemical Engineering and Processing: Process Intensification*, *43*, 339–346.

Lee, M., Shin, J., and Lee, S. (2011) Manage risks with dividing-wall column installations. *Hydrocarbon Processing*, *90*, 59–62.

Lei, Z., Li, C., Li, Y., and Chen, B. (2004) Separation of acetic acid and water by complex extractive distillation. *Separation & Purification Technology*, *36*, 131–138.

Lei, Z., Zou, Z., Dai, C. *et al.* (2011) Synthesis of dimethyl ether (DME) by catalytic distillation. *Chemical Engineering Science*, *66*, 3195–3203.

Long, N.V.D., Lee, S., and Lee, M. (2010) Design and optimization of a dividing wall column for debottlenecking of the acetic acid purification process. *Chemical Engineering and Processing*, *49*, 825–835.

Long, N.V.D. and Lee, M. (2011) Reduce costs and energy consumption of deethanizing and depropanizing fractionation steps in NGL recovery process. *Computer Aided Chemical Engineering*, *29*, 1658–1662.

Midori, S., Zheng, S.N., and Yamada, I. (2000) Analysis of divided-wall column for extractive distillation. *Kagaku Kogaku Ronbunshu*, *26*, 627–632.

Midori, S., Zheng, S.N., and Yamada, I. (2001) Azeotropic distillation process with vertical divided-wall column. *Kagaku Kogaku Ronbunshu*, *27*, 756–760.

Mueller, I. and Kenig, E.Y. (2007) Reactive distillation in a dividing wall column – rate-based modeling and simulation. *Industrial & Engineering Chemistry Research*, *46*, 3709–3719.

Mueller, I., Pech, C., Bhatia, D., and Kenig, E.Y. (2007) Rate-based analysis of reactive distillation sequences with different degrees of integration. *Chemical Engineering Science*, *62*, 7327–7335.

Mueller, I. and Kenig, E.Y. (2010) Numerical investigation of the reactive dividing wall column exemplified by methyl acetate hydrolysis. *Chemie Ingenieur Technik*, *82*, 2109–2118.

Muller, M. and Hubsch, U. (2005) Dimethyl ether, in *Ullmann's Encyclopedia of Industrial Chemistry*, 7th edn, Wiley-VCH Verlag GmbH, Weinheim.

Niggemann, G., Hiller, C., and Fieg, G. (2010) Experimental and theoretical studies of a dividing-wall column used for the recovery of high-purity products. *Industrial & Engineering Chemistry Research*, *49*, 6566–6577.

Olujić, Ž., Kaibel, B., Jansen, H. *et al.* (2003) Distillation column internals/configurations for process intensification. *Chemical and Biochemical Engineering Quarterly*, *17*, 301–309.

Olujić, Ž., Jödecke, M., Shilkin, A. *et al.* (2009) Equipment improvement trends in distillation. *Chemical Engineering and Processing: Process Intensification*, *48*, 1089–1104.

Parkinson, G. (2005) Distillation: New wrinkles for an age-old technology. *Chemical Engineering Progress*, *101*, 10–19.

Parkinson, G. (2007) Dividing-wall columns find greater appeal. *Chemical Engineering Progress*, *103*, 8–11.

Pendergast, J.G., Vickery, D., Au-Yeung, P., and Anderson, J. (2009) Consider dividing wall columns. *Chemical Processing*, *72*, 23–27.

Premkumar, R. and Rangaiah, G.P. (2009) Retrofitting conventional column systems to dividing-wall columns. *Chemical Engineering Research and Design*, *87*, 47–60.

Rangaiah, G.P., Ooi, E.L., and Premkumar, R. (2009) A simplified procedure for quick design of dividing-wall columns for industrial applications. *Chemical Product and Process Modeling*, *4* (1), Article 7. doi: 10.2202/1934-2659.1265.

Ruffert, G. and Olf, G. (2004) Extraktivdestillation - erfolg mit neuen entwicklungsmethoden. *Chemie Technik*, *3*, 86–88.

Sander, S., Flisch, C., Geissler, E. *et al.* (2007) Methyl acetate hydrolysis in a reactive divided wall column. *Chemical Engineering Research and Design*, *85*, 149–154.

Sandoval-Vergara, R., Barroso-Muñoz, F.O., Hernández-Escoto, H. *et al.* (2008) Implementation of a reactive dividing wall distillation column in a pilot plant. *Computer Aided Chemical Engineering*, *25*, 229–234.

Sattler, K. and Feindt, H.J. (1995) *Thermal Separation Processes*, VCH, Weinheim.

Schönbucher, A. (2002) *Thermische Verfahrenstechnik*, Springer, New York.

Schultz, M.A., Stewart, D.G., Harris, J.M. *et al.* (2002) Reduce costs with dividing-wall columns. *Chemical Engineering Progress*, *98*, 64–71.

Schultz, M.A., O'Brien, D.E., Hoehn, R.K. *et al.* (2006) Innovative flowschemes using dividing wall columns. *Computer Aided Chemical Engineering*, *21*, 695–700.

Slade, B., Stober, B., and Simpson, D. (2006) Dividing wall column revamp optimises mixed xylenes production, in *Distillation & Absorption 2006 (Symposium)*, IChemE, Rugby, pp. 1–10.

Stichlmair, J.G. and Herguijuela, J.R. (1992) Separation regions and processes of zeotropic and azeotropic ternary distillation. *AIChE Journal*, *38*, 1523–1535.

Sun, L.Y., Qi, C.X., Li, J., and Li, Q.S. (2011) Design of catalytic divided wall column. *Advanced Materials Research*, *219–220*, 1589–1592.

Sun, L.Y., Qi, C.X., Li, J., and Li, Q.S. (2011) Control of catalytic divided wall column. *Advanced Materials Research*, *225–226*, 496–499.

Vane, L.M. (2008) Separation technologies for the recovery and dehydration of alcohols from fermentation broths. *Biofuels, Bioproducts and Biorefining*, *2*, 553–588.

Wang, S.J., Huang, H.P., and Yu, C.C. (2011) Design and control of an ideal reactive divided-wall distillation process. *Asia-Pacific Journal of Chemical Engineering*, *6*, 357–368.

Ward, O.P. and Singh, A. (2002) Bioethanol technology: Developments and perspectives. *Advances in Applied Microbiology*, *51*, 53–80.

Yildirim, O., Kiss, A.A., and Kenig, E.Y. (2011) Dividing wall columns in chemical process industry: A review on current activities. *Separation and Purification Technology*, *80*, 403–417.

7

Heat Pump Assisted Distillation

7.1 INTRODUCTION

Despite the many well-known benefits of distillation and its widespread use, one major drawback is the significant energy requirements, since distillation can generate more than 50% of plant operating cost. To solve this problem, several technologies were proposed to reduce the energy requirements of distillation, with potential energy savings typically in the range 20–50% (Harmsen, 2010; Yildirim, Kiss, and Kenig, 2011; Kiss, Flores Landaeta and Infante Ferreira, 2012). Distillation has a relatively low thermodynamic efficiency (Araujo, Brito, and Vasconcelos, 2007), requiring the input of high quality energy in the reboiler to perform the separation task. At the same time, a similar amount of heat at lower temperature is rejected in the condenser. Several heat pump concepts have been proposed to upgrade that discharged energy and thus reduce the consumption of valuable utilities. Heat pump (HP) systems can be used to upgrade the low quality energy in the condenser to drive the reboiler of the column. The vapor compression (VC), thermal and mechanical vapor recompression (TVR and MVR) technologies are used to upgrade the heat by compressing the vapor distillate or a working fluid (Annakou and Mizsey, 1995; Fonyo and Benko, 1998; McMullan, 2003). Compression–resorption heat pumps (CRHPs) and absorption heat pumps (AHPs) increase the energy efficiency by means of absorption equilibrium (Mučić, 1989). Owing to the higher temperature lifts, the thermo-acoustic heat pump (TAHP) has a broader applicability range (Bruinsma and Spoelstra,

Advanced Distillation Technologies: Design, Control and Applications, First Edition.
Anton Alexandru Kiss.

2010), while the internally heat integrated distillation column (HIDiC) enhances both the heat and mass transfer, especially when structured packing is used also for exchanging heat (Bruinsma *et al.*, 2012).

The evaluation of the wide range of heat pump assisted distillation technologies and the appropriate selection for each application is a topic of great concern for separation experts. Alternatives for heat recovery from a distillation column were already proposed in the early 1950s (Freshwater, 1951; Kniel, 1952). Subsequently, the oil crisis of recent decades has motivated extensive research on energy efficiency in distillation. Several proposals were made to reduce the energy requirements of distillation, including the application of different heat pump technologies and configurations, as well as heat and thermal integration (Mah, Nicholas, and Wodnik, 1977; Linnhoff, Dunford, and Smith, 1983; Omideyi, Kasprzycki, and Watson, 1984. A high number of alternatives and configurations led to comparison studies with a single case study, such as that described by Meszaros and Meili (1994) for butane/isobutene separation. However, the scope of that paper was very limited, since it only included recompression and bottom flashing schemes. A first selection scheme also incorporating these options—but involving more complex calculations and equations—was developed earlier by Omideyi, Kasprzycki, and Watson (1984). Subsequent research simplified the selection guide for multiple heat pumps technologies, including also MVR, AHP, and TVR (Fonyo and Mizsey, 1994). More recently, Bor and Infante Ferreira (2011) evaluated the performance of selected heat pumps as a function of the required temperature lift, to provide guidelines for their selection in any application. The temperature lift can be related to the temperature difference between the heat sources and sinks, namely, the condenser and reboiler in the case of distillation, which in turn is determined by the product cuts that are separated between the top and bottom of the column. However, the previous research did not include other key distillation technologies that could outperform the heat pumps schemes analyzed, such as, for example HIDiC or cyclic distillation (de Rijke, 2007; Bruinsma *et al.*, 2012; Maleta *et al.*, 2011).

This chapter describes the available heat pump technologies. Moreover, it presents a novel selection scheme based on an extensive literature survey, taking into account the most promising energy efficient distillation technologies that presently feature shorter implementation times. Only the key aspects of the overall efficiency were analyzed: boiling points differences (ΔT_b), or the temperature lift ($\Delta T_{lift} = \Delta T_b + \Delta T_{df}$) to upgrade heat from the source accounting for the driving force (Wallas, 1990), the nature of the components involved, operating pressure of the system, product distribution and purity specifications, reboiler temperatures (T_{reb}) and duties (Q_{reb}), as well as the relative volatility of the components (α_{ij}). The efficiency indicators include the (ideal) coefficient of performance (COP) for heat pumps, which relates to the total operating cost (TOC) as well as the total annual cost (TAC) and the payback time (PBT). The following sections provide a comprehensive literature overview of the energy efficient technologies considered here and the newly proposed selection schemes for multicomponent and binary distillation.

Each and every energy efficient distillation technology yields its maximum savings only at given specific conditions (Meszaros and Meili, 1994; Harmsen, 2010; Bor and Infante Ferreira, 2011). Although there is a large amount of literature concerning the different solutions addressing the energy requirements of distillation, most of these systems are applied in different separation tasks, thus complicating the comparison and subsequent selection of the most suitable options. Currently, this task is assigned to dedicated experts, consuming valuable time and resources. To solve this problem, a practical scheme for the selection of energy efficient distillation technologies was developed (Kiss, Flores Landaeta and Infante Ferreira, 2012). The result is a simple evaluation scheme that allows quick and easy selection of the most suitable energy efficient distillation technologies. Ultimately, the application of the proposed scheme aims to speed up the design phase of more sustainable distillation processes.

7.2 WORKING PRINCIPLE

A heat pump (HP) is a machine or device that moves heat from one location (the "source") to another location (the "sink" or "heat sink") using mechanical work (Figure 7.1). Most HP technologies move heat from a low temperature heat source to a higher temperature heat sink. The most common examples are refrigerators, air conditioners, and reversible-cycle heat pumps for providing thermal comfort. Heat pumps can also operate in reverse, providing heat. Heat pumps can be also considered as a heat engine that is operating in reverse. One common type of heat pump works by exploiting the physical properties of an evaporating and condensing fluid known as a refrigerant (Silberstein, 2002). The following types of heat pumps are available:

- Mechanically driven heat pump (assumptions: 40% electricity generated efficiency, 50% HP Carnot efficiency, $T_L = 50\,°C$, $100\,°C$, $150\,°C$);

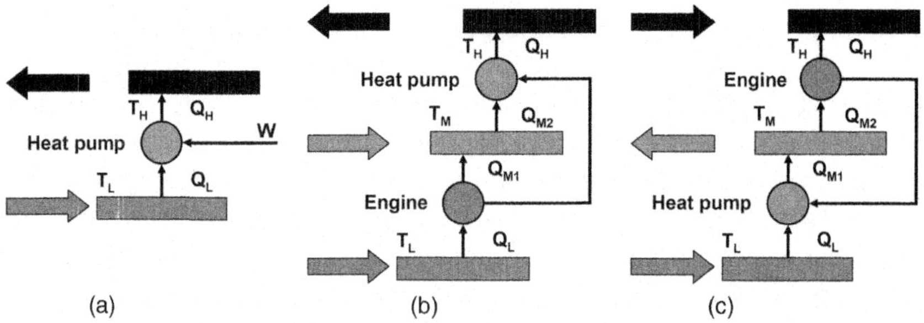

Figure 7.1 Mechanically driven heat pump (a), heat driven heat transformer (b), and heat driven heat pump (c)

- heat driven heat transformer (assumptions: conversion in heat pump and engine 50% of Carnot efficiency, T_M (50, 100, 150 °C, T_L (20 °C);
- heat driven heat pump (assumptions: heat available at T_H (212 and 300 °C, conversion in heat pump and engine 50% of Carnot efficiency, T_L (50, 100 °C).

7.3 VAPOR (RE)COMPRESSION

Figure 7.2 shows the main types of vapor compression and vapor recompression technologies (Kiss, Flores Landaeta and Infante Ferreira, 2012): vapor compression and mechanical or thermal vapor recompression.

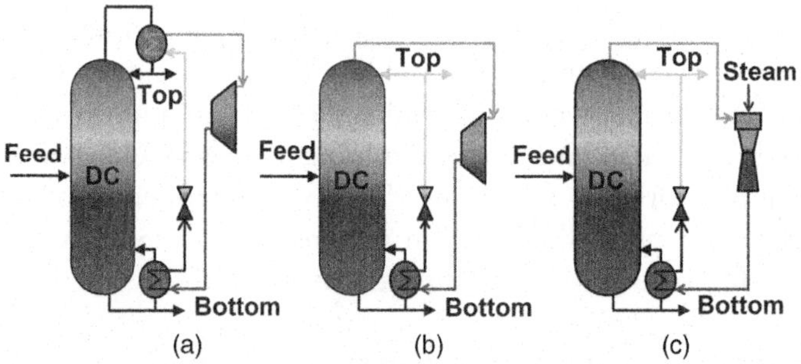

Figure 7.2 Vapor compression (a), mechanical vapor recompression (b), and thermal vapor recompression (c)

7.3.1 Vapor Compression

Vapor compression (VC) is a classical heat pump design proven already at industrial scale. It uses a specific fluid as heat transfer medium that runs between the heat source and sink through a pipeline (Annakou and Mizsey, 1995). A compressor is installed in between to provide the required work input, while a flash valve closes the cycle (Oliveira, Parise, and Pitanga Marques, 2002). Since all these elements are external to the distillation process, the distillation column does not require major modifications, except for the adjustments in heat exchangers for the change in utilities (Meszaros and Meili, 1994). VC is particularly beneficial when dealing with corrosive or fouling compounds. However, the design is very dependent on the ability of the heat transfer fluid to meet stringent operational, environmental and safety requirements (Göktun, 1995). For most applications, there is no adequate alternative—and even when an acceptable heat transfer fluid is found, the energy savings are not always translated into overall economic savings (Omideyi *et al.*, 1985). On one side, the compressors are very expensive and hard-to-maintain equipment, while in contrast, the equipment and inefficiencies in obtaining mechanical energy (work) impact heavily the final energy bill.

7.3.2 Mechanical Vapor Recompression

Mechanical vapor recompression (MVR) is a state-of-the-art industrial system for binary distillation, being widely applied in the separation of close boiling components (Fonyo and Mizsey, 1994). In such a system, the top vapor is used as heat transfer medium, being fed directly to the compressor. Accordingly, the heat pump also performs the function of the condenser, thus saving one heat exchanger as compared to the classic alternative (Olujić *et al.*, 2006, 2009). Moreover, it avoids the need to cool the heat transfer fluid below the boiling point of the top product—an issue of importance in the VC scheme for heat transfer purposes. Notably, MVR features slightly higher efficiency and lower investment costs than VC (Wang *et al.*, 2011b). However, MVR does not tackle directly the main drawback of VC: the economics involved in the compressor usage. Similarly, the distillate still has to meet at least the operational requirements for the heat transfer medium, not to mention the criteria for safe and economic compressor operations (Campbell *et al.*, 2008). All these constraints severely limit the application window of MVR technology.

7.3.3 Thermal Vapor Recompression

Thermal vapor recompression (TVR) is a particular variant of MVR, in which the compressor is replaced by a steam ejector as work input mechanism (McMullan, 2003). In terms of the advantages of the steam ejector, TVR has been widely implemented in industry (Feng and Berntsson, 1997). The steam ejector uses the Venturi effect to obtain mechanical energy from steam injection into a special variable diameter pipeline (Fonyo and Benko, 1998). This makes TVR a robust design with reduced capital and maintenance expenditures, as there are no rotating pieces involved (Perry and Green, 1999). However, the steam ejector has a relatively low efficiency in converting mechanical energy. Moreover, the design of the steam ejector is crucial in achieving economical operation (El-Dessouky *et al.*, 2002). There are wide changes in steam consumption even at small deviations from the optimal operating point. Notably, the steam input is mixed with the distillate to generate the required pressure (Chen and Sun, 1997). Clearly, as steam is being added to the vapor distillate, the applications of TVR are mostly for systems producing water as top product. In theory, the motive fluid for the ejector can be (part of) the distillate flow, which could be boiled and used to pressurize the vapor to the pressure level required in the reboiler. Nevertheless, such applications are rarely encountered due to the potential heat transfer losses. An alternative to MVR/TVR is the self-heat recuperation technology described by Matsuda *et al.* (2011). However, the addition of two compressors leads to unacceptable payback times. Chemical heat pumps were also proposed, but they involve the addition of endothermic and exothermic chemical reactors, rendering them unfeasible economically (Chung *et al.*, 1997; Wang, Zhang, and Wang, 2008).

7.4 ABSORPTION–RESORPTION HEAT PUMPS

7.4.1 Absorption Heat Pump

An *absorption heat pump* (AHP) considers thermochemical conversion to enhance operational efficiency. In this case absorption pairs are used as heat transfer fluids, such as, for example, ammonia and water or lithium bromide and water. The AHP is a well-known cycle (Bor and Infante Ferreira, 2011), widely applied in refrigeration—although there are also standalone (pilot plant) implementations of AHPs suitable for distillation or multistage evaporation processes (Wang, Zhang, and Wang, 2008;

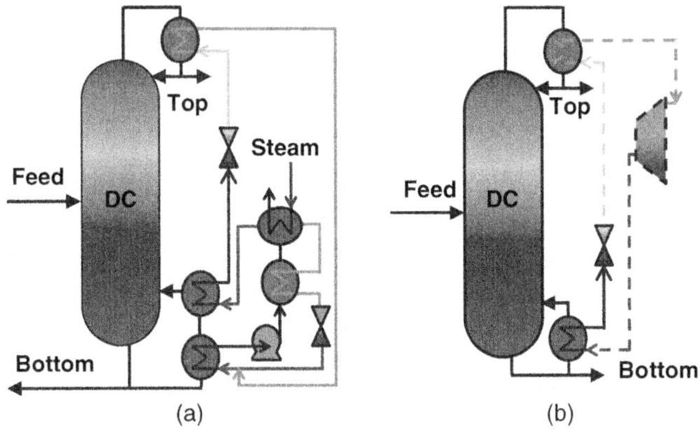

Figure 7.3 Absorption heat pump (a) and compression–resorption heat pump (b)

Wang and Lior, 2011). As shown in Figure 7.3a , the heart of the AHP is the steam driven desorber that separates the absorption pair (Tufano, 1997; Kiss, Flores Landaeta and Infante Ferreira, 2012). One of the components condenses first in the reboiler of the distillation column, and then it is flashed to cool down the condenser. Afterwards, it is mixed with the other component from the regenerator, to deliver heat in the reboiler through exothermic absorption (Ziegler, 2002). The resulting liquid is then pressurized and heated to displace the equilibrium and the whole cycle is repeated. AHP is preferred as it avoids the inconveniences and expenses of using a compressor as driver for the heat pump. However, the requirement of five heat exchangers gives AHP a very pricy installation cost and, therefore, long payback times (Díez *et al.*, 2009).

7.4.2 Compression–Resorption Heat Pump

A *compression–resorption heat pump* (CRHP)—also called hybrid heat pump—is a recent approach used to take advantage of thermochemical sorption processes. CHRP can achieve high temperature levels and lifts, with relatively high COP. CHRP uses the VC scheme in which the working fluid is replaced by an absorption pair (Bor and Infante Ferreira, 2011). When the vapor zeotropic mixture approaches the reboiler, the condensation and absorption process run at the same time, thus enhancing the heat transfer. This gives CRHP enhanced overall efficiency and reduced energy requirements—a critical issue for economic operations in wider temperature ranges (Nordtvedt, 2005). After the reboiler, the rich mixture is then flashed to take heat from the condenser, evaporating both components from the working pair. Evaporation in the desorber is incomplete so that solution recirculation between the desorber and the resorber and/or wet compression is possible (Zamfirescu *et al.*, 2004). When all the solution is recirculated, the cycle is called the Osenbruck cycle. The term hybrid wet compression (HWC) is used for cycles where all the solution is sent to the compressor, avoiding the use of a solution pump. Nonetheless, solving the issue of wet compression is of crucial importance for the wider implementation of CRHP. At present, the issue is still under research, with successful pilot plant implementations of standalone CRHP already in place (Taboada and Infante Ferreira, 2008).

The advantages of CHRP are related to the use of environmentally friendly refrigerants—such as water, ammonia, and CO_2—that can significantly contribute to the improved performance of heating processes. Specifically for industrial heating processes, CHRP allows energy performance gains of more than 20% as compared to VC. The use of a mixture allows lower pressure levels, and condensation and evaporation at gliding temperatures—which can result in higher efficiency. Wet compression has the effect of suppressing vapor superheating, and it can also improve the heat pump efficiency—if the technical problems surrounding it are solved. Ammonia–water mixtures can be used as efficient working fluids in CHRP, showing several advantages: (i) higher COP because of the use of a non-isothermal phase transition of the mixture in the heat exchangers at constant pressure; (ii) the mixture allows the achievement of high temperature operation at relatively low operating pressures; (iii) the cycle can be designed to show a temperature glide in the resorber that corresponds to the temperature glide of the industrial flow that has to be heated; and (iv) for specific operating conditions the cycle performance is significantly better than for the VC cycle (Bor and Infante Ferreira, 2011).

7.5 THERMO-ACOUSTIC HEAT PUMP

Thermo-acoustic (TA) relates to the physical phenomenon that a temperature difference can create and amplify a sound wave and vice versa that a sound wave is able to create a temperature difference. A sound wave is associated with changes in pressure, temperature, and density of the medium through which the sound wave propagates. In addition, the medium itself is moved around an equilibrium position. An acoustic wave is brought into interaction with a porous structure with a much higher heat capacity compared to the propagation. This porous structure acts as a kind of heat storage–regenerator (Bruinsma and Spoelstra, 2010).

Within thermo-acoustics a distinction is made between a thermo-acoustic engine or prime mover (TA-engine) and a thermo-acoustic heat pump (TA-heat pump). The first relates to a device creating an acoustic wave by a temperature difference while in the second an acoustic wave is used to create a temperature difference (Spoelstra, 2007, 2008).

In a *thermo-acoustic engine,* when a temperature gradient is imposed across a regenerator (porous structure) by, for example, a cold and a hot heat exchanger, an acoustic wave passes from the cold side and an acoustic cycle takes place with a parcel of gas (Figure 7.4):

- *Compression:* The gas is being compressed by the passing pressure wave. Since the gas is in very close thermal contact with the regenerator, the temperature stays the same locally.
- *Heating:* Successively the gas parcel is moved to a hotter part of the regenerator. Since the temperature over there is higher than the gas parcel, the gas is heated.
- *Expansion:* Then the pressure wave that first compressed the gas parcel is now expanding it. Again, the gas is not cooled here, due to the close thermal contact with the regenerator.
- *Cooling:* The gas parcel is moved back to its original position, and it is still hotter than the structure (regenerator), resulting in heat transfer from the gas to the structure.

During this cycle the gas is being compressed at low temperature, while expansion takes place at high temperature (Figure 7.4). This means that work is performed on the gas. The effect of this work is that the pressure amplitude of the sound wave is increased. The thermodynamic cycle just

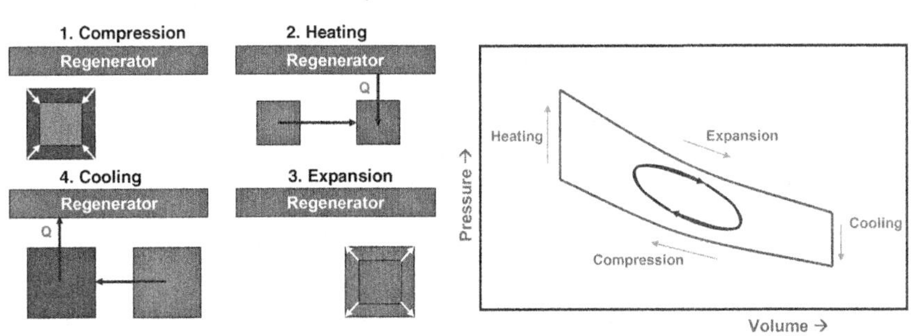

Figure 7.4 Working principle of the thermo-acoustic cycle

described resembles the well-known Stirling cycle. The acoustic wave has the function of both pistons normally present in a Stirling engine. In this way it is possible to create and amplify a sound wave by a temperature difference. The thermal energy is converted into acoustic energy that can be regarded as a kind of mechanical energy (Spoelstra, 2007, 2008). The reverse process of the above cycle happens in a TA-heat pump. The thermodynamic cycle is run in the reverse direction, meaning that acoustic energy is used to pump heat from a lower temperature to a higher temperature level.

The *thermo-acoustic heat pump* (TAHP) is the frontrunner in using a different mechanism for the work input. Although it is a relatively new technology, the proof-of-principle stage has been successfully completed—with scaling up currently the subject intense research efforts (Bruinsma and Spoelstra, 2010; Spoelstra, 2007, 2008; Tijani *et al.*, 2011). The main reason is that TAHP features a wide applicability range—much larger than the heat pumps previously mentioned. Basically, TAHP is a thermo-acoustic device that uses high-amplitude sound waves to pump heat from one place to another.

Figure 7.5 illustrates the application of a TAHP to a distillation column as well as the working principle (Kiss, Flores Landaeta and Infante Ferreira, 2012). The thermo-acoustic device consists of heat exchangers, a resonator, and a regenerator (on traveling wave devices) or stack (on standing wave devices). Depending on the type of engine, a driver or loudspeaker might be used as well to generate sound waves. To limit the space used, an electric driver (linear motor) generates the acoustic power cased inside a resonator, with the temperature lifts being determined by the size of resonator, as well as the properties and pressure of the acoustic medium (Gardner and Howard, 2009). The resonator, housing the TA

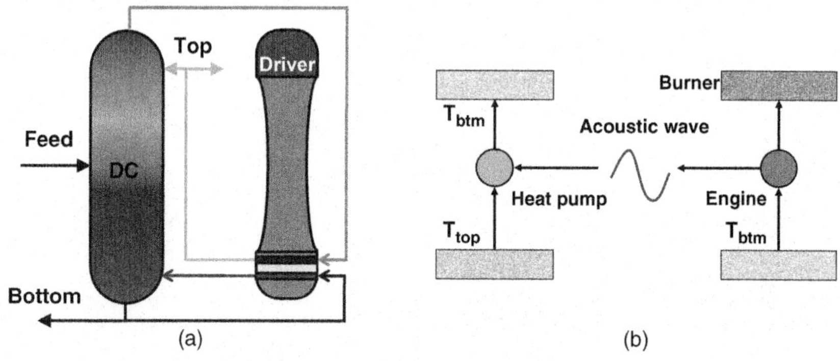

Figure 7.5 Thermo-acoustic heat pump applied to distillation (a) and schematics (b)

engine and the TA heat pump, determines the operating frequency, acts as a pressure vessel, and transports the acoustic power between the components. Note that there is no concern about the noise levels, since these are similar to current industrial standards (e.g., below 85 db).

Since the medium used is a gas (air or a noble gas such as helium), the system complies with safety and environmental concerns, while virtually having no limitations in its applications—it can be applied down to cryogenic temperatures, with temperature lifts of 100 °C or more. The reliability of a TAHP is considered to be very high, as it has basically no moving parts for the thermodynamic cycle when a waste heat driven system is used—although there is a moving part in the case of a linear motor. Other moving parts deal with the supply and removal of the heat to and from the system.

A TAHP is flexible with respect to changes in temperatures and powers. If the amount of waste heat reduces, all powers within the thermo-acoustic system will drop accordingly and vice versa, at the cost of slightly decreased efficiency of the system. Changes in temperature are accommodated in a similar way. The acoustic response time of the system is very fast, therefore the overall system response will be determined by the thermal inertia. This system can easily be started, since it already uses high-temperature heat to drive the system. The cost estimates are based on material quantities that are necessary to realize the system—mainly stainless steel for the resonator, heat exchangers, and regenerator. Alternative materials like aluminum or copper could be used for the heat exchangers. The CapEx is currently estimated at k€150–250 per MW heat input, but expected to reduce.

Thermo-acoustic systems can be scaled by using dimensionless numbers. Based on these numbers, scaling rules can be obtained that relate systems of different sizes, using different working media and working pressure. However, note that heat transfer and heat losses will not scale according to these rules.

Although the working principle of thermo-acoustic technology is rather complex, the practical implementation is quite simple. This offers great advantages with respect to the economic feasibility of this technology, as well as additional benefits such as:

- No moving parts for the thermodynamic cycle, so very reliable and providing a long life span;
- environmentally friendly working medium, such as, for example, air or noble gas;
- the use of air or a noble gas as working medium offers a large window of applications because there are no phase transitions;

- use of simple materials with no special requirements, which are commercially available in large quantities and, therefore, are relatively inexpensive;
- a large variety of applications can be covered on the same technology base.

7.6 OTHER HEAT PUMPS

7.6.1 Stirling Cycle

The Stirling cycle is a thermodynamic cycle that describes the general class of Stirling devices. The cycle is the same as most other heat cycles in that there are four main processes: (i) compression, (ii) heat addition, (iii) expansion and (iv) heat removal. However, these processes are not discrete, but rather the transitions overlap. The idealized Stirling cycle consists of four thermodynamic processes acting on the working fluid (Sier, 1999).

- *Isothermal expansion:* the expansion-space is heated externally, and the gas undergoes near-isothermal expansion;
- *isovolumetric or isochoric heat-removal:* the gas is passed through the regenerator, thus cooling the gas, and transferring heat to the regenerator for use in the next cycle;
- *isothermal compression:* the compression space is intercooled, so the gas undergoes near-isothermal compression;
- *isovolumetric or isochoric heat-addition:* the compressed air flows back through the regenerator and picks-up heat on the way to the heated expansion space.

The cycle is reversible, meaning that if supplied with mechanical power it can function as a heat pump for heating or refrigeration & cryogenic cooling. The cycle is defined as a closed-cycle regenerative cycle with a gaseous working fluid. The term *closed-cycle* means the working fluid is permanently contained within the thermodynamic system. This also categorizes the engine device as an external heat engine. *Regenerative* refers to the use of an internal heat exchanger called a regenerator, which increases the device's thermal efficiency. The Stirling engine is currently exciting interest as the core component of micro combined heat and power (CHP) units, in which it is more efficient and safer than any other comparable steam engine.

7.6.2 Vuilleumier Cycle

The Vuilleumier cycle is a thermodynamic cycle with main applications in low-temperature cooling. The cycle follows the route: hot side → displacer → ambient → displacer → cold side. The entire device is built inside a single pressure vessel, which cycles up and down in pressure once per cycle of the displacers, with all parts of the system at the same pressure at any given time (Carlsen, 1989).

In some respects, the Vuilleumier cycle resembles a Stirling cycle or engine, being like a non-kinematic duplex Stirling omitting the shared piston—it has two displacers connected mechanically, as compared to one in the Stirling cycle. The coupling maintains the appropriate phase difference, and the hot displacer is larger than the cold displacer. As the displacers are not pistons, they do no work and, therefore, no work is required to operate the cycle, in an ideal case. However, friction and other losses mean that some work is still required in reality. Remarkably, devices operating on this cycle are able to produce temperatures of 77 K without pre-cooling, and as low as 15 K using liquid nitrogen to pre-cool (for a heat flow of 1 W).

7.6.3 Brayton Cycle

The Brayton cycle is a thermodynamic cycle that describes the workings of the gas turbine engine, the basis of the jet engine and others. This is also sometimes known as the Joule cycle. The ideal Brayton cycle consists of (Gilmour, 1994):

- Isentropic process—ambient air is drawn into the compressor, where it is pressurized;
- isobaric process—the compressed air then runs through a combustion chamber, where fuel is burned, heating that air (at constant pressure), since the chamber is open to flow in and out;
- isentropic process—the heated, pressurized air then gives up its energy, expanding through a (series of) turbine(s); some of the work extracted by the turbine is used to drive the compressor;
- Isobaric process—heat rejection (in the atmosphere).

Since neither the compression nor the expansion can be truly isentropic, losses through the compressor and the expander represent sources of unavoidable working inefficiencies. In general, increasing the compression ratio is the most direct way to increase the overall power output of a Brayton system. The actual Brayton cycle consist of these processes: adiabatic process (compression), isobaric process (heat addition), adiabatic process (expansion) and isobaric process (heat rejection). There are two types of Brayton cycles, open to the atmosphere and using internal combustion chamber or closed and using a heat exchanger.

7.6.4 Malone Cycle

The Malone cycle is similar to the Brayton cycle, but the Malone engine has check valves, no displacer, and pulsating unidirectional flow instead of reciprocating flow (Lucke, 1902; Gilmour, 1994). The characteristics of this cycle are: large thermal expansion, low compressibility (small volume changes), large heat capacity (low flow rates), and large heat transfer. The Malone engine is a liquid-based engine, the working medium being a liquid near its critical point (e.g., high pressure liquid water). The engine used high temperature water as its working fluid, and was thus referred to as the *hot water engine*. In independent testing published in *Los Alamos Science* (1993), the design showed a similar efficiency to a gasoline engine—about 27%.

7.6.5 Solid–Sorption cycle

The solid-sorption cycle has a working principle based on the reversible sorption reaction between a gas/vapor and a solid (porous) material. Note that the sorption principle is already widely applied in separation and purification processes such as pressure swing adsorption (PSA) and temperature swing adsorption (TSA). The key characteristics of the solid-sorption cycle are physical adsorption (onto surface), chemical absorption (into bulk material), temperature lifts achieved by "thermal" compression, batch processes, and no moving parts. Some of the known working pairs are salt–ammonia, salts–water, carbon–ammonia, carbon–methanol, metal–hydrogen, silica gel–water, zeolite–water—such as, for example:

$$MgCl_2 \cdot 2NH_3(s) + 4NH_3(g) \leftrightarrow MgCl_2 \cdot 6NH_3(s)$$

Table 7.1 conveniently compares the different gas–liquid and solid-sorption cycles. The following notation is used: COP—coefficient of performance, PER—primary energy requirement, HP—heat pumps,

Table 7.1 Head-to-head comparison of gas–liquid cycles

Gas–liquid cycle parameter	Thermoacoustic	Stirling	Vuilleumier	Brayton	Malone
Drive:	Mechanically, thermally	Mechanically, thermally	Thermally	Mechanically	Mechanically, thermally
Operating window:	Wide range	Wide range	Wide range	Wide range	Depending on working medium
Temp. lift (°C):	> 100	> 100	> 100	> 100	~30
Development stage:	PoP (waste heat) Demonstration	Large exp. with engines and coolers	20 kW system for residential applications	Developed for niche applications	Development stopped
Performance:	COP = 3.9–5.5 Mc COP = 1.3–2.5 HP COP = 0.2–0.3 HT	COP = 4.5 Mec COP = 1.4 HP COP = 0.3 HT	COP = 2.0 HP COP = 0.3 HT	COP = 1.2 Mec	N/A
Reliability:	High	Moving parts Sealing is important	Moving parts No tight seals	Moving parts Known components	Moving parts
Solid–sorption cycle parameter	**Salt-ammonia**	**Carbon–ammonia**	**Metal–hydrogen**	**Silica gel–water**	**Zeolite-water**
Drive:	Thermally	Thermally	Thermally Hybrid/comp	Thermally	Thermally
Operating window:	Limited, salts dependent	Limited Hybrid systems	Limited, metal dependent	Limited 0–150 °C	Limited 0–250 °C
Temperature lift (°C):	> 100	~ 60	> 50	~40	> 50
Development stage:	Early stage	PoC for cooling	Early stage	Commercially available for cooling	Demo unit domestic heating 10 kW
Performance:	COP = 1.7 HP COP = 0.4 HT	Suitable only for cooling	COP = 1.7 HP COP = 0.4 HT	COP = N/A HP COP = 0.4 HT	COP = 1.6 HP COP = 0.2 HT
Reliability:	No moving parts but stability is an issue	No moving parts but stability is an issue	No moving parts but stability is an issue	No moving parts. Maintain vacuum conditions	No moving parts. Maintain vacuum conditions

HT—heat transformer, or heat transfer, PoP—proof of principle, PoC—proof of concept, N/A—not applicable, or information not available.

7.7 HEAT-INTEGRATED DISTILLATION COLUMN

The *heat integrated distillation column* (HIDiC) is the most radical approach of heat pump design, making use of internal heat-integration (Nakaiwa *et al.*, 1997; Gadalla *et al.*, 2006; Bruinsma *et al.*, 2012). Instead of using a single point heat source and sink, the whole rectifying section of a distillation column becomes the heat source, while the stripping part of the distillation column acts as a heat sink (Figure 7.6) (Matsuda *et al.*, 2010; Kiss, Flores Landaeta and Infante Ferreira, 2012). The problem of different sizes for rectifying and stripping sections can be relatively easy tackled by using one of the many alternative HIDiC configurations, as described in Chapter 8. This internal heat-integration widely enhances the reachable coefficient of performance (COP), because the required temperature difference for heat transfer is kept low with gliding temperatures across both parts (Campbell *et al.*, 2008). According to Bruinsma and Spoelstra (2010), the COP is defined as the ratio between the amount of heat upgraded (Q_b) and the heat pump energy requirements (W): COP = $(Q_b/W) \leq (T_{reb}/\Delta T_b)$. The work input is provided by a compressor installed at the top outlet of the stripper section, while the heat pump

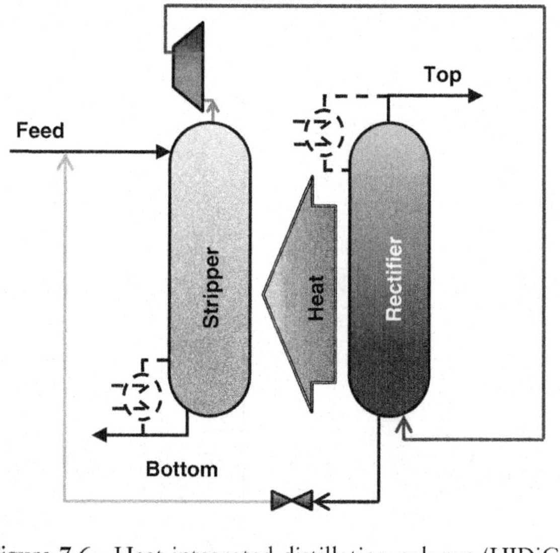

Figure 7.6 Heat-integrated distillation column (HIDiC)

cycle is closed by the valve flashing the liquid bottom outlet of the rectifier section. The HIDiC success relies actually on good hardware performance for both heat and mass transfer tasks at the same time (Nakaiwa *et al.*, 2003). Currently, there is only a pre-commercial 15 kton yr^{-1} implementation of HIDiC in Japan (Huang *et al.*, 2008).

7.8 TECHNOLOGY SELECTION SCHEME

The main aim of this section is to provide design guidelines, allowing any process engineer to narrow down significantly the number of suitable energy efficient options for a given separation task at early design stages. Currently, the whole technology selection process is carried out by distillation experts. When designing new processes—before approaching the selection schemes—one must ensure that distillation is indeed the technique providing greatest efficiency. Moreover, heat integration possibilities within the process, plant, or with other distillation columns should be considered in parallel (Linnhoff, Dunford, and Smith, 1983). The use of the selection schemes allows significant time savings for separation experts, since they are required to evaluate only the final decision through rigorous simulations, while the engineering departments and plant managers could qualitatively evaluate the suggestions provided by the scheme, and raise awareness about the implementation of energy efficient solutions. The ultimate goal is that this practical selection scheme becomes a major contributor in the path towards more sustainable distilling processes (Kiss, Flores Landaeta and Infante Ferreira; 2012).

In principle, the selection scheme can be used for new designs or retrofit applications, but the suitable proposals vary for each case. For binary distillation, the scheme offers at least two technology options for each concluding condition, one of them being also suited for retrofits. Those solutions that make use of components external to the distillation process (e.g., VC, MVR, TVR, AHP, CRHP, and TAHP) can be used in either new designs or retrofit. Note that the tray-integrated versions of the AHP and CRHP can only be applied in new designs. The solutions that make use of components internal to the distillation process (e.g., HIDiC and CyDist) are used mainly in new designs. However, cyclic distillation can also be used for retrofit by changing the operating mode and internals of existing (binary) distillation columns. For multicomponent separations, DWC and Kaibel columns are mainly applied in new-built plants, but retrofits have also been made possible using the unfixed dividing-wall technology from Julius Montz (Olujić *et al.*, 2009; Yildirim, Kiss, and Kenig, 2011).

7.8.1 Energy Efficient Distillation Technologies

In addition to heat pump assisted distillation, there are other advanced technologies able to reduce the energy requirements, such as, for example, multi-effect distillation (MED). Another example is cyclic distillation (CyDist) that reduces the energy demand by enhancing the separation efficiency via pseudo-steady-state operation based on separate phase movement (Gaska and Cannon, 1961; Maleta *et al.*, 2011). This has great potential to bring new life to existing distillation columns, by revamping them with new internals and a periodic mode of operation. For multi-component distillation, the Kaibel column and dividing-wall column (DWC) technologies allow significant energy savings by avoiding the remixing effects (Yildirim, Kiss, and Kenig, 2011). Another alternative is the multi-partitioned DWC—equivalent to an extended Petlyuk setup. Note, however, that any of these technologies can provide energy savings of 20–50% as compared to conventional distillation, but only when certain conditions are met—so a specific operating range is applicable to all of them.

The selection scheme takes into account the heat pumps that are in an advanced implementation stage in industry—most of them being already applied at industrial scale, while others are at least at the stage of working pilot plant or prototype. Moreover, along with the selected heat pumps, other energy efficient distillation technologies must be considered in the selection scheme, such as cyclic distillation and thermally coupled distillation columns (e.g., Petlyuk, DWC, Kaibel column).

Cyclic distillation (CyDist) is not a heat pump system, but it has emerged as another important trend for improving distillation performance: enhancing the separation efficiency through pseudo-steady-state operation based on separate phase movement (SPM)—achieving up to 50% energy savings (Gaska and Cannon, 1961; Maleta *et al.*, 2011). Figure 7.7 shows the key two steps (vapor period followed by a liquid period) that repeat periodically in a pseudo-steady-state operation (Kiss, Flores Landaeta and Infante Ferreira, 2012). CyDist uses special internals—that are quite robust and similar to bubble caps trays—to avoid the simultaneous vapor and liquid flows across the distillation column (Maleta *et al.*, 2011). This way, liquid back-mixing is reduced and the separation efficiency is drastically enhanced. CyDist is already implemented in the food industry for concentrating ethanol (e.g., 20 m^3 per day). For instance, a cyclic distillation column with 15 trays provides the same separation performance as a conventional column with 50 trays (Matsubara, Watanabe, and Kurimoto, 1985). However, the main obstacle for the widespread implementation of CyDist is the pseudo-steady-state operation, which requires special

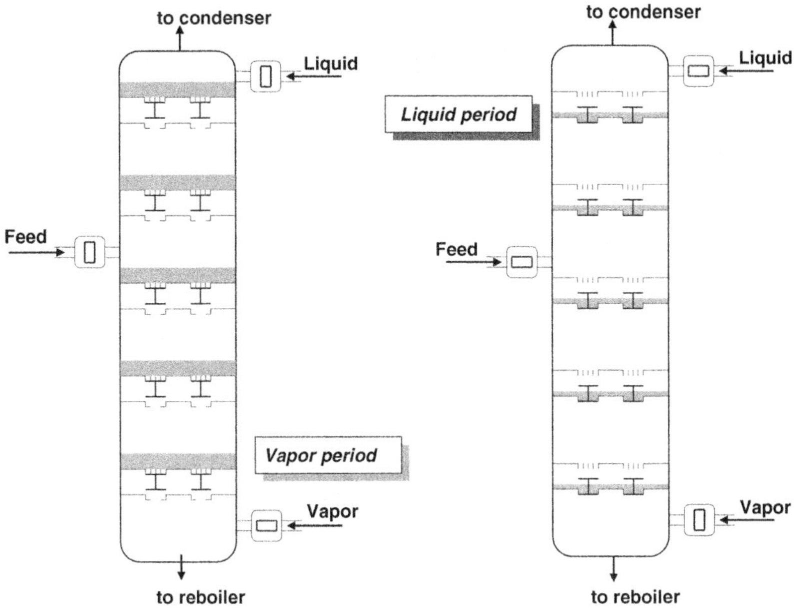

Figure 7.7 Schematics of a cyclic distillation system

operator training and additional safety measurements (Sommerfeld *et al.*, 1966). More details about cyclic distillation are presented in Chapter 9.

Multiple effect distillation (MED) is derived from a particular variant of heat integrated processes. For binary distillation it consists of two or more distillation columns, one running at high pressure and other at lower pressure. The system is arranged in such way that the high-pressure condenser acts as the low-pressure reboiler (Engelien and Skogestad, 2005; Zhang, Shengrong, and Feng, 2010). In that way, the heat rejected in the first stage is taken into the second one. The working principle is widely applied in evaporation processes (Al-Shammiri and Safar, 1999). Reported industrial applications in the distillation field include mainly aqueous separations, such as acetic acid/anhydride or methanol/water separations (Zhang, Shengrong, and Feng, 2010; Campbell *et al.*, 2008). Many applications have been also suggested for multicomponent applications with each pressure stage performing one separation task, but the competitive advantages in that field are lost in favor of dividing-wall based technologies (Chung *et al.*, 1997; Engelien and Skogestad, 2005). MED systems feature high investment costs by using at least two columns, one of which operated at high pressure. Additionally, these systems are rather unstable and usually require complex control structures (Engelien, Larsson, and Skogestad, 2003; Bansal *et al.*, 2000). Therefore, the application of multi-effect distillation is economically advised only in energy-intense separation tasks with high boiling point differences.

Membrane distillation (MD) is especially useful for applications involving bulk water removal, such as seawater and brackish water desalination, process water treatment, and water purification. MD is a thermally-driven separation process that is possible due to phase change. A hydrophobic membrane is used as a barrier for the liquid phase, while letting the (water) vapor phase pass through its pores. The partial vapor pressure difference—related to a temperature difference—gives the driving force of the process. The most common technologies are direct contact (DCMD), air gap (AGMD), vacuum (VMD), sweeping gas (SWGMD), and vacuum multi-effect (V-MEMD). Owing to the niche applications, MD technology should be considered only on a case-by-case basis.

HiGee distillation uses the rotating packed bed (RPB) concept in a high-gravity (100–500*g*) technology that emerged several decades ago (Ramshaw, 1983; Rao, Bhowal, and Goswami, 2004) claiming HETP values as low as 1–2 cm, about 3–6 times higher throughput, and a volume reduction of 2–3 orders of magnitude lower than that of conventional packed columns. Although many commercial applications of HiGee are known in absorption, stripping, and reactive precipitation, very few commercial applications in distillation have been reported so far (Rao, Bhowal, and Goswami, 2004; Wang *et al.*, 2011). A key reason is that several problems such as the dynamic seal, middle feed, liquid distributor, and the multi-rotor configuration were not properly addressed. To successfully solve these problems, a novel kind of HiGee device was recently proposed and developed by Wang *et al.* (2011)—the so-called rotating zigzag bed (RZB) that contains a unique rotor. Remarkably, the RZB fills the gap in HiGee distillation and it has the potential for a bright future (Wang *et al.*, 2011).

Modern distillation technologies for multicomponent separations are based on the Petlyuk configuration. The *dividing-wall column* (DWC) (Figure 7.8a) (Kiss *et al.*, 2013) is a practical implementation of the Petlyuk setup, in which the prefractionator and the main column are built in the same shell but separated by a vertical wall inside the DWC (Dejanović, Matijašević, and Olujić, 2010; Dejanović *et al.*, 2011a,b). The specific features of a DWC allows it to achieve 25–30% in energy savings, by avoiding remixing effects (Kiss and Bildea, 2011; Yildirim, Kiss, and Kenig, 2011). In addition, DWC also provides a low CapEx and plant footprint. Moreover, the advantages of DWC can be extended to extractive, azeotropic, and reactive separations (Kiss and Suszwalak, 2012; Kiss *et al.*, 2012a). Most research efforts are currently focused on

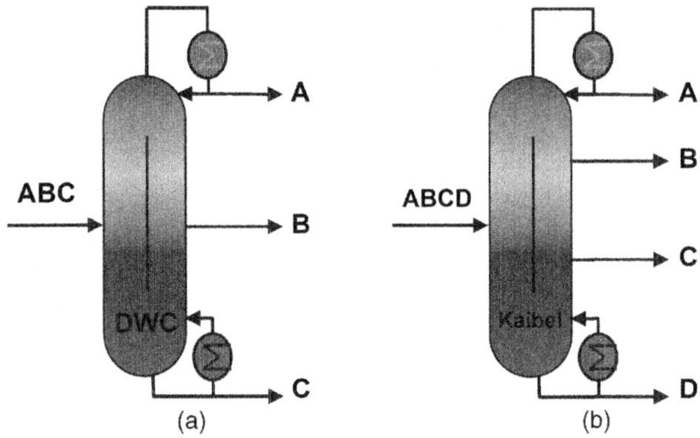

Figure 7.8 Dividing-wall column (a) and Kaibel column (b)

expanding the applications of DWC. Note that the DWC operation is limited to a single pressure operation in the column. As a result, in the case of wide boiling components, the economic performance can be seriously jeopardized by the requirement of non-conventional utilities (Yildirim, Kiss, and Kenig, 2011). Previous chapters cover this topic in more detail.

The *Kaibel column* is similar to a DWC. It extends the approach of the Petlyuk setup to a column with four product streams (Figure 7.8b) (Kiss, Flores Landaeta and Infante Ferreira, 2012). The prefractionator of a Kaibel column is designed to perform the sharp separation between the two middle boiling products (Halvorsen *et al.*, 2011). Since the prefractionator does not operate at the minimum energy requirement, the Kaibel column is slightly less efficient thermodynamically than a DWC but this is economically compensated by the fact that it can separate more products in one unit. However, the efficiency gains do not always pay-off the added issues in design and operability. Nevertheless, the energy savings, the relatively robust design, and good operability make the Kaibel column an important option in multicomponent distillation (Dejanović *et al.*, 2011). An alternative to the Kaibel column is the multi-partitioned DWC—in fact the thermodynamic equivalent of an extended Petlyuk configuration—but this setup has only been studied theoretically, with no practical implementation being reported so far (Dejanović, Matijašević, and Olujić, 2010).

7.8.2 Multicomponent Separations

Figure 7.9 shows the selection scheme proposed for multicomponent separations (Kiss, Flores Landaeta and Infante Ferreira, 2012). The

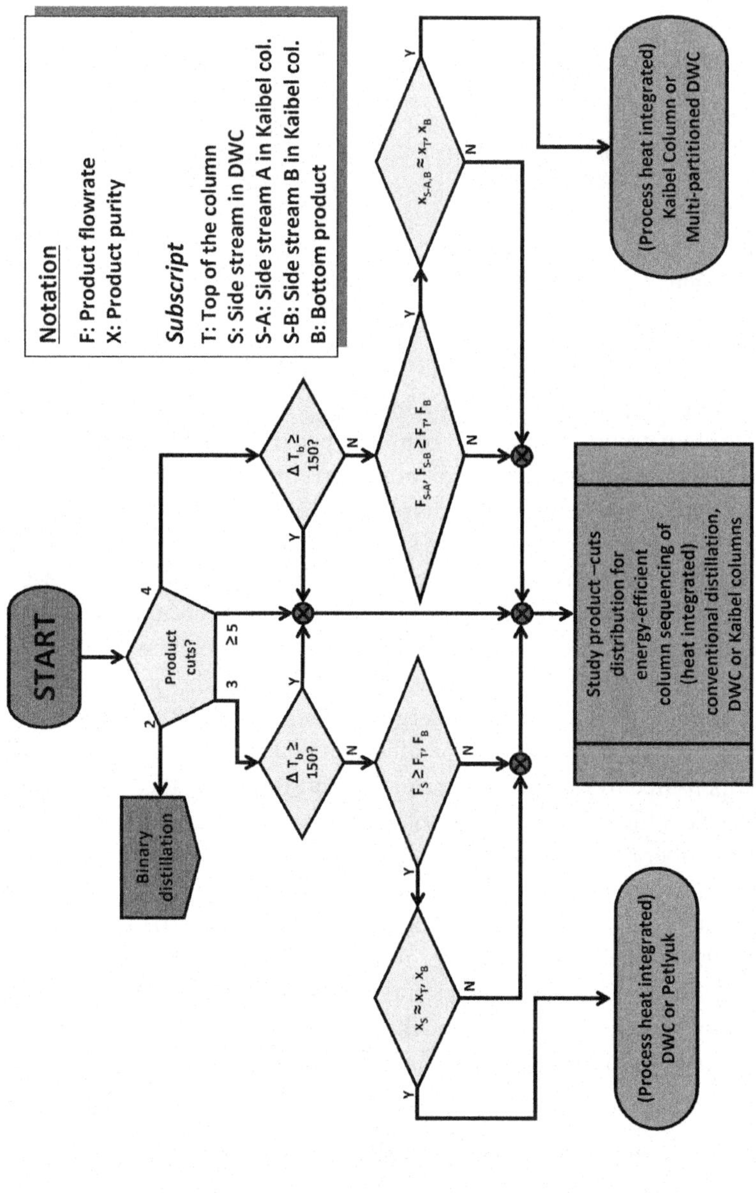

Figure 7.9 Main selection scheme for energy efficient distillation technologies

systems are primarily classified according to the type of separation tasks performed. Additionally—only in the case of Figure 7.9 (Kiss, Flores Landaeta and Infante Ferreira, 2012)—the selection criteria include also the boiling points, flow rates and purity specifications of the number of components (or product cuts) to be separated. The type of separation is primarily reviewed to select the most efficient technology in each case. As explained before, the DWC concept—and its thermodynamically equivalent Petlyuk configuration—was designed to handle three products, while the Kaibel column and multi-partitioned DWC were conceived with four or more products in mind. Thus, setting the number of components as the first verification criteria is straightforward. The following criteria for technology selection involve the operational parameters that determine the suitability of a DWC or Kaibel column in any situation. The main information about DWC and Kaibel column applications and case studies is collected and organized in Table 7.2 —which for convenience displays only the most relevant cases or interesting applications (Kiss, Flores Landaeta and Infante Ferreira, 2012). Over 40 entries were analyzed, but only those achieving energy and/or economic savings higher than 20% were actually considered in this study. The preferred applicability range is obtained from the evaluation of the operating conditions for those cases with considerable energy or economic savings. The review of Yildirim, Kiss, and Kenig (2011) provides a more comprehensive list of DWC applications (see also Chapter 6).

An important parameter for the operability of DWC and Kaibel columns is the temperature span across the column—in fact the difference in boiling points between the top and bottom product (ΔT_b). As these columns operate at a single pressure, a reasonable ΔT_b ensures their operability with common utilities: cooling water and steam. Based on the data reported in Table 7.2 (Kiss, Flores Landaeta and Infante Ferreira, 2012) a practical estimate is $\Delta T_b = 150\,K$. Note that considering heating oil as a common utility is questionable, but its use certainly allows additional 50 K higher ΔT_b (Douglas, 1988). On the other hand, a 30 K reduction is expected for processes operating in warmer regions. A lower operational limit of $\Delta T_b = 80\,K$ was also chosen based on practical reasons: the lower the ΔT_b the higher the energy requirements. Those energy requirements can be translated into taller columns or higher reboiler duties. Considering that DWC or Kaibel columns involve separations of three or four components, the column length for each separation is limited by practical criteria as well. Thus, the load on the reboiler might be even higher. For those cases, efficient solutions for binary distillation such as HIDiC or cyclic distillation could provide much higher energy efficiency.

Table 7.2 Overview of case studies, experiments, and industrial implementation of distillation technologies for multicomponent separations

Technology	Condition of separation task	Performance information	Application	Reference
Kaibel	Equimolar feed of methanol, ethanol, 1-propanol, 1-butanol.	Ideal savings are 33%, simulated savings >26%, compared to the conventional three column case	Case study	Ghadrdan, Halvorsen, and Skogestadad (2011)
Kaibel	Separation of platforming products into four products: C_5–C_6 (T_b 40 °C, P 2.7 bar), benzene rich (T_b 115 °C, P 3.2 bar), toluene (T_b 111 °C, P 1 bar), heavies (T_b 164 °C, P 1.6 bar)	Kaibel column at top pressure 2.2 bar affords 42% energy savings compared to conventional three-column design with different pressures. This translates into ~40% total annual cost saving per year	Case study	Dejanović *et al.* (2011)
DWC	Separation of platforming products into three products: C_5–C_6 (Key: 2-methylpentane), benzene rich (key: benzene), heavies (key: toluene)	DWC operating at 2.7 bar requires ~43% less energy than two conventional columns at 1.7/2.7 bar, respectively. Total annual costs are ~47% less for the DWC option compared with the conventional option	Case study	Dejanović, Matijašević, and Olujić (2010)
DWC	Separation of reforming products: lights (key: benzene), int. (key: toluene), heavies (key: xylenes)	DWC scheme achieves 13% energy savings and 4% investment cost reduction compared to conventional two-column sequence. operating at < 2 bar	Case study	Lee, Kim, and Hwang (2004)

DWC	Mixed xylenes taken as side-product.	DWC revamp offers: either 50% energy savings or increased capacity and 25% energy savings; operating pressure: ~1.5 bar	Industrial implementation	Slade, Stober, and Simpson (2006)
DWC	Retrieval of C_{11}–C_{13} "heart cut" from crude kerosene	In UOP Process, DWC yields 30% energy savings with 28% lower capital cost than conventional alternative	Industrial implementation	Dejanović, Matijašević, and Olujić (2010)
DWC	Hexane industrial process: light key: *n*-pentane, intermediate key: *n*-hexane, heavy key: *n*-heptane.	DWC energy savings are up to 34% compared to conventional two-column sequence	Case study	Kim, Hwang, and Nakaiwa (2004)
DWC	Test different feed fractions for high purity (~99%) separation of *n*-hexanol, *n*-octanol, and *n*-decanol.	Characterization of DWC performance with different separation tasks. Minimum purity under adverse conditions: 98%.	Pilot plant implementation	Niggemann, Hiller, and Fieg (2010)
DWC	Extraction of aromatics (benzene) using Morphylane technology	Reported 36% energy savings, 30% less equipment with 40% less plant footprint for extractive DWC alternative	Industrial implementation	Kolbe and Wenzel (2003)
DWC	Extraction of ethanol from water with ethylene glycol (E-DWC) and pentane (A-DWC); operating at atmospheric pressure	A-DWC yields 20% energy savings compared to conventional azeotropic distillation, while E-DWC delivers 10% energy savings compared to an optimized conventional extractive technology. In general, extractive process is less energy intensive than the azeotropic one	Case study	Kiss and Suszwalak (2012)

Moreover, recent reviews state that energy savings of the DWC and Kaibel columns are higher at larger side-stream flow rates and high purities, with about equal difference in volatilities between the three key components—more heuristic rules are available in the open literature (Kaibel, 1987; Aspiron and Kaibel, 2010; Dejanović, Matijašević, and Olujić, 2010; Yildirim, Kiss, and Kenig, 2011).

7.8.3 Binary Distillation

The maze of choices for binary distillation technologies is illustrated in Figure 7.10 (Kiss, Flores Landaeta and Infante Ferreira, 2012). Most technologies are grouped as heat pump assisted distillation (HPAD). Consequently, the selection criteria for binary distillation include operating pressure, nature of the components with respect to corrosiveness and fouling, boiling point differences (ΔT_b), temperature lift (ΔT_{lift}), reboiler duty (Q_{reb}), and temperature level (T_{reb}) as well as the relative volatility between components (α_{i-j}). The analysis applied to obtain these criteria and their values is similar to the one used in multicomponent distillation. In this case, the reviewed data includes more than 70 technology applications and case studies. The technology should provide about 20–50% savings in order to be considered for future application under the conditions reported by the literature. After the initial screening, the data was organized considering the relevant details of the application/case study and the energy/economic savings obtained in each particular case.

Table 7.3 shows the technologies used and the most important studies collected in the literature survey (Kiss, Flores Landaeta and Infante Ferreira, 2012). Mainly two classes of binary mixtures have been reported—(relatively) small hydrocarbons and water–alcohols—but these cover a quite large range of industrial processes. The technology selection criteria and the limiting values between similar situations were drawn from Table 7.3. In some cases, particularly at low ΔT_b, only the top three technologies providing the maximum savings were selected. For the readers' convenience, Table 7.4 presents a more detailed comparison of the efficiency indicators for various distillation technologies (Kiss, Flores Landaeta and Infante Ferreira, 2012): coefficient of performance (COP) and energy savings (%) that directly relate to the total operating cost (TOC) as well as to the CO_2 emissions, reduction of the total annual cost (ΔTAC), and the payback time (PBT). The most important criteria for selection are the difference in boiling points (ΔT_b) or the temperature lift (ΔT_{lift}). Basically, ΔT_{lift} adds to ΔT_b the required driving force for heat

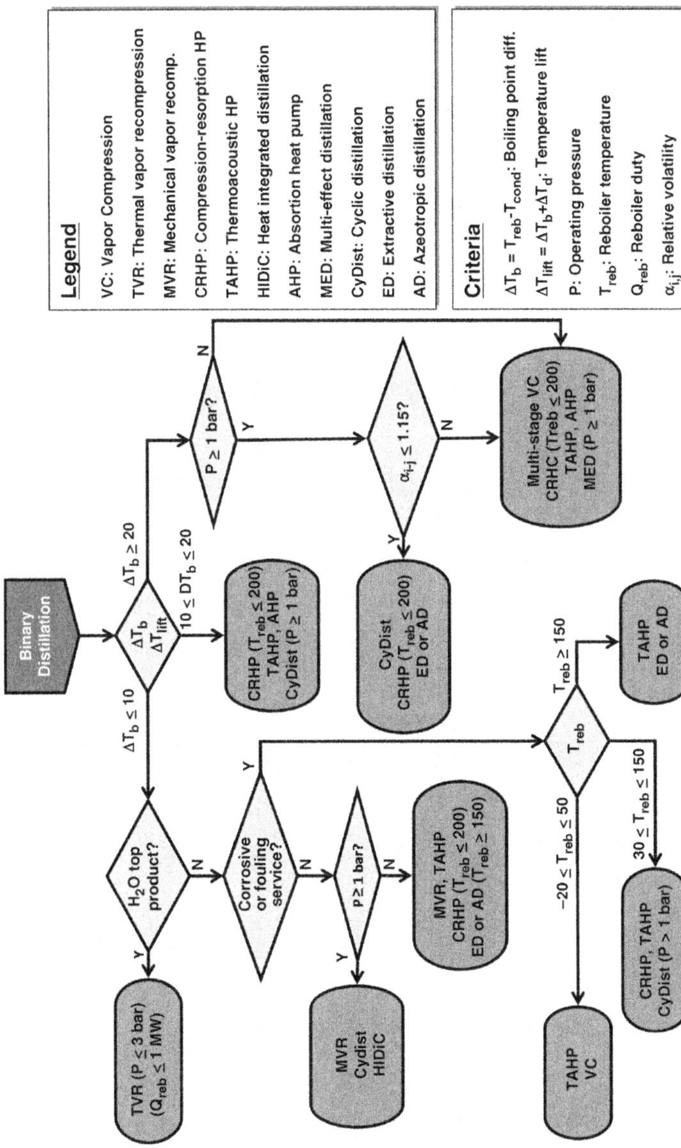

Figure 7.10 Selection scheme for energy efficient binary distillation technologies (T in °C)

Table 7.3 Overview of case studies, experiments, and industrial implementation of energy efficient distillation technology for binary separations

Technology	Condition of separation task	Performance information	Application	Reference
VC, MVR, B = bottom flashing, column integration	C_4 splitters: $P \approx 5$ bar. $\Delta T_b \approx 10$ K. Feed upgrade from 19% to 99.1% 1-butene	MVR allows higher energy & cost savings (~50% lower TAC) than column integrated schemes, but its CapEx is almost twice as high. VC savings are similar to MVR with higher energy costs. 36% energy savings from column integration	Case study	Meszaros and Meili (1994)
VC	Ethanol recovery from fertilization product. Key components: ethanol–water	VC with R114 working fluid yields $3 \leq \text{COP} \leq 4$ at $50\,\text{K} \leq \Delta T_{lift} \leq 64\,\text{K}$. Changing to ethanol as working fluid, total energy savings are up to 50%.	Case study	Omideyi *et al.* (1985)
VC	Theoretically constructed cases	VC with R114 working fluid yields $\text{COP} = 3.2$ at $\Delta T_b = 67\,\text{K}$	Case study	Wallin, Franck, and Berntsson (1990)
MVR	Propane–propylene separation (P between 4 and 16 bar)	Single compressor—optimal pressure: 9 bar Double compressor—optimal pressure: 10 bar Both feature ~37% annual cost savings and ~60% less CO_2 emissions. Double compressor has better operability	Case study	Annakou and Mizsey (1995)
MVR AHP	1. Butane–isobutane separation ($\Delta T_b \approx 15\,\text{K}$).	1. MVR yields $\text{COP} \approx 6.44$ and 29% savings in utilities requirements in	Case study	Fonyo and Benko (1998)

	2. Acetic acid dehydration ($\Delta T_b \approx 55$ K).	case 1. AHP yields COP ≈ 1.87 and 41% utilities savings. AHP lowest payback time. 2. MVR yields COP ≈ 2.63 and AHP yields COP ≈ 1.67. None gives utilities saving. Payback time is negative for all cases.		
MVR AHP	*n*-Butane–i-butane separation: $\Delta T_b \approx 10$ K @ $P = 710$ kPa	MVR yields 33% energy savings and 9% less TAC. AHP incorrectly implemented Both compared to conventional distillation	Case study	Díez *et al.* (2009)
TVR	Evaporators, steam-stripping, peroxide distillation, and other aqueous separations	Usual COP ≈ 1.1 with $\Delta T_{lift} < 40$ K. Lowest investment cost per kW output	Industrial implementation	Feng and Berntsson, 1997; Perry and Green (1999)
CRHP	Heat stream upgrading	The system delivers 1000 kW with $\Delta T_b = 20$ K and $\Delta T_{lift} = 25$ K gives COP ≈ 9.1	Pilot plant implementation	Mučić (1989)
CRHP, MVR	Benzene–toluene distillation	Diabatic CRHP distillation design yields COP ≈ 8 with ΔT_{lift} max 40 K. Conventional MVR yields COP ≈ 7	Case study	Taboada and Infante Ferreira (2008)
TAHP	Heat stream upgrading	5 kW burner driven TAHP with argon yields $\Delta T_{lift} = 40$ K, delivering 200 W acoustic power at 100 °C	Pilot plant implementation	Spoelstra (2008)
TAHP	Benzene–toluene distillation	Electrical driven TAHP yields COP ≈ 2	Case study	Tijani *et al.* (2011)
CyDist	Ethanol recovery from fertilization product. Key components: ethanol–water	Reduce losses from 5% to 0.5% of plant capacity. Compared to a similar case in conventional process, separation efficiency increases 200–300%. Energy savings of up to 50%	Industrial implementation	Maleta *et al.* (2011)

Table 7.3 (*Continued*)

Technology	Condition of separation task	Performance information	Application	Reference
CyDist	Methylcyclohexane–heptane atmospheric separation.	Plate efficiencies of up to 130% were achieved. Protruded packing is 45% more efficient than Raschig rings.	Proof of principle implementation	McWhirter and Cannon (1961)
CyDist	Methanol–water atmospheric distillation	CyDist requires 20% to 50% less vapor flow compared to conventional distillation.	Pilot plant implementation	Matsubara, Watanabe, and Kurimoto (1985)
HIDiC	Fractionation of C_4–C_6 mixture: Top: *n*-pentane; bottom: *i*-pentane $P = 190/101$ bar.	HIDiC achieves 26% energy savings compared with conventional existing column, while requiring ~22% more energy than minimum reflux distillation.	Pre-commercial implementation	Huang *et al.* (2008)
HIDiC	Cyclohexane/*n*-heptane equimolar feed, 95% product purity. $P = 1.1$ bar. $\Delta T_b \approx 17$. Pressure ratios from 1.4 to 1.7	Primary energy savings range from 81% at 1.7 pressure ratio to 69.7% at 1.4, compared to a conventional distillation process. Research also included performance from proof-of-principle experimental study	100 ktpa case study	Bruinsma *et al.* (2012)
HIDiC, MVR	Propane–propylene separation. Conventional column: $P = 18$ bar MVR: $P = 10$ bar HIDiC: $P = 18/13$ or $18/15$ bar	MVR industrial implementation gives COP ≈ 7.4 and 37% TAC savings against conventional column ($\Delta T_{lift} \approx 24.9$ K). HIDiC case study yields COP ≈ 10 and 25% TAC savings against MVR with $\Delta T_b \approx 10.9$ K.	Case study/ industrial implementation	Olujić *et al.* (2003, 2006)

HIDiC MVR	1. Methanol–water separation ($P = 2.6/1.3$ bar for HIDiC) 2. Ethylbenzene–styrene separation ($P = 0.24$ bar for MVR)	1. Top HIDiC design yields 84% energy savings, but only 9% lower TAC 2. Bottom HIDiC design better. MVR pressure ratio: 2.8, while HIDiC 2.2. MVR affords 49% energy savings, while HIDiC gives 53%. However, MVR savings in TAC are 35%, while HIDiC are 28%. All compared to conventional distillation	Case study	Gadalla *et al.* (2007); Gadalla (2009)
HIDiC MVR	Acetic acid dehydration; vacuum operating pressure.	82% energy savings with MVR and 23% with HIDiC compared to heat integrated columns	Case study	Campbell *et al.* (2008)
MED	Methanol–water (HP = 5–8 bar; LP = 1.4 bar)	A five-column multi-effect distillation scheme achieves 33.6% energy savings compared to the existing four-column multi-effect distillation	Case study/ industrial implementation	Zhang, Shengrong, and Feng (2010)
MED	CH_3SiCl_3-$(CH_3)_2SiCl_2$ (HP = 2 bar; LP = 1 bar)	Energy savings of 43.67% for multi-effect distillation compared to conventional distillation process	Case study	Wang and Wang (2012)

Table 7.4 Comparison of efficiency indicators for various energy efficient technologies

Technique	Separation task	ΔT_b (K)	ΔT_{lift} (K)	COP	PES (%)	ΔTAC (%)	Remarks	References
AHP	Butane–isobutane	15	25	1.9	47	−42	PBT = 0.9	Fonyo and Benko (1998)
AHP	Water–acetic acid	55	65	1.7	40	−12	PBT = 7.1	Fonyo and Benko (1998)
CHRP	Steam production	10	30	9.1	78	—	—	Mučić (1989)
CRHP	Benzene–toluene	30	<30	8.3	76	—	—	Taboada and Infante Ferreira (2008)
CyDist	Ethanol–water	20	N/a	—	<50	—	Azeotrope	Maleta *et al.* (2011)
CyDist	Methylcyclohexane–*n*-heptane	20	N/a	—	~45	—	Low rel. volatility	McWhirter and Cannon (1961)
CyDist	Methanol–water	35	N/a	—	>15 but <80	—	—	Matsubara (1985)
HIDIC	*n*-Pentane–cyclopentane	9	20	10.3	71	—	—	Huang *et al.* (2008)
HIDIC	Cyclohexane–*n*-heptane	17	33	9.6	75	−62	PBT = 1.8	Bruinsma *et al.* (2012)
HIDIC	Propane–propylene	7.7	10.9	10.0	80	−72	PBT = 1.2	Olujić *et al.* (2006)
HIDIC	Methanol–water	35	55	12.9	41		—	Gadalla (2009)
HIDIC	Benzene–toluene	30		4.0	46	−46	PBT = 0.9	Gadalla *et al.* (2007)
HIDIC	Acetic acid dehydration	35	55	3.2	27	—	—	Campbell *et al.* (2008)
MVR	1-Butene–*n*-butane	10		5.5	63	−60	PBT = 0.6	Meszaros and Meili (1994)
MVR	Propane–propylene	7.6	11.8	—			−60% CO_2	Annakou and Mizsey (1995)
MVR	Benzene–toluene	30	35	6.7	70	—	—	Taboada and Infante Ferreira (2008)
MVR	Butane–isobutane	15	20	6.4	69	−63	PBT = 1.1	Fonyo and Benko (1998)
MVR	Water–acetic acid	55	60	2.6	23	+65	PBT = 31.8	Fonyo and Benko (1998)
MVR	Butane–isobutane	12.9	35.8	7.2	72	−64	PBT = 0.3	Díez *et al.* (2009)

MVR	Propane–propylene	7.7	24.9	7.5	73	−65	PBT = 1.4	Olujić *et al.* (2006)
MVR	Acetic acid dehydration	35	55	8.4	50	—	—	Campbell *et al.* (2008)
MVR	Methanol–water	35	40	6.6	69	−67	PBT = 0.7	Feng and Berntsson (1997)
TAHP	Methanol–water	35	45	2.0	0	—	—	Tijani *et al.* (2011)
TAHP	Methanol–water	35	55	2.3	14	—	—	Tijani *et al.* (2011)
TVR	Methanol–water	35	40	1.1	23	−19	PBT = 1.5	Feng and Berntsson (1997)
VC	Methanol–water	35	45	3.0	33	−29	PBT = 2.5	Feng and Berntsson (1997)
VC	1-Butene–*n*-butane	10	20	6.2	67	−57	PBT = 1.6	Meszaros and Meili (1994)
VC	Methanol–water	35	45	4.1	50	—	—	Omideyi *et al.* (1985)
VC	Methanol–water	35	55	3.1	34	—	—	Omideyi *et al.* (1985)
VC	Methanol–water	35	43	5.0	60	−57	PBT = 0.9	Wallin, Franck, and Berntsson (1990)
Multi-VC (2st)	Water-acetic acid	55	65	1.8	−10	+99	PBT = negative	Fonyo and Benko (1998)
MED	Methanol–water	35	N/a	—	36	+32	—	Zhang, Shengrong, and Feng (2010)

Note: primary energy savings (PESs) are the yearly savings of primary energy in comparison to the primary energy requirements of a conventional column; PBT is the simple payback time in years (assuming a plant life time of ten years; primary energy cost = €30 MWh^{-1}; electrical energy cost = € 65 MWh^{-1}; boiler efficiency = 0.85; efficiency of electricity conversion = 0.42; investments as proposed in papers); negative ΔTAC means savings; ΔT_b is given at the pressure reported in the reference.

transfer between the working fluids (ΔT_{df}): $\Delta T_{lift} = \Delta T_b + \Delta T_{df}$. According to common practices, the driving force is $\Delta T_{df} =$ 5–20 K (Wallas, 1990): <5 K for HIDIC, 5–10 K for MVR or TVR, 10 K or lower for CHRP since glide is used, or 10–20 K for others, depending on design.

Thus, considering their values, ΔT_{lift} and ΔT_b can be considered to be more or less the same criteria. The importance of this criterion lies on two key facts. On one hand, it is well known that distillation energy requirements are higher when the separation involves close boiling products (Wallin, Franck, and Berntsson, 1990; Fonyo and Benko, 1998; Bruinsma and Spoelstra, 2010). On the other hand, the efficiency of the selected technologies greatly depends on the ΔT_b, particularly for heat pumps (Fonyo and Mizsey, 1994; Feng and Berntsson, 1997; de Rijke, 2007). Thus, at lower ΔT_b, technology efficiency and conventional energy requirements are higher. Consequently, the largest number of options is tailored to lower ΔT_b. Similarly, most distillation processes in industry fall in the middle range of ΔT_b. Thus, significant developments in that field are expected from massive implementation of energy efficient solutions (Harmsen, 2010). Last but not least, a high ΔT_b involves huge amounts of work for upgrading heat at the industrial scale, as well as a relatively high T_{reb} ($\geq$250 °C). Thus, solutions with high investment costs or even the introduction of mass separating agents are preferred options for reducing the huge energy requirements of conventional distillation.

Another criterion with high impact is the operating pressure. As shown in Table 7.3 (Kiss, Flores Landaeta and Infante Ferreira, 2012), the performance of some technologies changes drastically when the operating pressure drops below atmospheric pressure. The effect is especially important when the top product is used as working fluid (strong connection between the distillation column and the heat pump), and the work input is provided through a compressor. Moreover, the internals of radical designs such as HIDiC and CyDist impose certain pressure drops levels that are unsuitable for vacuum applications (Nakaiwa *et al.*, 1997; Maleta *et al.*, 2011).

When the required products are corrosive or fouling, it is economically wise to avoid contact of these substances with the rotating equipment. Moreover, the equipment internal design for any technology considered should withstand such treatments. Thus, the heat pumps that use working fluids/pairs as heat transfer media with minimal changes in the conventional column design are preferred. Consequently, it is required to evaluate the reboiler temperature to select the most efficient option. Similarly, the robust design of CyDist internals may allow its use with corrosive substances at high operating pressure, but further research is needed on the effect of the moving valve parts (Maleta and Maleta, 2010; Lita, Bildea, and Kiss, 2012).

Finally, there are certain technologies that require additional selection criteria. They were treated as niche applications, as beneficial performance was particularly reported on limited conditions. An example is the limitation on the reboiler duty (Q_{reb}) imposed by the steam ejector efficiency on TVR (Fonyo and Benko, 1998). Similarly, the strong dependence of VC on the heat transfer fluid properties limits the ranges of T_{reb} suitable for operation. VC is recommended only for T_{reb} matched by the working fluid meeting the strong environmental, safety, and performance requirements. Last but not least, the non-ideal behavior of certain components is accounted in the scheme. In such cases, the relative volatility (α_{i-j}) changes drastically with the composition profile across the distillation column. Thus, flexible technologies capable of handling difficult separations are preferred (e.g., AD or ED). Keep in mind that the selection scheme only narrows-down the options for achieving an energy efficient process—so the final choice should be made based on rigorous simulations.

7.8.4 Selected Scheme Applications

The physical properties of the components involved should always be available before using the scheme. This is usually not a problem, since that data is also required at other stages of process research and design. Although the first variable checked in the scheme is the difference in boiling points, it is advisable to verify first other operational constrains—as the boiling points depend on the chosen operating pressure. Sometimes, selection of the operating pressure relies on the particular properties of the components involved, such as corrosiveness, fouling, and so on. However, in most cases, the use of common utilities (steam, cooling water, etc.) determines the selection of the operating pressure. As mentioned already, the use of non-common utilities can seriously jeopardize the economic operation of the distillation column. Therefore, it is advisable to gather the property data prior to approaching the scheme. After the data gathering process, the technology selection scheme can be used. Some possible situations are illustrated in the following paragraphs.

The separation of *methanol–water* systems involves components that are neither corrosive nor fouling (Haynes, 2011). The boiling point for each component can be calculated at atmospheric pressure as a first estimate to check whether common utilities can be used. The boiling point difference at atmospheric pressure is about 35 °C, with the top temperature of about 65 °C (methanol) and the bottom temperature of 100 °C (water). With that data in mind, the selection scheme can be applied. The first criterion verified is the difference in boiling points, which in this particular case fits in the area of wide boiling components ($\Delta T_b \geq 20\,^{\circ}C$). The next check is the operating pressure, the preferred value being atmospheric pressure. The final verification is the presence of non-ideal VLE behavior. As the thermodynamic behavior reveals no azeotropes, the preferred options are VC, CRHP, TAHP, and AHP. Comparing the result of the scheme with further in-depth analysis of the literature in Table 7.4, the largest reported primary energy savings for methanol–water separation reported so far is for CRHP (74%, obtained by slight extrapolation of existing cases), MVR (69%), VC (46%, average of values from different sources), and AHP (43%, interpolated between existing cases). The data available for the TAHP concerns a laboratory prototype and shows low primary energy savings (14%). The reduction in total annual cost (TAC) ranks the best solutions as MVR (67%), VC (43%), and AHP (27%). There is no published data available about the CHRP and TAHP. Concerning the payback time, the shortest one is reported for MVR (0.7 years), closely followed by VC (0.9 years) and with a significant difference for TVR (1.5 years). The payback time of AHP will be much larger (4.0 years). Payback times of 3.0 years for CHRP, 4.4 years for VCHP and 5.6 years for AHP are obtained using data from Bor and Infante Ferreira (2011). These payback times are based on installed costs and cannot be compared directly with the payback times from the other sources, based on equipment costs only.

Another example of a revamp on binary separations could be the separation of propane–propylene (*PP-splitting*). These gases are separated in existing plants by distillation at about 10 atm without major concerns about corrosiveness or fouling problems (Haynes, 2011). The boiling point of each component is calculated to verify which energy efficient technologies could reduce the current energy requirements. At 10 atm, propane boils at ~27 °C, while propylene boils at ~20 °C. This results in a low ΔT_b ($\leq 10\,^{\circ}C$). Applying the separation scheme with the complete information, and considering that water is not the top distillate product, the TVR option can be ruled out. Following the selected path the operating pressure is higher than atmospheric, hence three viable options are suggested. Out of these options, the state-of-the-art system for propane–propylene distillation is MVR, but interesting results were obtained in simulation studies for HIDiC (Olujić *et al.*, 2003, 2006), while the potential of CyDist is now ready to unfold in follow up studies (Lita, Bildea, and Kiss, 2012).

7.9 CONCLUDING REMARKS

Heat pump assisted distillation technologies were proposed and developed in order to upgrade the low-level energy and re-use it in the process, thus reducing the consumption of valuable utilities. With the exception of solid-sorption systems, the heat pumps presented here have no intrinsic (temperature) limitations and, therefore, the application range is quite large within the chemical process industry. Temperature lifts of up to 30–100 °C are feasible, leading to significant savings in the primary energy requirements (PER) of up to 80% (Kiss, Flores Landaeta and Infante Ferreira, 2012).

Moreover, the technology selection scheme presented here provides valuable guidelines in the design of energy efficient distillation processes. Although the main focus is on heat pumps, other energy efficient technologies were also included for completeness. Using the scheme, process designers can effectively narrow down the number of technology alternatives that can deliver significant energy savings considering the particular process conditions. The selection scheme features several key criteria for the conditions required to successfully apply each technology.

- VC, MVR, and TVR usually allow higher savings for low T_{reb} and low ΔT_{lift} (<20 °C);
- TAHP and CRHP can be applied from low to middle ΔT_{lift} (in the range of 20–50 °C);
- technologies such as HIDiC and cyclic distillation are applicable only at higher pressures, mainly for non corrosive compounds, low ΔT_{lift} and relative volatility closed to unity;
- for multicomponent distillation, DWC and Kaibel distillation columns are the preferred option for ΔT_b lower than 80 K and high purity side stream requirements.

REFERENCES

Al-Shammiri, M. and Safar, M. (1999) Multi-effect distillation plants: state of the art. *Desalination*, *126*, 45–59.

Annakou, O. and Mizsey, P. (1995) Rigorous investigation of heat pump assisted distillation. *Heat Recovery Systems & CHP*, *15* (3), 241–247.

Araujo, A.B., Brito, R.P., and Vasconcelos, L.S. (2007) Exergetic analysis of distillation processes –A case study. *Energy*, *32*, 1185–1193.

Aspiron, N. and Kaibel, G. (2010) Dividing wall columns: Fundamentals and recent advances. *Chemical Engineering and Processing*, *49*, 139–146.

Bansal, V., Ross, R., Perkins, J.D., and Pistikopoulos, E.N. (2000) The interaction of design and control: double-effect distillation. *Journal of Process Control*, *10*, 219–227.

Bor, D.M.van de. and Infante Ferreira, C.A. Quick selection of heat pump types and optimization of loss mechanisms. Presented at the 10th IEA Heat Pump Conference, May 16–19 (2011), Tokyo, Japan.

Bruinsma, D. and Spoelstra, S. (2010) Heat pumps in distillation, in *Proceedings from Distillation & Absorption 2010: conference proceedings, 12–15 September, Eindhoven, The Netherlands* (ed. A.B. Hahn, H. Kooijman, and A. Górak) University of Eindhoven, pp. 21–28.

Bruinsma, O.S.L., Krikken, T., Cot, J. *et al.* (2012) The structured heat integrated distillation column. *Chemical Engineering Research and Design*, *90*, 458–470.

Campbell, J.C., Wigala, K.R., van Brunt, V., and Kline, R.S. (2008) Comparison of energy usage for the vacuum separation of acetic acid/acetic anhydride using an internally heat integrated distillation column (HIDiC). *Separation Science and Technology*, *43* (9), 2269–2297.

Carlsen, H. (1989) Development of a gas fired Vuilleumier heat pump for residential heating, in *Proceedings of the 24th IntersocietyEnergy Conversion Engineering Conference, IECEC-89*, IEEE, Piscataway, pp. 2257–2263.

Chen, Y.M. and Sun, Ch.Y. (1997) Experimental study of the performance characteristics of a steam-ejector refrigeration system. *Experimental Thermal and Fluid Science*, *15*, 384–394.

Chun, H.D., Yi, J.S., Yeo, Y.K. *et al.* (1997) Optimization of fractionation operation in the benzene-toluene-xylene plant. *Korean Journal of Chemical Engineering*, *14* (5), 325–333.

Chung, Y., Kim, B.J., Yeo, Y.K., and Song, H.K. (1997) Optimal design of a chemical heat pump using the 2-propanol/acetone/hydrogen system. *Energy*, *22*, 525–536.

Dejanović, I., Matijašević, Lj., and Olujić, Ž. (2010) Dividing wall column – A breakthrough towards sustainable distilling. *Chemical Engineering and Processing*, *49* (6), 559–580.

Dejanović, I., Matijašević, Lj., Halvorsen, I.J. *et al.* (2011a) Designing four-product dividing wall columns for separation of a multicomponent aromatics mixture. *Chemical Engineering Research and Design*, *89* (8), 1155–1167.

Dejanović, I., Matijašević, Lj., Jansen, H., and Olujić, Ž. (2011b) Designing a packed dividing wall column for an aromatics processing plant. *Industrial & Engineering Chemistry Research*, *50*, 5680–5692.

Díez, E., Langston, P., Ovejero, G., and Dolores Romero, M. (2009) Economic feasibility of heat pumps in distillation to reduce energy use. *Applied Thermal Engineering*, *29*, 1216–1223.

Douglas, J.M. (1988) *Conceptual Design of Chemical Processes*, McGraw-Hill, New York.

El-Dessouky, H., Ettouney, H., Alatiqi, I., and Al-Nuwaibit, G. (2002) Evaluation of steam jet ejectors. *Chemical Engineering and Processing*, *41*, 551–561.

Engelien, H.K., Larsson, T., and Skogestad, S. (2003) Implementation of optimal operation for heat integrated distillation columns. *IChemE Transactions*, *81*, 227–281.

Engelien, H.K. and Skogestad, S. (2005) Multi-effect distillation applied to an industrial case study. *Chemical Engineering and Processing*, *44*, 819–826.

Feng, X. and Berntsson, T. (1997) Critical COP for an economically feasible industrial heat-pump application. *Applied Thermal Energy*, *17* (1), 93–101.

Fonyo, Z. and Mizsey, P. (1994) Economic applications of heat pumps in integrated distillation systems. *Heat Recovery Systems & CHP*, *14* (3), 249–263.

Fonyo, Z. and Benko, N. (1998) Comparison of various heat pump assisted distillation configurations. *IChemE Transactions*, *76*, A, 348–360.

Freshwater, D.C. (1951) Thermal economy in distillation. *Transactions of the Institution of Chemical Engineers*, *29*, 149–160.

Gadalla, M., Olujić, Z., de Rijkeb, A., and Jansens, P.J. (2006) Reducing CO_2 emissions of internally heat-integrated distillation columns for separation of close boiling mixtures. *Energy*, *31*, 2409–2417.

Gadalla, M., Jiménez, L., Olujić, Ž., and Jansens, P.J. (2007) A thermo-hydraulic approach to conceptual design of an internally heat-integrated distillation column (i-HIDiC). *Computers and Chemical Engineering*, *31*, 1346–1354.

Gadalla, M.A. (2009) Internal heat integrated distillation columns (iHIDiCs) - New systematic design methodology. *Chemical Engineering Research and Design*, *87*, 1658–1666.

Gardner, D.L. and Howard, C.Q. (2009) Waste-heat-driven thermo-acoustic engine and refrigerator. Presented at the Acoustics Seminar, November 23rd–25th, Adelaide, Australia.

Gaska, R.A. and Cannon, M.R. (1961) Controlled cycling improves various processes. *Industrial and Engineering Chemistry*, *53*, 630–631.

Ghadrdan, M., Halvorsen, I.J., and Skogestadad, S. (2011) Optimal operation of Kaibel distillation columns. *Chemical Engineering Research and Design*, *89*, 1382–1391.

Gilmour, T.A. (1994) Malone–Brayton cycle engine/heat pump, Patent Department of the Navy, Washington DC, 07/1994.

Göktun, S. (1995) Selection of working fluids for high-temperature heat pumps. *Energy*, *20* (7), 623–625.

Halvorsen, I.J., Skogestad, S., Dejanović, I. *et al.* (2011) Multi-product dividing wall columns: a simple and effective assessment and conceptual design procedure. *Chemical Engineering Transactions*, *25*, 611–616.

Harmsen, J. (2010) Process intensification in the petrochemicals industry: Drivers and hurdles for commercial implementation. *Chemical Engineering and Processing*, *49*, 70–73.

Haynes, W.M. (ed.) (2011) *CRC Handbook of Chemistry and Physics*, 92nd edition, CRC Press, Boca Raton.

Huang, K., Iwakabe, K., Nakaiwa, M. *et al.* (2008) Consider heat integration to improve separation performance. *Hydrocarbon Processing*, pp. 101–108.

Kaibel, G. (1987) Distillation columns with vertical partitions. *Chemical Engineering and Technology*, *10*, 92–98.

Kiss, A.A. and Bildea, C.S. (2011) A control perspective on process intensification in dividing-wall columns. *Chemical Engineering and Processing*, *50*, 281–292.

Kiss, A.A., Segovia-Hernandez, J.G., Bildea, C.S. *et al.* (2012a) Reactive DWC leading the way to FAME and fortune. *Fuel*, *95*, 352–359.

Kiss, A.A., Ignat, R.M., Flores Landaeta, S.J., and de Haan, A.B. (2013) Intensified process separation of aromatics powered by Kaibel and dividing-wall columns. *Chemical Engineering and Processing: Process Intensification*. Article in press. doi: 10.1016/j.cep.2012.06.010

Kiss, A.A., Flores Landaeta, S.J., and Infante Ferreira, C.A. (2012) Towards energy efficient distillation technologies – Making the right choice. *Energy*, *47*, 531–542.

Kiss, A.A. and Suszwalak, D.J.-P.C. (2012) Enhanced bioethanol dehydration by extractive and azeotropic distillation in dividing-wall column. *Separation and Purification Technology*, *86*, 70–78.

Kiss, A.A. and Suszwalak, D.J.-P.C. (2012) Innovative dimethyl ether synthesis in a reactive dividing-wall column. *Computers & Chemical Engineering*, *38*, 74–81.

Kim, Y.H., Hwang, K.S., and Nakaiwa, M. (2004) Design of a fully thermally coupled distillation column for hexane process using a semi-rigorous model. *Korean Journal of Chemical Engineering*, *21* (6), 1098–1102.

Kniel, L. (1952) Fractional distillation, US Patent No. 2619814.

Kolbe, B. and Wenzel, S. (2003) Novel distillation concepts using one-shell columns. *Chemical Engineering and Processing*, *43*, 339–346.

Lee, J.Y., Kim, Y.H., and Hwang, K.S. (2004) Application of a fully thermally coupled distillation column for fractionation process in naphtha reforming plant. *Chemical Engineering and Processing*, *43*, 495–501.

Linnhoff, B., Dunford, H., and Smith, R. (1983) Heat integration of distillation columns into overall processes. *Chemical Engineering Science*, *38* (8), 1175–1183.

Lita, I., Bildea, C.S., and Kiss, A.A. (2012) Modeling, design and control of cyclic distillation systems. Presented at the 20th International Congress of Chemical and Process Engineering (CHISA), 25–29 August 2012, Prague, Czech Republic, Paper 1119.

Lucke, C.E. (1902) The heat engine problem. *Transactions of the American Society of Mechanical Engineers*, ASME Paper No. 923, pp. 231–232.

Mah, R.S.H., Nicholas, J.J., Jr., and Wodnik, R.B. (1977) Distillation with secondary reflux and vaporization: A comparative evaluation. *AIChE Journal*, *23* (5), 651–658.

Maleta, B.V. and Maleta, O. (2010) Mass-exchange contact device, US Patent Application No. 2010/0219060A1.

Maleta, V.N., Kiss, A.A., Taran, V.M., and Maleta, B.V. (2011) Understanding process intensification in cyclic distillation systems. *Chemical Engineering and Processing: Process Intensification*, *50*, 655–664.

Malone, J. (1993) John Malone and the Invention of Liquid-based Engines, *Los Alamos Science* (21) http://www.fas.org/sgp/othergov/doe/lanl/pubs/00326023.pdf.

Matsubara, M., Watanabe, N., and Kurimoto, H. (1985) Binary periodic distillation scheme with enhanced energy conservation - II: Experiment. *Chemical Engineering Science*, *40*, 755–758.

Matsuda, K., Iwakabe, K., Ohmori, T., and Nakaiwa, M. (2010) Dynamic behavior of an internally heat-integrated distillation column (HIDiC). *Chemical Engineering Transactions*, *21*, 127–132.

Matsuda, K., Kawazuishi, K., Kansha, Y. *et al.* (2011) Advanced energy saving in distillation process with self-heat recuperation technology. *Energy*, *36*, 4640–4645.

McMullan, A. (2003) Industrial heat pumps for steam and fuel savings (DOE/GO-102003-1735), U.S. Department of Energy, www1.eere.energy.gov/industry/bestpractices/ pdfs/heatpump.pdf – URL last accessed on 24.01.2013.

McWhirter, J.R. and Cannon, M.R. (1961) Controlled cycling distillation in a packed-plate column. *Industrial and Engineering Chemistry*, *53* (8), 632–634.

Meszaros, I. and Meili, A. (1994) 1-Butene separation processes with heat pump assisted distillation. *Heat Recovery Systems & CHP*, *14* (3), 315–322.

Mučić, V. (1989) Resorption compression heat pump with solution circuit for steam generation using waste heat of industry as heat source. *Heat Pump Centre Newsletter*, *7/1*, 14–15.

Nakaiwa, M., Huang, K., Owa, M. *et al.* (1997) Energy savings in heat-integrated distillation columns. *Energy*, *22*, 621–625.

Nakaiwa, M., Huang, K., Endo, A. *et al.* (2003) Internally heat-integrated distillation columns: A review. *Chemical Engineering Research and Design*, *81*, 162–177.

Niggemann, G., Hiller, C., and Fieg, G. (2010) Experimental and theoretical studies of a dividing-wall column used for the recovery of high-purity products. *Industrial and Engineering Chemistry Research*, *49*, 6566–6577.

Nordtvedt, S.T. (2005) Experimental and theoretical study of a compression/absorption heat pump with ammonia/water as working fluid, PhD Thesis, NTNU, Norway.

Oliveira, S.B.M., Parise, J.A.R., and Pitanga Marques, R. (2002) Modeling of an ethanol–water distillation column assisted by an external heat pump. *International Journal on Energy Research*, *26*, 1055–1072.

Olujić, Ž., Fakhri, F., de Rijke, A. *et al.* (2003) Internal heat integration – the key to an energy-conserving distillation column. *Journal of Chemical Technology and Biotechnology*, *78*, 241–248.

Olujić, Ž., Sun, L., de Rijke, A., and Jansens, P.J. (2006) Conceptual design of an internally heat integrated propylene-propane splitter. *Energy*, *31*, 3083–3096.

Olujić, Ž., Jödecke, M., Shilkin, A. *et al.* (2009) Equipment improvement trends in distillation. *Chemical Engineering and Processing*, *48*, 1089–1104.

Omideyi, T.O., Kasprzycki, J., and Watson, F.A. (1984) The economics of heat pump assisted distillation systems-i. a design and economic model. *Heat Recovery Systems*, *4* (3), 187–200.

Omideyi, T.O., Parande, M.G., Supranto, S. *et al.* (1985) The economics of heat pump assisted distillation systems. *Heat Recovery Systems & CHP.*, *5* (6), 511–518.

Perry, R.H. and Green, D.W. (eds) (1999) *Perry's Chemical Engineering Handbook*, 7th edn, McGraw-Hill, New York.

Ramshaw, C. (1983) HiGee distillation - An example of process intensification. *Chemical Engineer – London*, *389*, 13–14.

Rao, D.P., Bhowal, A., and Goswami, P.S. (2004) Process intensification in rotating packed beds (HiGee): An appraisal. *Industrial and Engineering Chemistry Research*, *43*, 1150–1162.

Rijke, A. (2007) Development of a concentric internally heat integrated distillation column (HIDiC), PhD thesis, TU Delft.

Sier, R. (1999) *Hot Air Caloric and Stirling engines. Vol. 1, A History*, L.A. Mair, Chelmsford.

Silberstein E., Heat pumps, Delmar Cengage Learning, 2002.

Slade, B., Stober, B., and Simpson, D. (2006) Dividing wall column revamp optimizes mixed xylenes production, *Distillation and Absorption 2006, IChemE Symposium Series*, No. 152, IChemE, Rugby, 1–10.

Sommerfeld, J.T., Schrodt, V.N., Parisot, P.E., and Chien, H.H. (1966) Studies of controlled cyclic distillation: I. Computer simulations and the analogy with conventional operation. *Separation Science and Technology*, *1* (2), 245–279.

Spoelstra, S. (2007) Innovative heat pump technology. *NPT Procestechnologie*, *14e*, 14–15.

Spoelstra, S. (2008) Saving energy in distillation with thermoacoustic heat pumps, ECN Brochure B-07-008.

Taboada, R. and Infante Ferreira, C.A. (2008) Compression resorption cycles in distillation columns. Presented at the International Refrigeration and Air Conditioning Conference at Purdue, USA, Paper 912.

Tijani, M.E.H., Vanapalli, S., Spoelstra, S., and Lycklama à Nijeholt, J.A. (2011) Electrically driven thermoacoustic heat pump. Presented at the 10th IEA Heat Pump Conference, June 27–August 31, Tokyo, Japan.

Tufano, V. (1997) Heat recovery in distillation by means of absorption heat pumps and heat transformers. *Applied Thermal Engineering*, 2, 171–178.

Wallas, S. (1990) *Chemical Process Equipment Selection and Design*, Butterworth-Heinemann, Boston.

Wallin, E., Franck, P.A., and Berntsson, T. (1990) Heat pumps in industrial processes – An optimization methodology. *Heat Recovery Systems & CHP*, *10* (4), 437–446.

Wang, Ch., Zhang, P., and Wang, R. (2008) Review of recent patents on chemical heat pump. *Recent Patents on Engineering*, 2 (3), 208–216.

Wang, G.Q., Xu, Z.C., and Ji, J.B. (2011a) Progress on Higee distillation - Introduction to a new device and its industrial applications. *Chemical Engineering Research and Design*, *89*, 1434–1442.

Wang, P., Li, W., Quan, S., and Xu, J. (2011b) Application of heat pump distillation in a gas separation device. Presented at the 2011 International Conference on Computer Distributed Control and Intelligent Environmental Monitoring, CDCIEM 2011, 19–20 February, China, pp. 145–147.

Wang, W. and Wang, W. (2012) Energy saving in methyl-chloro-silane distillation. Presented at the 2012 International Conference on Computer Distributed Control and Intelligent Environmental Monitoring, CDCIEM 2012, Zhangjiajie, Hunan, China, 5–6 March.

Wang, Y. and Lior, N. (2011) Thermo-economic analysis of a low-temperature multi-effect thermal desalination system coupled with an absorption heat pump. *Energy*, *36*, 3878–3887.

Yildirim, O., Kiss, A.A., and Kenig, E.Y. (2011) Dividing wall columns in chemical process industry: a review on current activities. *Separation and Purification Technology*, *80*, 403–417.

Zamfirescu, C., Nannan, N., Marin, M., and Infante Ferreira, C.A. (2004) Oil free two phase ammonia (water) compressor, Report K-336, NOVEM-BSE-NEO, TUDelft.

Zhang, J., Shengrong, L., and Feng, X. (2010) A novel multi-effect methanol distillation process. *Chemical Engineering and Processing*, *49*, 1031–1037.

Ziegler, F. (2002) State of the art in sorption heat pumping and cooling technologies. *International Journal of Refrigeration*, *25*, 450–459.

8

Heat-Integrated Distillation Column

8.1 INTRODUCTION

Although distillation is by far the most applied separation technology, its major drawback is the degradation of energy associated with the temperature difference between the reboiler and condenser of a distillation column – which leads to a low thermodynamic efficiency of typically around 10% (Humphrey, Seibert, and Koort, 1991; de Rijke, 2007). Consequently, several technologies have been developed and implemented during recent decades to improve the thermal efficiency of distillation processes, such as thermally coupled columns (e.g., Petlyuk or DWC) or heat pumps assisted distillation. For example, vapor recompression (VRC) is a proven technology used to improve the energy efficiency of a distillation column and it is already applied industrially for the separation of close-boiling mixtures.

Heat-integrated distillation column (HIDiC) is the most radical approach of a heat pump design, making use of internal heat-integration (Nakaiwa *et al.*, 1997; Gadalla *et al.*, 2006; de Rijke, 2007; Bruinsma *et al.*, 2012). Instead of using a single point heat source and sink, the whole rectifying section of a distillation column becomes the heat source, while the stripping part of the distillation column acts as a heat sink (Matsuda *et al.*, 2011; Matsuda, Iwakabe, and Nakaiwa, 2012). The work input is provided by a compressor installed at the top outlet of the stripper section, while the heat pump cycle is closed by a valve flashing the liquid bottom outlet of the rectifier section. Basically, the stripping

Advanced Distillation Technologies: Design, Control and Applications, First Edition.
Anton Alexandru Kiss.

section is physically separated from the rectifying section and the heat is transferred inside the distillation column – from the rectifying to the stripping section – due to the increased operating pressure of the rectifying section by means of a compressor (de Rijke, 2007). Figure 8.1 illustrates both the mechanical vapor recompression (MVR) and HIDiC technologies as well as the differences between them (Kiss, Flores Landaeta, and Infante Ferreira, 2012). Remarkably, an internally heat-integrated distillation column (HIDiC) combines the advantages of vapor recompression and the diabatic operation to reduce the energy requirements of a single distillation column. The theoretical advantage of HIDiC over MVR systems is that in a HIDiC configuration the compressor operates only over the stripping section, and not the whole column. For this reason, a HIDiC should operate at a lower compression ratio and thus a lower compressor duty as compared to MVR (de Rijke, 2007). Nevertheless, as described hereafter, the actual success of HIDiC relies on very good hardware performance for both heat and mass transfer tasks at the same time (Nakaiwa *et al.*, 2003).

Notably, HIDiC technology offers several very attractive features that keep stimulating the researchers to pursue its implementation in the chemical process industry (Nakaiwa *et al.*, 2003):

- *High energy efficiency:* The highest degree of internal heat integration within HIDiC offers typically higher energy efficiency as compared with conventional distillation or other types of heat-integrated distillation columns, such as heat pump assisted distillation columns. The arrangement of the internal heat transfer area for every paired stage provides a large number of extra degrees of

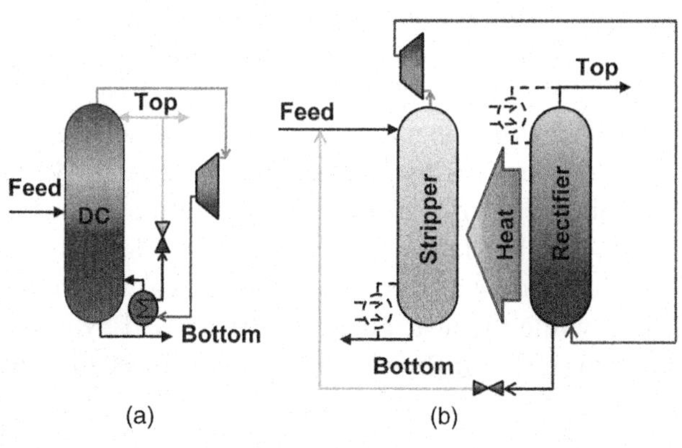

Figure 8.1 Mechanical vapor recompression (MVR) (a) and HIDiC configuration (b)

freedom for the process design, thus making it possible to utilize the maximum potential energy savings within a distillation column. This was confirmed by numerous simulations and bench-scale experiments (Nakaiwa *et al.*, 2003; de Rijke, 2007).
- *Zero external reflux and boil-up operation:* The common practice in distillation is to use a condenser and a reboiler to generate external reflux and boil-up for operation. For HIDiC, the internal heat integration plays this role and generates these two flows internally, such that neither a condenser nor a reboiler is necessary, in principle (Nakaiwa *et al.*, 2003; Matsuda, Iwakabe, and Nakaiwa, 2012).
- *Enhanced potential of internal heat-integration techniques:* Using internal heat integration is very effective in improving process energy efficiency and it can find many applications within distillation processes, such as multicomponent separation, batch distillation, pressure-swing distillation, and even reactive distillation. Moreover, internal heat integration can be considered between two distillation columns that may have no direct connections at all.

Following modern trends towards more sustainable processes, HIDiC technology has developed into a very popular topic that has received much scientific attention in recent years, as reported by numerous papers of various research groups – the focus ranging from thermodynamic analysis, mass and heat transfer mechanism, to process design, operation and applications (Huang *et al.*, 1996a,b; Huang *et al.*, 2007, 2008; Nakaiwa *et al.*, 1997, 2001, 2003; Olujić *et al.*, 2003, 2006, 2008; Gadalla *et al.*, 2005; Gadalla *et al.*, 2006, 2007a,b; de Rijke, 2007; Gadalla, 2009; Fukushima, Kano, and Hasebe, 2006; Schmal *et al.*, 2006; Mane and Jana 2010; Cabrera-Ruiz, Jimenez-Gutierrez, and Segovia-Hernandez, 2011; Kansha, Kishimoto, and Tsutsumi, 2011; Pulido *et al.*, 2011; Shenvi, Herron, and Agrawal, 2011; Suphanit, 2011; Kiran, Jana, and Samanta, 2012; Bruinsma *et al.*, 2012; Matsuda, Iwakabe, and Nakaiwa, 2012).

8.2 WORKING PRINCIPLE

The working principle of HIDiC can be explained by making a comparison with the conventional adiabatic distillation and the vapor recompression heat pump (Kiss, Flores Landaeta, and Infante Ferreira, 2012; Flores Landaeta and Kiss, 2012). Note that in contrast to the multi-effect distillation (MED) methods, vapor recompression is primarily a method for enhancing the thermal efficiency and economy of single columns

(Freshwater, 1951, 1961; Null, 1976). Vapor recompression (Figure 8.1a) is especially effective because it tackles the thermal inefficiency problem (i.e., introduction and removal of heat to and from the distillation column) directly at the source: the pressure and temperature of the overhead vapor is upgraded – by means of a compressor – to a higher level, such that it can be used as heat source for the reboiler (de Rijke, 2007). Although vapor recompression is a very useful technique for saving energy, VRC is rather capital intensive and, hence, it is in practice economical only for close boiling mixtures – where small compression ratios and duties are required due to a small temperature difference between the top and bottom of the column. This should not be surprising as the potential for energy saving is largest in the case of separating a close-boiling mixture, where high reflux ratios and large reboiler duties are required for separation (Null, 1976; Nakaiwa *et al.*, 2003; de Rijke, 2007).

Diabatic distillation is an outstanding concept for energy savings in a single distillation column (Le Goff, Cachot, and Rivero, 1996; Schaller *et al.*, 2005; de Rijke, 2007). Compared to the classical *adiabatic distillation,* a diabatic column replaces the reboiler and the condenser – usually connected to the bottom and at the top of the column – with a condenser and reboiler integrated in the rectification and stripping sections, respectively (Figure 8.2 a). Owing to the gradual supply and removal of heat along the stripping and rectification sections, respectively, diabatic distillation uses in a more efficient way the heat of condensation and evaporation. Heat transfer takes place at a lower temperature difference, which implies smaller exergy losses associated with the heat transfer. Nonetheless, diabatic distillation as such has not been implemented in practice so far due to some practical drawbacks (de Rijke, 2007).

The effect of internal heat transfer on the distillation process can be conveniently evaluated using the McCabe–Thiele diagram, as plotted in Figure 8.2b. In classic distillation, the operating lines have specific slopes – their ratio being equal to the ratio of the molar flow rates of liquid and vapor flowing counter-currently in the rectification and stripping sections of an adiabatic column. However, the two straight lines are replaced in a diabatic column by a continuous operating curve that is parallel to the equilibrium curve. The distance between the operating and the equilibrium curve gives a measure for the irreversibility of the process. In the limit case when these curves overlap, the exergy losses are zero and the distillation process is perfectly reversible. However, a finite difference must be maintained in practice, as bringing these two curves together – while reducing the exergy losses to a minimum – would result in an inoperable column. Consequently, the theoretical potential for maximum

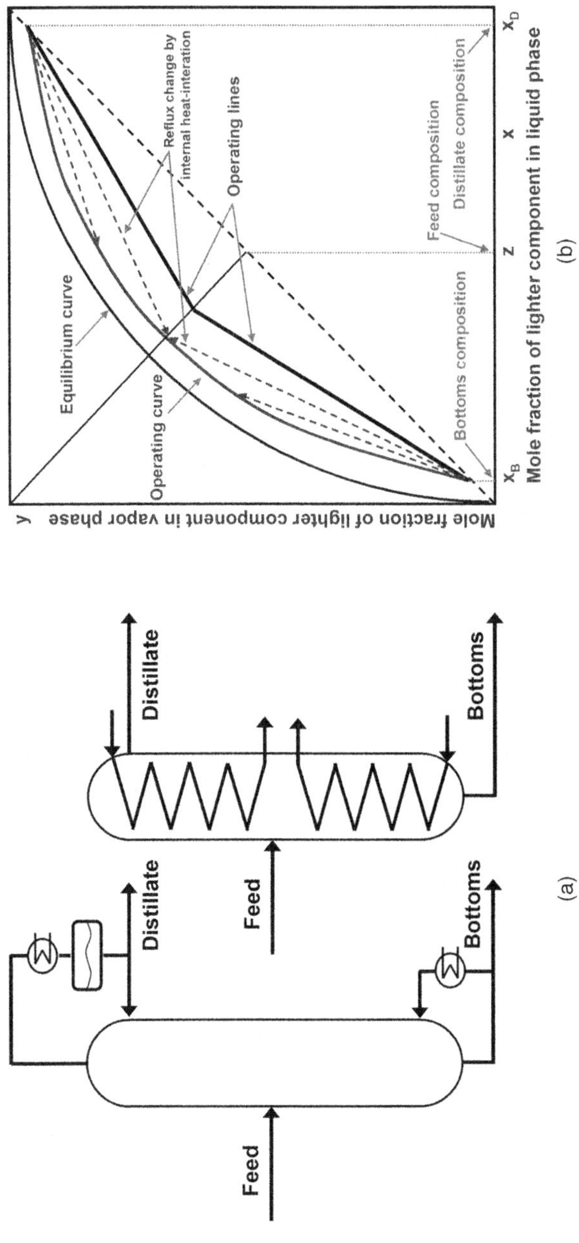

Figure 8.2 Conventional and diabatic distillation columns: schematic representation (a) and McCabe–Thiele diagram (b)

energy savings cannot be exploited, and a tradeoff is required between the operating reflux and the number of stages – just as in adiabatic distillation. A key drawback of this concept is that it requires the use of a special heating media in the reboiler, and the utilities leaving the condenser and reboiler are at such temperatures that they can hardly be re-utilized in the process (de Rijke, 2007).

A *heat-integrated distillation column* (HIDiC) is another possibility for internal heat integration that combines the advantages of vapor recompression and diabatic distillation in a column. In a HIDiC, heat is transferred from the hotter rectification section (operated at higher pressure) to the colder stripping section (Figure 8.1b), thus leading to a gradual evaporation and condensation along the length of the stripping and rectifying section, respectively (Nakaiwa *et al.*, 2003; de Rijke, 2007; Kiss, Flores Landaeta, and Infante Ferreira, 2012). Note that the problem of different sizes for rectifying and stripping sections can be tackled relatively simple by using one of the alternative HIDiC configurations (Figure 8.3) (Kiss, Flores Landaeta, and Infante Ferreira, 2012). Thus, a small stripper can be connected only to the top or bottom part of a

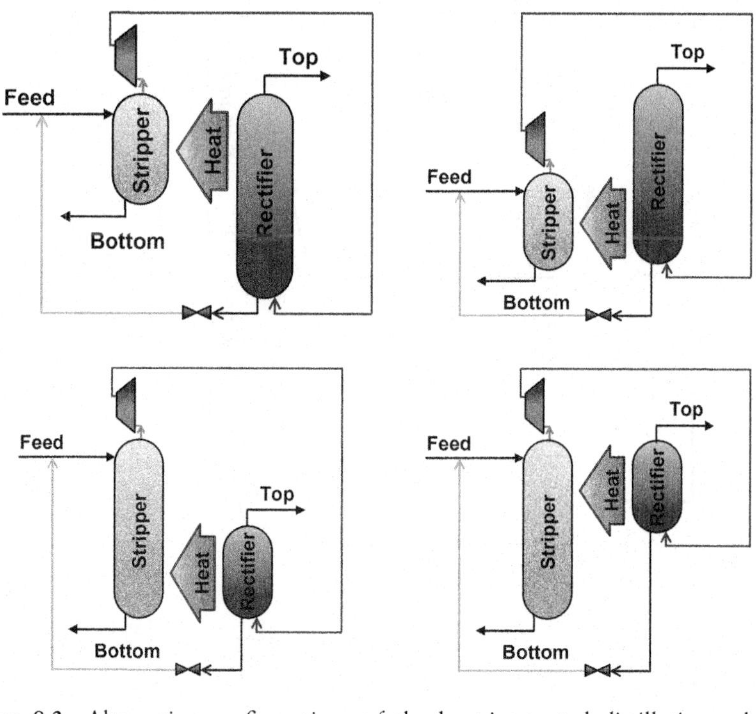

Figure 8.3 Alternative configurations of the heat-integrated distillation column (HIDiC)

rectifier, and similarly a small rectifier can be connected only to the top or bottom part of a stripper. Sometimes, it is actually beneficial to modify the general process configuration and intentionally break the symmetry between the rectifying and stripping sections to improve significantly the performance of process operations – but this comes at the expense of a certain degree of loss in energy efficiency (Nakaiwa *et al.*, 2003; de Rijke, 2007; Matsuda, Iwakabe, and Nakaiwa, 2012).

To realize a temperature driving force, the vapor leaving the top of the stripping section is directly recompressed such that the rectifying section is operated at a higher pressure than the stripping section. While VRC requires the heat pump to operate over the complete temperature difference that exists in the system, the main advantage of HIDiC is its ability to operate at a lower temperature difference and compression ratio as compared to a conventional heat pump system. Theoretically, a HIDiC can operate even without reboiler and condenser. Moreover, the height of a HIDiC could be much lower – in the best case just half the normal column height (de Rijke, 2007).

8.3 THERMODYNAMIC ANALYSIS

An extensive thermodynamic analysis of the distillation processes was first applied by Freshwater (1951, 1961), who also suggested transferring heat from the rectifying to the stripping section of a single distillation column, to reduce the temperature lift to be overcome by a heat pump. The idea was worked out systematically by Flower and Jackson (1964) who showed the advantages of this technique by simulation studies based on the second law of thermodynamics. The HIDiC concept was thoroughly evaluated later under the name *secondary reflux and vaporization* (SRV) by Mah, Nicholas, and Wodnik (1977) and Fitzmorris and Mah (1980). In the following years, several patents related to the equipment were filed (Haselden, 1977; Seader, 1980; Govind, 1986, 1987). The name HIDiC used nowadays was introduced by Nakaiwa (1988), who performed simulation studies and pilot plant experiments that proved substantial energy savings of HIDiC as compared with its conventional counterpart. Subsequently, the possibility of operating even without either a reboiler or a condenser, in a so-called ideal HIDiC, was indicated by Takamatsu, Nakaiwa, and Nakanishi (1996), while Aso *et al.* (1998) proposed and patented a new type known as shell & tube HIDiC. Nakaiwa *et al.* (2003) provide more details in a broad literature review of their work in the area of HIDiC technology.

In a classic distillation column, heat is supplied in the reboiler and removed in the condenser. Owing to the temperature difference between the heating and cooling media used in the reboiler and condenser, respectively, the separation of components is always accompanied by a degradation of energy, even when heat losses are neglected. To realize a closer approach to a reversible distillation column, one could use inter-coolers and inter-heaters, instead of a separate reboiler and condenser. The basic idea here is to distribute the addition and removal of heat from the distillation process more uniformly along the length of the column. The thermodynamic advantages of a more equal partition of energy introduction and removal from a distillation column were reported by several authors (Haselden, 1958; Flower and Jackson 1964; Le Goff, Cachot, and Rivero, 1996; Rivero, 2001; de Koeijer, 2002; de Rijke, 2007). Actually, the *equipartition of entropy production* was introduced as a general concept by Tondeur and Kvaalen (1987), who suggested that the *optimal design* (in view of the second law) is the one where the entropy production is uniformly distributed among the process variables. The equipartition of entropy production was also proposed as a new design principle that is applied in both a quasi-reversible or diabatic distillation column and a HIDiC. In a HIDiC setup, this design principle is combined with a heat pump operating over a temperature range that is narrower than that of a conventional heat pump system. This combination of benefits leads to a very powerful, possibly ultimate design of a distillation column with respect to energy savings (de Rijke, 2007). Note that the thermodynamic efficiency of an ideally heat-integrated distillation column was discussed at a greater length by Liu and Qian (2000) and Nakaiwa *et al.* (2003).

For a separation process the minimum amount of work (W_{min} expressed in J s^{-1}) required to make a complete separation is given by the following equation:

$$W_{min} = F(\Delta H - T\Delta S) \tag{8.1}$$

where F is the feed flow-rate (kmol s^{-1}), ΔH (kJ mol^{-1}) is the change in enthalpy, T (K) is the temperature and ΔS (kJ $mol^{-1}\cdot K^{-1}$) is the change in entropy. For an ideal mixture of n components the minimum work for separation can be expressed as follows (de Rijke, 2007):

$$W_{min} = FRT\sum_{i=1}^{n} x_i \ln x_i \tag{8.2}$$

where R (8.3145 J $mol^{-1}\cdot K^{-1}$) is the ideal gas constant, T (K) is the temperature of the mixture and x_i (−) is the mol fraction of component i

in the feed. Note that $W_{\min}$ is the thermodynamic minimum, independent of any particular process. Actual processes operate with finite driving forces that are irreversible and consequently use more energy than the thermodynamic minimum. For a conventional distillation column, the minimum energy ($Q_{\min,\text{conv}}$ expressed in J s^{-1}) required to completely separate an ideal binary mixture when the feed is saturated liquid can be reduced to the following expression (Keller and Humphrey, 1997):

$$Q_{min,\text{conv}} = F\Delta H_{\text{vap,B}}\left[\frac{1}{\alpha_{12}-1} + x_{\text{F},1}\right] \qquad (8.3)$$

where $\Delta H_{\text{vap,B}}$ (kJ kg^{-1}) is the heat of vaporization of the bottom product, α_{12} (–) is the relative volatility of the components, and $x_{\text{F},1}$ (–) is the molar fraction of component 1 (one) in the feed.

The maximum thermodynamic efficiency ($E_{\max}$) is defined as the minimum work for separation ($W_{\min}$) divided by the minimum energy required for a separation process ($Q_{\min}$). Therefore, in the case of conventional distillation of an ideal system, the following relationship holds true:

$$E_{\max,\text{conv}} = \frac{W_{min}}{Q_{\min,\text{conv}}} = \frac{RT\sum_{i=1}^{n} x_i \ln x_i}{\Delta H_{\text{vap,B}}\left[\frac{1}{\alpha_{12}-1} + x_{\text{F},1}\right]} \qquad (8.4)$$

Accordingly, the maximum thermodynamic efficiency for conventional distillation is very low, only around 10% (Keller and Humphrey, 1997; Olujić *et al.*, 2003). The relative volatility (α_{12}) is in fact the most important parameter, as a relative volatility close to unity leads to a very high reflux rate and consequently to high energy requirements in a conventional distillation column. Therefore, systems with low relative volatility are the best candidates for energy savings in distillation. As a general rule, these close boiling systems are mainly binary or nearly binary mixtures (de Rijke, 2007).

For a binary mixture separated in an ideal HIDiC – operating without reboiler and condenser and with the feed as saturated liquid – the separation can be completely driven by the shaft work of a compressor. The compressor duty (expressed in J s^{-1}) is equal to the energy required by the HIDiC:

$$Q_{\text{HIDiC}} = W_{\text{comp}} = VRT_{\text{in}}\frac{K}{K-1}\left[\left(\frac{P_{\text{out}}}{P_{\text{in}}}\right)^{\frac{K-1}{K}} - 1\right] \qquad (8.5)$$

where V (kmol s^{-1}) is the molar vapor flow-rate leaving the top of the column, $K = C_P/C_V$ (–) is the ratio of the specific heat capacity at constant pressure (C_P) to that at constant volume (C_V), P_{in} (kPa) is the absolute inlet pressure (i.e., pressure of the stripping section), and P_{out} (kPa) is the absolute outlet pressure (i.e., pressure of the rectifying section). Consequently, the thermodynamic efficiency of the HIDiC for the separation of an ideal system (E_{HIDiC}) may be expressed as follows (de Rijke, 2007):

$$E_{HIDiC} = \frac{FRT \sum_{i=1}^{2} x_i \ln x_i}{VRT_{in} \frac{K}{K-1} \left[\left(\frac{P_{out}}{P_{in}} \right)^{\frac{K-1}{K}} - 1 \right]} \tag{8.6}$$

The compressor duty – determining the energy efficiency of HIDiC – depends on two key variables: the pressure ratio (P_{out}/P_{in}) and the vapor flow rate entering the compressor (V) – with this vapor flow rate also being influenced by the partial vaporization of the high pressure liquid that is fed back from the rectifying section. Consequently, to maximize the thermodynamic efficiency of a HIDiC, the compression ratio and the vapor load should be minimized (de Rijke, 2007).

8.4 POTENTIAL ENERGY SAVINGS

To approach the potential energy savings of a HIDiC, one must consider both concepts of the partial and ideal heat-integrated distillation column (Nakaiwa *et al.*, 2003; de Rijke, 2007).

8.4.1 Partial Heat Integrated Distillation Column (p-HIDiC)

In contrast to an ideal HIDiC – which can operate without a reboiler or condenser and where the separation is completely driven by the shaft work of a compressor (Takamatsu, Nakaiwa, and Nakanishi, 1996) – in a partial HIDiC only part of the heat is transferred internally from the rectifying to the stripping section, so both an external reboiler and condenser are still required (Mah, Nicholas, and Wodnik, 1977; Seader, 1980; Seader and Baer, 1984). Mah, Nicholas, and Wodnik (1977) suggested a *secondary reflux and vaporization* (SRV) column that uses vapor recompression between the top vapor and the bottom reboiler to enhance the thermal economy of the partial HIDiC. A primary reflux is obtained in the conventional way by feeding part of the condensed top product back in the column, while the secondary reflux is achieved by internal condensation in the rectifying section due to the heat transfer between the rectifying and stripping sections. As a consequence, the concept of Mah, Nicholas, and Wodnik (1977) requires two compressors. The first is the SRV compressor that compresses the vapor leaving the stripping section into the bottom of the rectifying section, and the second is the MVR compressor that compresses the vapor leaving the top of the column into the bottom reboiler. The binary mixtures investigated by Mah, Nicholas, and Wodnik (1977) were ethylene–ethane, propylene–propane, and *trans*-2-butene and *cis*-2-butene – with the first one emerging as the most promising application (Table 8.1) (de Rijke, 2007). Furthermore, Seader (1980) also reported large energy savings of 56% for the system ethylene–ethane in a partial HIDiC operated without an extra vapor recompression compressor.

8.4.2 Ideal Heat Integrated Distillation Column (i-HIDiC)

An ideal HIDiC can operate even without a reboiler or condenser and the separation is completely driven by the shaft work of a compressor (Takamatsu, Nakaiwa, and Nakanishi, 1996). The system benzene–toluene separated in an ideal HIDiC was reported in several simulation studies (Nakaiwa *et al.*, 1997, 2001, 2003; Liu and Qian 2000) and validated experimentally in a pilot plant (Naito *et al.*, 2000). Energy

Table 8.1 Reported energy savings of a partial or ideal HIDiC for various binary systems

Model system	α_{12}	Pressure ratio ($P_{out}(P_{in})$	Energy req. HIDiC versus conventional	Reference
Ethylene–ethane	1.80	3.0	−54% (p-HIDiC)	Mah *et al.* (1997)
Ethylene–ethane	1.80	3.0	−56% (p-HIDiC)	Seader *et al.* (1980)
trans-2-Butene–*cis* 2-butene	1.08	2.4	−15% (p-HIDiC)	Mah *et al.* (1977)
Benzene–toluene	2.40	2.0	−60% (i-HIDiC)	Nakaiwa *et al.* (1997, 2001)
Benzene–toluene	2.40	2.1	−60% (i-HIDiC)	Jansens *et al.* (2001)
Benzene–toluene	2.40	3.0	−52% (i-HIDiC)	Liu and Qian. (2000)
Ethylbenzene–styrene	1.40	2.3	−80% (i-HIDiC)	Jansens *et al.* (2001)
Propylene–propane	1.15	1.4	−90% (i-HIDiC)	Jansens *et al.* (2001)

savings of up to 60% were reported for this model system – as shown in Table 8.1 (de Rijke, 2007) – though the number of stages was increased by a factor 2–3 as compared to a conventional column.

According to Jansens *et al.* (2001) and Olujić *et al.* (2003), HIDiC is indeed able to operate at a compression ratio lower than that of a heat pump and HIDiC is also especially attractive for close boiling mixtures. Benzene–toluene was taken again as reference system, but in this case with the same number of stages as the conventional column. From that simulation study it was concluded that a HIDiC PP-splitter can save up to 90% energy. Despite the conclusion of Mah, Nicholas, and Wodnik (1977) that SRV was not favorable for the separation of propylene–propane, it is worth noting that the compression ratio was lower (1.4) in the study of Jansens *et al.* (2001) and higher (2.2) in the work of Mah, Nicholas, and Wodnik (1977). From these studies, it can be concluded that HIDiC is especially feasible for separations with low relative volatility and tight product specifications. In the case of a sub-ambient condenser temperature, a heat pump can also eliminate the use of refrigerants (de Rijke, 2007).

8.5 DESIGN AND CONSTRUCTION OPTIONS

Although HIDiC is an attractive energy efficient distillation technology, it poses great difficulties in realizing an effective configuration in practice – hence good equipment design is the key to its industrial implementation (Olujić *et al.*, 2009). This section provides some design guidelines along with the design alternatives proposed so far in the literature (Nakaiwa *et al.*, 2003; de Rijke, 2007).

Several similarities exist between HIDiC and conventional distillation columns, although their process configurations look rather different (Nakaiwa *et al.*, 2003). Notably, in the case of HIDiC there are two extreme operating conditions: the maximum and the minimum reflux ratio operating mode. These extreme operating conditions are very closely related to the pressure difference between the rectifying and the stripping sections ($P_R - P_S$). The maximum reflux ratio operation is where the internal heat integration is so intensive that the operation curve overlaps the 45° diagonal line (thus requiring the minimum number of stages), while the minimum reflux ratio operation corresponds to the situation where the operating curve is tangent to the equilibrium curve (thus requiring an infinite number of stages). Accordingly, one can easily understand why the pressure difference is a key variable for the design of HIDiC (Nakaiwa *et al.*, 2003).

A systematic design procedure for HIDiC was proposed by Gadalla (2009), who developed a design hierarchy that includes two phases of design: thermodynamic and hydraulics. The design procedure is applied using simulation-based design methods. In thermodynamic design, the temperature profiles for column sections are used as a design tool to guide the designers, while the hydraulic capacities of stages are analyzed to determine the maximum physical space area available for heat exchange. Hence, feasibility regions for heat integration and hydraulic design are identified (Gadalla, 2009).

Prior to any design of HIDiC, conventional distillation calculations are performed – while in some cases heat pump data is also available. Based on the available conventional calculations, a design hierarchy is used for guiding the designer to perform a systematic design and to provide valuable insights for optimum results. The design process starts by simulating a conventional column or heat pump configuration for the given design problem. The required (input) data for the simulation design are, typically, the feed flow rate and conditions, feed composition, product and separation requirements, column pressure, and the pressure drop. The simulated design (output) parameters are the number of stages in each column section, reboiler duty, condenser duty, and product compositions. In addition to these parameters, in the case of heat pump assisted distillation, the column configuration is simulated to calculate also the electicity consumption of the compressor and the working pressure ratio. According to the design hierarchy of Gadalla (2009), this leads first to a design of basic HIDiC that has no heat exchange or integration, and then a complete HIDiC is designed based on the basic design. The complete HIDiC setup is simulated by increasing (step-by-step) the level of heat transfer between the individual columns until the reboiler duty is reduced to minimum (p-HIDiC) or zero (i-HIDiC). The design step of a complete HIDiC is carried out through a two-step approach: first according to the thermodynamic capabilities of the design, and then based on the hydraulic capacity of stages. The final design is completed by an optimization procedure and improvements, such as addition of stages, flash drum, and gas turbine for more energy and power efficiency (Gadalla, 2009).

Typically, in the case of internal heat integration, the vapor flow increases in the stripping section due to an internal evaporation in the heat integrated part, while in the corresponding integrated part of the rectifying section the vapor flow decreases due to internal condensation. However, in the non-integrated part on both sides of the columns the vapor and liquid flow rates are almost unchanged. For reasons of energy saving, a low temperature difference between the rectifying and the stripping section is preferred for HIDiC but this leads to a rather large heat transfer area requirement per stage (Olujić *et al.*, 2003). Accordingly, the HIDiC design should allow for large heat transfer area inside the column. Because of the changing vapor flows in the column a very important prerequisite for a heat-integrated column design is that it allows changing area for vapor flow. This could be done by changing the diameter of the stripping section and rectifying section stepwise along the height of the column and/or by a variation of the tray layout (de Rijke, 2007; Gadalla, 2009).

Close boiling separations require large reflux ratios in order to generate enough liquid–vapor contact for the required product purity. In high-pressure applications, the column diameter is relatively small due to the increased density of the vapor. The combination of a large reflux ratio and a small column diameter leads to a large liquid flow rate, relative to the cross sectional area of the distillation column. Consequently, sieve or valve trays are applied where structured packing cannot handle large liquid loads. Until several years ago it was considered that a HIDiC design should allow mainly trays as distillation internals (de Rijke, 2007), but Bruinsma *et al.* (2012) showed also the possibility of efficiently using structured packing. This actually opens the way to novel applications, such as reactive HIDiC – with a study on r-HIDiC already reported by Pulido *et al.* (2011) for the synthesis of *tert*-amyl methyl ether (TAME). The most important constructive solutions proposed so far for HIDiC are described hereafter (Nakaiwa *et al.*, 2003; de Rijke, 2007).

8.5.1 Inter-coupled Distillation Columns

In early designs, two column shells were connected via heat exchangers (Figure 8.4) (Haselden, 1977). This design accounts for the different heat requirement in each stage, but it also implies significant heat loses and additional equipment costs. The performance of this design is heavily linked to the heat transfer means used. The basic concept is based on two separate parallel columns interconnected on every distillation tray by piping. The tubular heat transfer bodies are submerged in the liquid (vapor mixture on the trays. This concept is applicable for tray columns and allows a flexible, but relatively small, heat exchange area per stage. A key drawback of this design is the large amount of piping between the adjacent columns, which leads to extra investment costs and extra heat losses to the environment (de Rijke, 2007). Beggs (2002) introduced a

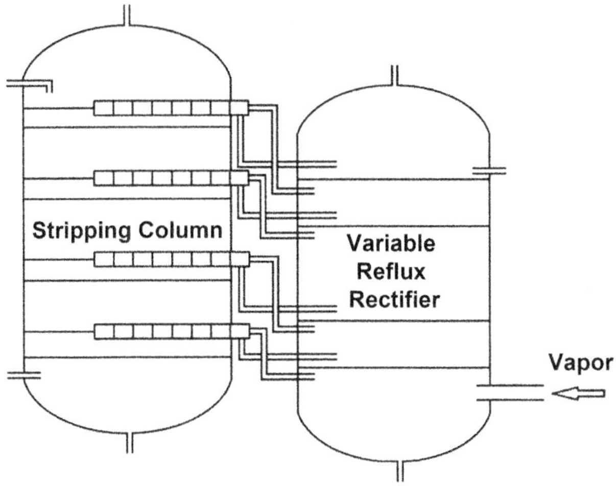

Figure 8.4 Two-column HIDiC design

separate stripping and rectifying section that is not connected at every tray but uses a single heat transfer loop – meaning much less tubing than in the concept of Haselden (1977). An external pump is required for transport of the heat transfer fluid that is heated along the length of the rectifying section and subsequently cooled along the length of the stripping section. Special heat transfer trays with serpentine ducts inside the tray are used, which implies that the total heat transfer area is restricted to the tray surface. Based on experimental measurements, overall heat transfer coefficients in the range 2200–4000 $W\,m^{-2}\,K^{-1}$ were reported for the system water–methanol (Kaeser and Pritchard, 2003).

Several researchers have proposed also the use of side heat exchangers connecting two separate columns (Nakaiwa *et al.*, 2003). These systems imply a trade-off between efficiency and complexity, as the heat transfer between stages is reduced according to specific requirements. This solution can be implemented with standard distillation hardware in a revamping operation. However, a detailed economical analysis is needed to account for the additional equipment costs and heat loses.

8.5.2 Distillation Column with Partition Wall

A patent was filed by Seader *et al.* (1980) on a column with two semi-cylindrical sections, where the heat transfer is realized by pipes mounted through the wall and trays of the stripping section. Figure 8.5a illustrates

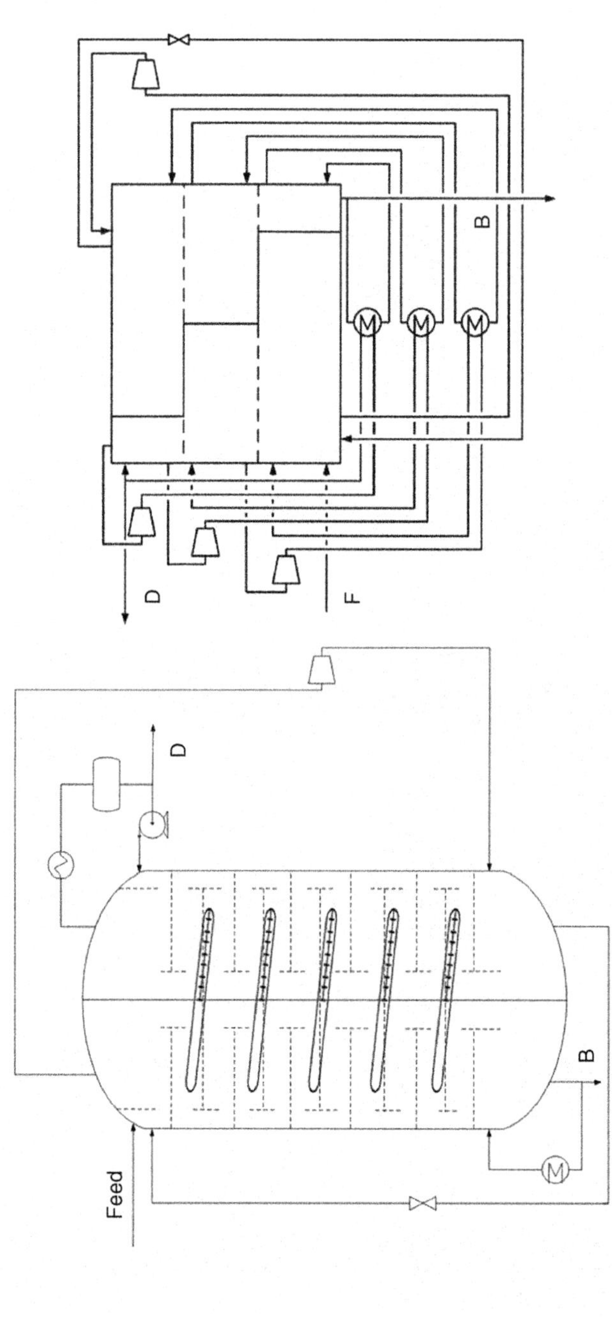

Figure 8.5 (a) Single shell HIDiC—detail of opposite downcomers design (b) Honeywell single shell design with heat transfer between opposite stages

the proposal of a single column shell with a dividing wall (Seader, 1980). This design allows heat transfer through the wall by the liquid on both sides of the downcomers, and it reduces the heat loses and the requirement for heat transfer means. Effective heat transfer over small temperature differences is possible due to the high internal heat transfer coefficients and negligible pressure drop along the heat pipes (Beggs, 2002). A special heat pipe fluid is needed, which means the introduction of extra components in the process. The concept is applicable for tray columns and allows flexible heat transfer area per stage. Seader and Baer (1984) carried out heat transfer measurements at bench-scale in a column equipped with heat pipes – using finned copper tubes with water as working fluid. Pure water was distilled in both the low- and high-pressure distillation columns. Heat transfer coefficients of 4700 $W\,m^{-2}\,K^{-1}$ were reported, based on the bare tube area at temperature driving forces of 12.4–22.7 K. On the practical side, note that heat transfer coefficients for hydrocarbon mixtures are considerably lower than the values reported for pure water. Moreover, an interesting aspect of intra-column heat transfer is the effect of heat transfer on the mass transfer efficiency of the distillation tray, which was not taken into account by Seader *et al.*

Recently, Honeywell proposed a similar column without heat transfer means (Figure 8.5b) (Sechrist, 2009). This design accounts for the different vapor loads in each stage by an off-center wall. Moreover, heat is transferred between opposite stages in side-heat exchangers. This solution lowers the pressure step required, but it adds additional equipment costs. The heat transfer means an increase in the pressure drop, and liquid bypassing can be an issue for the heat transfer across the wall. Note that the use of packing is discouraged in these proposed designs (Sechrist, 2009).

8.5.3 Concentric Distillation Column

A concentric HIDiC, in which the annular stripping section is configured around the rectifying section, was proposed by Govind (1986, 1987) and Glenchur and Govind (1987). Basically, the use of concentric columns reduces the heat loses and slightly enhances the heat transfer performance (Figure 8.6) (Govind, 1986). In principle, this concentric configuration has ideal properties with respect to the heat transfer because the heat in the rectifying section cannot leak to the environment – in other words, the heat has no other way to go but to the stripping section. This concept was proposed for tray columns, with the heat transfer area restricted to the

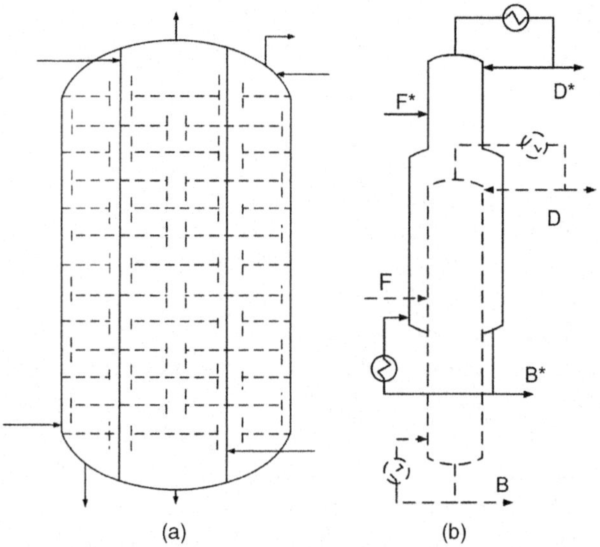

Figure 8.6 Concentric HIDiC design (a) and multistage distillation arrangement (b) in concentric columns

area of the column wall. Consequently, the heat transfer area is relatively small – insufficient in most cases – and the design does not allow any flexibility for changing the heat transfer area per stage (de Rijke, 2007).

Notably, this solution incorporates the downcomer scheme with two-pass trays in the inner column and single pass in the outer tower. Moreover, it accounts for the differences in size between the rectifying and stripping sections and it can even be used in multistage distillation (Figure 8.6b) (Govind, 1986). The overall heat transfer coefficient for the concentric tubes varies with the vapor load, in the range 600–800 $W\,m^{-2}\,K^{-1}$, and the height of an equivalent theoretical plate (HETP) is rather high, up to 4 m when using packing (Noda *et al.*, 2000). Nonetheless, the concentric tubes design still lacks the required flexibility for efficient heat transfer performance. Therefore, the use of modular heat transfer means is recommended to increase the heat transfer area, and also variable tube diameter can be used to accommodate better the changing vapor loads (de Rijke, 2007).

8.5.4 Concentric Column with Heat Panels

A concentric column equipped with heat panels – described in a subsequent section – was patented by de Graauw *et al.* (2003). In this case, the distillation internals can be trays or random/structured packing. Heat transfer bodies can be placed in either the stripping or rectifying section. In the bottom and upper part of the HIDiC, heat transfer bodies are preferentially placed in the rectifying and stripping section, respectively (de Rijke, 2007). In contrast to the heat pipes in the design of Seader *et al.* (1980), the panels in this concept are in open connection with the other section of the distillation column. Hence panels placed in the stripping section are in open connection with the rectifying section such that vapor from the rectifying section can enter the panels, condense inside, and liquid can flow back to the rectifying section. On the outer surface of the panels present in the stripping section, simultaneous evaporation of liquid takes place. Obviously, this implies that the panels should be well wetted. When the heat transfer bodies are positioned in the rectifying section, the liquid from the stripping section enters the heat transfer element via the open connection, then it evaporates partially inside and the vapor flows back to the stripping section. On the outer surface, vapor present in the rectifying section will condense simultaneously. Note that the heat transfer panels should be wetted by froth and entrainment from a tray (de Rijke, 2007). In this concept, heat transfer internals can be placed in the downcomer or in the active area between two distillation trays, which results in a large and flexible heat transfer area. A strong advantage of this concept is that it allows the designer to exclude certain trays with low or negative temperature driving forces from the heat transfer process – in other words, no heat panels are installed on these trays.

The proposed HIDiC configurations are illustrated schematically in Figure 8.7a, while Figure 8.7b shows the placement of the heat transfer panels in the rectifying or stripping section for a concentric column (de Rijke, 2007). It is certainly worth noting the two main directions in the design of HIDiC: heat exchanging distillation columns and fractionating heat exchangers.

8.5.5 Shell & Tube Heat-Exchanger Column

The problem of limited heat transfer area – as encountered in the concentric column design – can be conveniently solved by adopting the concept of multiple tubes in a shell, as proposed by Aso *et al.* (1998). In this way, a fractionating heat exchanger is actually constructed rather than a heat-exchanging distillation column. This solution takes advantage of all the constructive know-how developed in the heat exchange industry. The design exhibits very good heat transfer properties

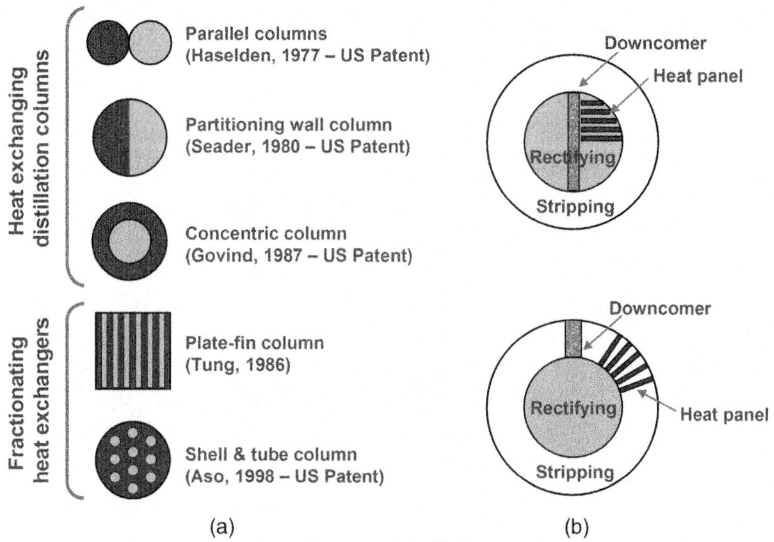

Figure 8.7 (a) Proposed configurations for a HIDiC (b) placement of the heat transfer panels in the rectifying or stripping section for a concentric column

and – similar to a concentric column – the heat can only be transferred to the stripping section that is configured around the tubes. Note that the shell and tube structure can enhance the heat transfer without using panels (Figure 8.8) (Aso *et al.*, 1998). In addition, this design allows a large and flexible heat transfer area but the excessive wall area given by the amount of tubes can be an issue when using packing as mass transfer means. For this configuration, the overall heat transfer coefficient exceeds $600\,W\,m^{-2}\,K^{-1}$ (Noda *et al.*, 2000). The design gives flexibility for area and hydraulics requirements, but it requires specialized maintenance procedures. A key drawback of the shell & tube HIDiC is that it is not suitable for trays in both the rectifying and stripping sections – due to the small diameter of the tubes and very complex tray design, respectively (de Rijke, 2007).

8.5.6 Plate-Fin Heat-Exchanger Column

The construction of a HIDiC from a plate-fin heat exchanger was suggested by Tung, Davis, and Mah, (1986), while the first patent applying a plate-fin design to HIDiC was filed later, in 2003. In a plate-fin device, alternate and adjacent vertical channels serve as strippers and rectifiers. Stripping and rectifying sections can be housed in

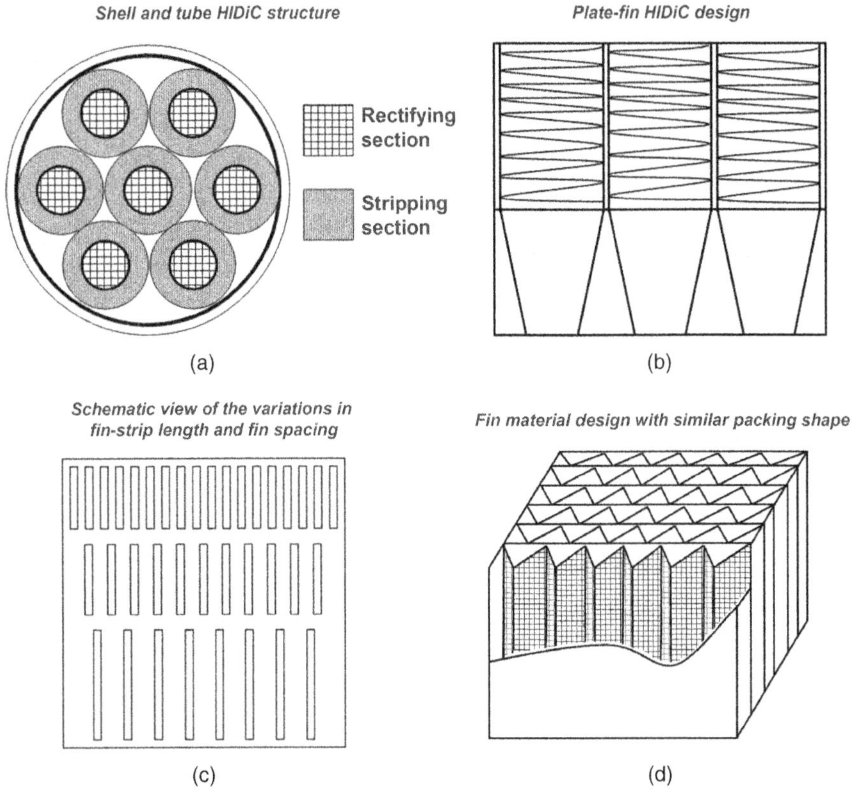

Figure 8.8 (a) Shell & tube HIDiC (b) plate-fin HIDiC and (c) variations in the fin length and spacing (d) fin design with similar packing shape

alternating fins, but liquid and vapor needs to be redistributed at the edges. Nevertheless, alternating plates can be used instead of fins to solve this problem. The variation in hydraulic loads can be accounted for with non-parallel plates or with variable fin lengths and spacing as in Figure 8.8b,c (Hugill, 2003). The former has less flexibility towards flooding criteria, but it is more robust than the latter. Moreover, the plate fin design might incorporate the mass transfer means. Figure 8.8d shows fin material shaped similarly to regular packing (Aitken, 1998; Sunder, Houghton, and Kuznetsov, 2004). In addition, regular fins might have a rough surface to enhance mass transfer. Consequently, every plate fin design can be adapted specifically for each application, making it comparatively expensive. Note that in this design the heat transfer area is large and flexible. By changing the channel and fin dimension, the heat transfer area can be changed along the length of the column. Difficulties arise in the (re)distribution of liquid and vapor over the set of parallel columns.

A model was developed (Tung, Davis and Mah, 1986) to predict the heat transfer rates and influence of internal heat transfer on the separation efficiency of a plate-fin device, and the model is in agreement with experimental data. Aitken (1998) filed a patent on a plate-heat exchanger for combined heat and mass transfer in which the space between the vertical plates is equipped with corrugated sheets similar to that of a structured packing. An alternative idea, which deviates more from the current plate fin heat exchanger designs, is to use non-parallel plates with a constant fin spacing (Hugill, 2003).

8.5.7 Heat Transfer Means

The simplest heat transfer means is to connect only physically the column sections. Thus, sealed metal tubes with fins can be used (Figure 8.9a) (Seader, 1980) – hence this system might be arranged for minimal pressure drop. The means were initially proposed for two-column designs, but they are used also for concentric configurations. These designs feature the lowest heat transfer performance, but they are simpler and less expensive than other alternatives. Moreover, a fluid media can be used to enhance the heat transfer efficiency – usually the same distilled substances are driven through the column for heat transfer purposes. There have been proposals for a closed heat transfer cycle in HIDiC (Haselden, 1977), but this can only increase the complexity and investment costs. The heat and mass transfer means might also be integrated in a single device to ensure the enhancement of heat exchange without significant effects on the mass transfer performance. However, there is still a lack of studies for these structures, with notable exceptions (Bruinsma *et al.*, 2012). Figure 8.9b illustrates the proposal of a combined heat and mass transfer tray (Legatski, 1954) that features horizontal pipes with fins, resembling a sieve tray – this design also includes a small column jacket. Figure 8.9c shows another proposal that has tubes on top of a typical sieve tray (Stage, 1974), with the tray being designed to immerse the heat transfer pipes in liquid. The pressure drop across the tray is reduced by having vapor holes. Overall heat transfer coefficients in the range 2200–4400 W $m^{-2} K^{-1}$ have been reported for these tray-shaped systems. Note that the trials were conducted for methanol–water separation, instead of hydrocarbon separations as in earlier reports.

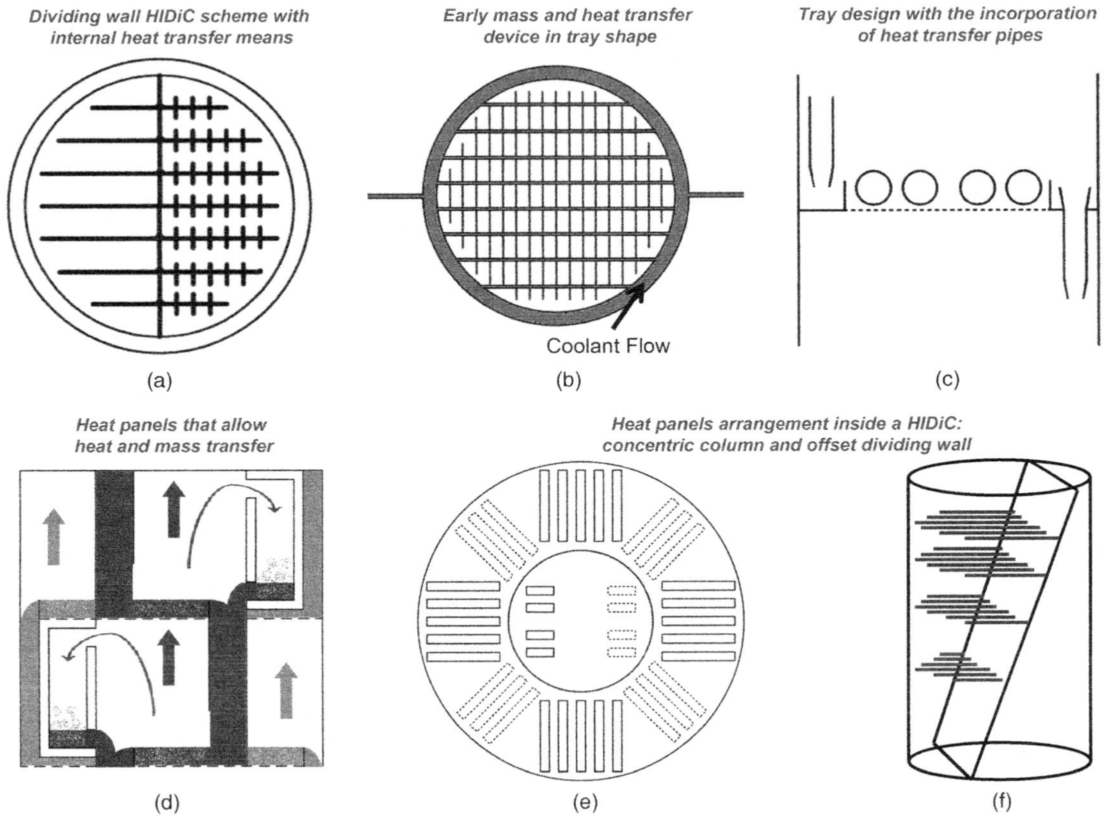

Figure 8.9 (a) Dividing-wall design with internal heat transfer means (b) early combined heat and mass transfer tray (c) tray design with heat transfer pipes (d)–(f) various heat panels arrangement inside a HIDiC

Several more complex structures were proposed to enhance the heat transfer inside a HIDiC – these designs being mostly intended for heat transfer. Figure 8.9d–f presents the column design configurations for these heat transfer means, while Figure 8.10 compiles the various heat transfer panels proposed (de Graauw *et al.*, 2003, 2008). Most designs have a minimized footprint in the flow direction to reduce the pressure drop, and feature extended bodies on the other section of the column – since enlarged bodies ensure reasonable heat transfer performance. The heat transfer area might be increased, but a lower pressure drop is also preferred for economic reasons. Figure 8.10 shows several types of arrangement for these panels inside the integrated columns (de Graauw *et al.*, 2003, 2008). The location of the panels follows the requirements for hydraulic and heat transfer area, and the installation of panels is avoided in stages with zero or negative temperature differences. Moreover, the number of panels can be increased in stages with higher driving force. Note that a single panel can transfer up to $1000\,\mathrm{W\,m^{-2}\,K^{-1}}$ (de Graauw *et al.*, 2008), and moreover tailor-made designs can be

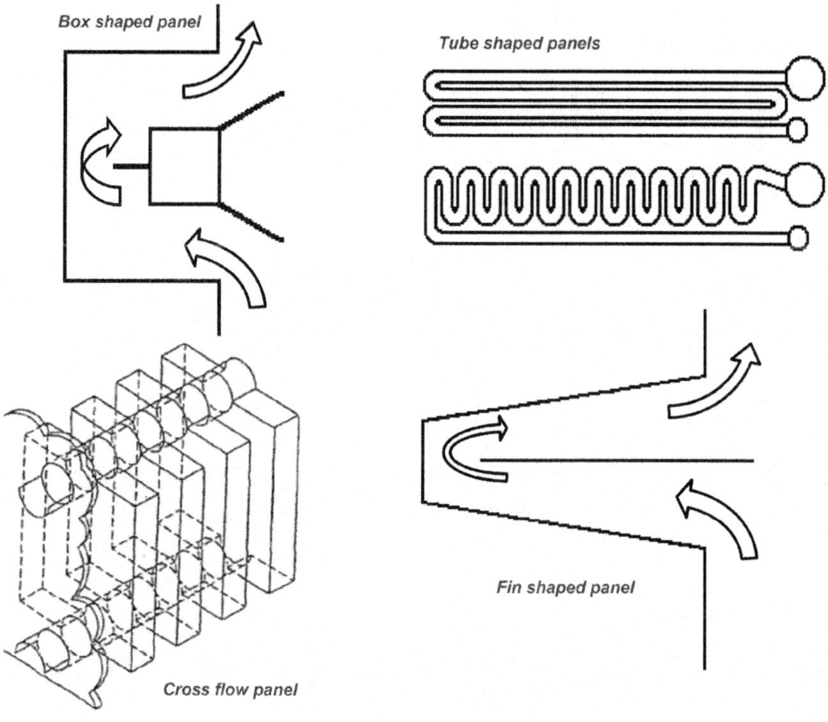

Figure 8.10 Different heat panel designs used in a HIDiC

realized that combine the required amount of panels. However, the final panel design should be carefully considered – although a low performance panel might be less expensive, the HIDiC requires a high number of panels, which greatly affects the column operation.

8.6 MODELING AND SIMULATION

The first model for HIDiC was described in the paper of Mah, Nicholas, and Wodnik (1977), and summarized by Nakaiwa (1988) with examples of binary separation problems. As the model is actually based on the tri-diagonal matrix method proposed by Wang and Henke (1966), the differences between the original one and the model for the HIDiC are briefly explained hereafter. To represent the amount of the heat exchanged between the rectifying and the stripping sections, the overall heat transfer coefficient (U) and area (A) must be introduced. Assuming a tray column for the HIDiC setup, the amount of the heat exchanged between the coupled stages (Q_{ex}) is expressed as follows:

$$Q_{\mathrm{ex},j} = U_j A_j \left(T_j - T_{j+N_{\mathrm{hex}}}\right) \tag{8.7}$$

where N_{hex} is the number of heat exchange stages and the subscript j denotes the j-th stage in the rectifying section. Treating $Q_{\mathrm{ex},j}$ as heat duty for a side cooler or reboiler, the tri-diagonal matrix for the mass and the energy balances is solved. If one takes $A = 0$ then $Q_{\mathrm{ex},j}$ also becomes zero and the terms for the heat exchange in the tri-diagonal matrix disappear. Thus, Nakaiwa's model for the HIDiC can also handle conventional columns without any corrections (Iwakabe *et al.*, 2006).

Various kinds of process models with different complexities have been developed so far for the HIDiC configurations. Although they are used for different purposes, they can reflect the general behavior of HIDiC. A very simple model is given below, where the principle of internal heat integration can be very well illustrated (Nakaiwa *et al.*, 2003):

$$V_{j+1}y_{j+1} - V_j y_j + L_{j-1}x_{j-1} - L_j x_j + \delta(j = f)Fz_{\mathrm{F}} = 0, \quad \text{for } j = 1 \ldots n \tag{8.8}$$

$$y_j = y_j(P_j, T_j, x_j), \quad \text{for } j = 1 \ldots n \tag{8.9}$$

$$L_1 = Q_1/\lambda_1 \text{ and } L_j = L_{j-1} + \frac{Q_j}{\lambda_j} + \Psi(j \geq f)Fq, \quad \text{for } j = 2 \ldots n \tag{8.10}$$

$$L_1 = Q_1/\lambda_1 \text{ and } L_j = L_{j-1} + \frac{Q_j}{\lambda_j} + \Psi(j \geq f)Fq, \quad \text{for } j = 2 \ldots n \quad (8.11)$$

$$V_n = -Q_n/\lambda_n \text{ and } V_{n-j} = V_{n-j+1} - \frac{Q_j}{\lambda_j} + \Psi(f \geq j)F(1-q), \text{ for } j = 1 \ldots n-1 \quad (8.12)$$

where j is the stage number (from 1 to n), L and V are the liquid and vapor flow-rate (mol h^{-1}), respectively, x_j and y_j are the liquid and vapor compositions, z_F is the mole fraction in the feed, q is the feed condition, T_j the temperature, P_j the pressure, Q the heat load, U the heat transfer coefficient, A the heat exchange area, δ is the Kronecker function, Ψ is the unit step function, and λ is the latent heat (kJ kmol^{-1}).

The T–H diagram of a distillation column – developed based on thermodynamic equilibrium operation – can indicate clearly the heating and cooling sections for a distillation column and thus guide the design of internal heat integration. For a binary distillation column, the rectifying part is a cooling section that must release a certain amount of heat to the environment, while its stripping part is a heating section that has to take in a certain amount of heat from the environment. These two coherent properties provide fundamentals for the consideration of internal heat integration within a conventional distillation column (Nakaiwa *et al.*, 2003). Therefore, it is reasonable in process development to reuse the heat available from the rectifying section to heat up the stripping section, while approaching more closely a reversible distillation operation. To achieve the internal heat integration between the rectifying section and the stripping section, one has to maintain sufficient temperature driving forces and this can be realized through heat pumps. With the assumption of numerous reversible heat pumps between corresponding stages of the rectifying and the stripping sections, the following equations can be obtained (Gokcen and Reddy, 1996):

$$\mathrm{d}Q_\mathrm{R} = \frac{T_\mathrm{R}}{T_\mathrm{S}\mathrm{d}Q_\mathrm{S}} \quad (8.13)$$

$$\mathrm{d}W = \frac{(T_\mathrm{S} - T_\mathrm{R})}{T_\mathrm{S}\mathrm{d}Q_\mathrm{S}} \quad (8.14)$$

The integration of these two equations leads to (Nakaiwa *et al.*, 2003; Matsuda, Iwakabe, and Nakaiwa, 2012):

$$\int_{T_C}^{T_F} \frac{dQ_R}{T_R} = \int_{T_F}^{T_B} \frac{dQ_S}{T_S} \tag{8.15}$$

$$W = \int_{T_F}^{T_B} \frac{T_S - T_R}{T_S} dQ_S = Q_S - Q_R \tag{8.16}$$

Clearly, under appropriate operating conditions, the distillation processes can become self-contained in heat utilization and the separation can be driven only by shaft work – as neither the condenser nor the reboiler appear to be necessary. Furthermore, the pressure elevation from the stripping section to the rectifying section and feed thermal condition must be determined carefully so as to satisfy the equation: $W = Q_S - Q_R$. This provides the fundamentals and guidelines for the development of a new model of internally heat-integrated distillation columns (HIDiC). Moreover, the internal heat integration causes the operation curve of the HIDiC to be exactly the same as the equilibrium one. Remarkably, HIDiC presents higher energy efficiency in process operation, as it needs less driving force than conventional distillation columns at the minimum reflux ratio operating condition, and considering the same mass transfer duty (Nakaiwa *et al.*, 2003).

In addition to being used in process design and operation studies, one potential application of the process models is to assist the process configuration development for the HIDiC. Model-based development of process configurations is thus highly desired, considering that the experimental-based development of process configurations is very expensive and time-consuming. This requires process models to be of great reliability and accuracy, and in general model validation through experimentation must be carried out to achieve these requirements (Nakaiwa *et al.*, 2003).

8.7 PROCESS DYNAMICS, CONTROL, AND OPERATION

One concern for applications of internal heat integration is the complicated process dynamics that may be introduced. When applying internal heat integration to chemical processes, it is better not to vary the original process dynamics drastically. In particular, it is important that the resultant process should be maintained as open-loop stable as possible – otherwise great efforts must be spent on process control of the resultant non-minimum phase system. Even so, process performances can degrade considerably (Nakaiwa *et al.*, 2003). For the HIDiC, several authors have already demonstrated by numerical evaluations and mathematical proof that the process is still open-loop stable despite the introduction of internal heat integration between the rectifying and stripping sections (Matsuda *et al.*, 2011; Matsuda, Iwakabe, and Nakaiwa, 2012). However, further adding the heat integration between the overhead product and the feed flows will turn the HIDiC into an open-loop integrated process that is no longer open-loop stable (Nakaiwa *et al.*, 2003). This can certainly cause extra difficulties in the process design and operation, while leading inevitably to deterioration of the control system performance. Therefore, a judgment must be made in the process development on whether to achieve the economics of internal heat integration at the expense of control system performances or vice versa.

Owing to the structure with no reboiler and no condenser, it is actually impossible to start up a HIDiC by itself. For this reason, the operation must be carried out by means of an external trim-condenser and an external trim-reboiler. A very important step towards a smooth process startup is to avoid inverse heat transfer from the stripping section to the rectifying section – otherwise there is a risk of operation problems, in addition to the extra consumption of valuable energy (Nakaiwa *et al.*, 2003). Moreover, due to the high degree of integration, it is necessary to synthesize special control systems for HIDiC. In the previous equation ($W = Q_S - Q_R$), the left-hand side is the energy of the compressor, which is closely related to the pressure difference between the rectifying and stripping sections ($P_R - P_S$), while the right-hand side is the imbalance of heat loads between the rectifying and the stripping sections. The thermal condition of the feed (q) is the dominant variable to change this heat imbalance. Clearly, the pressure difference ($P_R - P_S$) and the feed thermal condition (q) can be used as manipulated variables for the process operation. Figure 8.11a shows a schematic diagram for a control configuration of a HIDiC (Nakaiwa *et al.*, 2003). Both the overhead and bottom products can be maintained accurately at their desired steady-state values using this control structure. Moreover, the feasibility of this control structure and other alternatives was confirmed by extensive simulation studies (Huang *et al.*, 1996a,b; Huang *et al.*, 2007; Nakaiwa *et al.*, 2003; Matsuda, Iwakabe, and Nakaiwa, 2012).

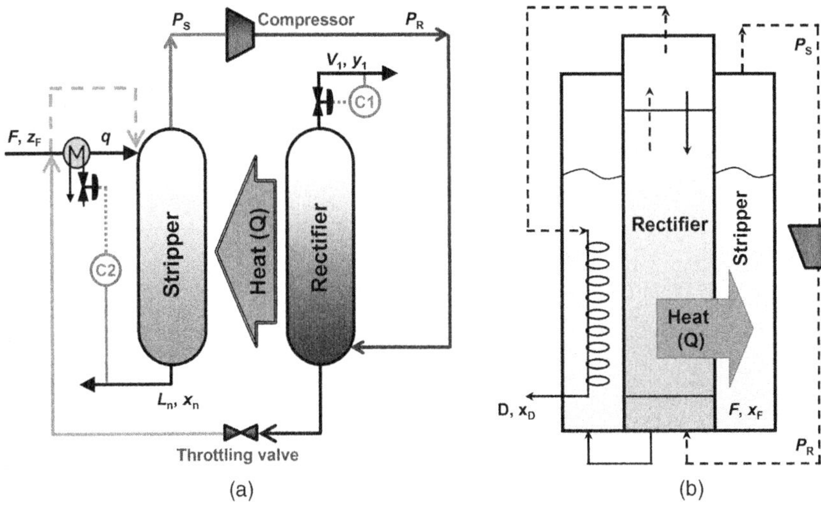

Figure 8.11 Working control strategy for HIDiC (a), and schematics of internally heat-integrated batch distillation column (b)

In a study focusing on the dynamics and control of HIDiC, Fukushima, Kano, and Hasebe (2006) showed that although HIDiC has a more complex structure and slower dynamics than a conventional column, the control performance of HIDiC is comparable to that of a conventional column as long as a suitable control system is designed. The distillate flow rate (D) is superior to the reflux flow rate (L) as a manipulated variable, because the dynamics of HIDiC becomes considerably faster by using D rather than L. To achieve high energy efficiency and controllability at the same time it was recommended to adopt the structure in which the feed is preheated by the distillate vapor product, while using only a condenser and no reboiler at all. The control performance of HIDiC becomes better as the heat transfer area decreases, and it becomes only marginally better as the pressure difference between the stripping section and the rectifying section decreases (Fukushima, Kano, and Hasebe, 2006).

The more recent predictions of the dynamic models developed by Process Systems Enterprise (2011) are in excellent agreement with the published measurements of Naito *et al.* (2000,) who reported a detailed experimental study of a bench-scale HIDiC operation. More remarkably, this match of results was achieved using purely predictive models without any parameter fitting. The start-up dynamic simulation quantified the pressure and heat exchange during the transition from cold and empty to full operation – this verified that the equipment could be started up as planned. Further runs could then be performed to minimize the start-up time to full production. It was also observed that the reboiler and condenser duties are substantially less than for the conventional column, while the additional compression energy requirement is relatively small (Process Systems Enterprise, 2011).

8.8 APPLICATIONS OF HIDIC

Despite the obvious potential, the very promising concept of HIDiC has not yet developed into successful full industrial-scale applications. The applications reported so far indicate that a properly designed HIDiC system can save up to 60% of the energy required for separation and, in addition, the condenser duty and its size can be also reduced by up to 60% (Nakaiwa *et al.*, 2003). As already described, the most suitable applications are those in which the difference in boiling of the two components to be separated is less than 25 °C. Nonetheless, HIDiC has also been proposed for the separation of binary mixtures with larger differences in boiling points. Table 8.2 provides a list of more recent case studies (pilot plant and industrial implementations) of HIDiC technology reported in the literature (Kiss, Flores Landaeta, and Infante Ferreira, 2012). Remarkable energy savings of over 80% were possible for various binary systems such as propane–propylene, methanol–water, or cyclohexane–(*n*)heptane.

An important part of the HIDiC project carried out in Japan is the cryogenic air separation in a HIDiC equipped with a plate-fin type flow path, with a low HETP (0.2–0.4 m) that is almost equivalent to that of regular packing. Experimental validation confirmed that the cryogenic air separation process can be operated at a lower pressure than that of the conventional process and, hence, a reduction in compressor duty is expected (Nakaiwa, 1988; Nakaiwa *et al.*, 2003; Matsuda, Iwakabe, and Nakaiwa, 2012).

Another HIDiC project was conducted in Europe, at Delft University of Technology and ECN (NL) in collaboration with other industrial partners (e.g., AkzoNobel, DSM, Shell, Sulzer Chemtech). An experimental setup was used to test the hydraulics and other characteristics of the HIDiC. They showed that significant energy savings of 90% can be achieved in a PP-splitter, and a conceptual design procedure of the HIDiC utilizing pinch analysis was proposed (Olujić *et al.*, 2003). Moreover, the reduction of CO_2 emission by using a HIDiC with no reboiler was estimated to be 83% as compared to a conventional distillation, and

Table 8.2 Reported case studies (pilot plant and industrial implementations) of HIDiC technology

Separation task	Performance information	References
Propane–propylene separation. Conventional column: $P = 18$ bar MVR: $P = 10$ bar HIDiC: $P = 18/13$ or 18 (15 bar	MVR industrial implementation gives COP ≈ 7.4 and 37% TAC savings against conventional column ($\Delta T_{lift} \approx 24.9$ K). HIDiC case study yields COP ≈ 10 and 25% TAC savings against MVR with $\Delta T_b \approx 10.9$ K	Olujić *et al.* (2003, 2006)
1-Methanol–water separation ($P = 2.6/1.3$ bar for HIDiC) 2-Ethylbenzene–styrene separation ($P = 0.24$ bar for MVR)	1. Top HIDiC design yields 84% energy savings, but only 9% lower TAC. 2. Bottom HIDiC design better. MVR pressure ratio: 2.8, while HIDiC 2.2. MVR gets 49% energy savings, while HIDiC gets 53%. However, MVR savings in TAC are 35%, while HIDiC are 28%. All compared to conventional distillation	Gadalla *et al.* (2007a,b); Gadalla (2009)
Acetic acid dehydration Vacuum operating pressure.	82% energy savings with MVR and 23% with HIDiC as compared to multi-effect distillation	Campbell *et al.* (2008)
Fractionation of C_4–C_6 mixture: Top: *n*-pentane; bottom: *i*-pentane $P = 190/101$ bar.	HIDiC achieves 26% energy savings as compared with conventional existing column, while requiring ~22% more energy than minimum reflux distillation	Huang *et al.* (2008a)
Cyclohexane–*n*-heptane equimolar feed, 95% product purity ($P = 1.1$ bar, $\Delta T_b \approx 17$ °C) Pressure ratios from 1.4 to 1.7	Primary energy savings range from 81% at 1.7 pressure ratio to 69.7% at 1.4, compared to conventional distillation process. Research also included performance from proof-of-principle experimental study	Bruinsma *et al.* (2012)

36% as compared to a heat pump (Gadalla *et al.*, 2006). Recently, Bruinsma *et al.* (2012) provided the experimental evidence needed to properly asses the techno-economic feasibility of HIDiC using structured internals to enhance both the heat and mass transfer. The plate-packing configuration using structured packing exhibited superior performance to that of a HIDIC based on the plate-fin heat exchanger. The mass and heat transfer efficiency increase pronouncedly with increasing throughput, but at the penalty of a higher pressure drop per stage. Simulation of an industrial scale plate-packing unit revealed that an even better performance can be obtained by increasing the volumetric thermal load via further optimization of internals (Bruinsma *et al.*, 2012). These results pave the way to novel applications of HIDiC (e.g., reactive distillation systems) and an easier industrial acceptance of this technology.

The principle of HIDiC can also be applied quite analogously to *batch distillation* columns, where internal heat integration between the rectifying section and reboiler can be devised – as shown by the schematics in Figure 8.11b (Nakaiwa *et al.*, 2003). Note that it was also demonstrated that the energy efficiency could be improved sharply as compared with its conventional counterparts. This could be attributed mainly to a reflux ratio that is much lower than the minimum one and consequently, a higher distillate rate of the overhead product (Nakaiwa *et al.*, 2003).

Separation of binary pressure-sensitive azeotropic mixtures through *pressure-swing distillation* (PSD) processes can also be facilitated with internal heat integration. A PSD process is one of the simplest and yet most economical techniques for separating binary azeotropes, providing that the azeotropic composition is sensitive enough to changes in the operating pressure. A PSD process consists of two columns, one operated at a relatively low pressure and temperature and the other at a high pressure and temperature. By this pressure elevation, the constraint imposed by the azeotropic point can be avoided and pure components can be produced from the high-pressure (HP) and low-pressure (LP) distillation columns, respectively. Figure 8.12 illustrates the working principle of PSD for the mixture water–tetrahydrofuran, which forms a pressure-dependent minimum boiling azeotrope. In the LP column the initial mixture (50% water, 50% THF) is separated into water and the azeotrope (~19% water, at 1 bar). By increasing the pressure in the HP column, the azeotropic composition changes (~38% water at 10 bar) and, hence, pure THF can be obtained in the bottom. The existence of HP and LP distillation columns also provides chances for considering internal heat integration within PSD processes. In such a heat integrated system, the rectifying section of the HP distillation column provides heat to the stripping section of the LP distillation column – and an application of HIDiC to the water–acetonitrile (ACN) system has already been reported (Nakaiwa *et al.*, 2001, 2003). However, note that the benefit of internal heat integration is closely related to the mixture to be processed. Therefore, as internal heat integration is generally more effective for close-boiling mixture separations, it is expected that internal heat integration within PSD processes will be applied most appropriately to such azeotropic mixture separations (Nakaiwa *et al.*, 2003).

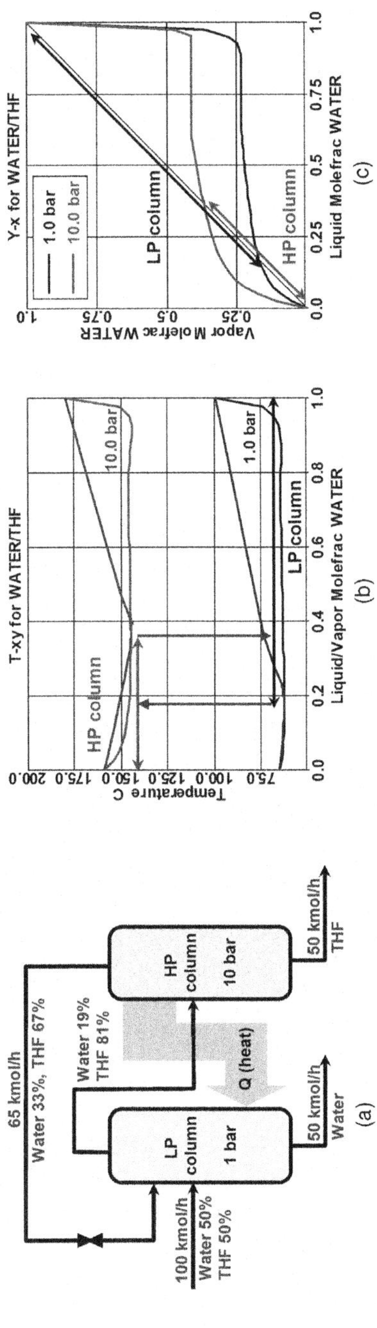

Figure 8.12 (a) Flow sheet of a pressure-swing distillation process for water–THF separation, (b) T–x,y diagram, and (c) y–x diagram for the mixture water–THF at pressures of 1 and 10 bar

The internal heat integration principle can also be applied to *different distillation columns* that may have no connections at all (Nakaiwa *et al.*, 2003). Providing that it is indeed feasible to consider internal heat integration between the two different distillation columns concerned, this can generally lead to more benefits in either energy utilization or capital investment than that based on condenser–reboiler-type heat integration configurations. Upon examination of a distillation train system (e.g., direct or indirect sequences) one can find many opportunities to consider this type of internal heat integration, which is even believed to hold higher potential for applications than the original HIDiC (Nakaiwa *et al.*, 2003).

8.9 CONCLUDING REMARKS

Heat-integrated distillation columns (HIDiC) claim among the highest energy savings in distillation. However, as with most heat pump assisted distillation processes, the capital investment costs are higher than those of conventional distillation – mainly due to the use of an additional compressor and a more complex configuration required for the enhanced heat transfer (de Rijke, 2007). In contrast to other heat pump assisted distillation columns, the HIDiC process involves internal heat integration between the whole rectifying and stripping sections and thus it has higher potential for energy savings. Both thermodynamics and bench-scale experimental evaluations proved that HIDiC holds much higher energy efficiency potential than conventional distillation columns for close-boiling binary mixture separations. Several recent simulation studies and experimental evaluations have further confirmed that the process can be operated very smoothly, with no special difficulties (Nakaiwa *et al.*, 2003; Fukushima, Kano, and Hasebe, 2006; de Rijke, 2007; Matsuda *et al.*, 2011; Matsuda, Iwakabe, and Nakaiwa, 2012; Process Systems Enterprise, 2011; Bruinsma *et al.*, 2012; Matsuda, Iwakabe, and Nakaiwa, 2012).

Several important issues must be considered during the HIDiC design, such as, for example, flexibility to changes in the operating conditions, influence of an impurity or a third component, and process dynamics and operation. These impose strict constraints on the energy efficiency that can be potentially achieved by a HIDiC system. Therefore, a trade-off between process design economics and process operation appears to be very important and it has to be carried out with great care (Nakaiwa *et al.*, 2003; de Rijke, 2007; Matsuda, Iwakabe, and Nakaiwa, 2012).

Notably, among the design and operating variables the compression ratio and the heat transfer coefficient and area (U and A) are the ones that significantly affect most the performance of a HIDiC. Consequently, an important research task during the recent years was the development of more appropriate configurations, including, for example, multi-tube and multi-shell, plate-type, or even structured-packing HIDiC (Nakaiwa *et al.*, 2003; de Rijke, 2007; Bruinsma *et al.*, 2012). Moreover, research and development on the HIDiC configurations has so far primarily been confined to the separations of binary mixtures. Considering that multicomponent mixture separations represent a major application of distillation columns, the development of a corresponding HIDiC technology (Iwakabe *et al.*, 2006; Kiran, Jana, and Samanta, 2012; Matsuda, Iwakabe, and Nakaiwa, 2012) is an important and extremely challenging topic for future work – especially if HIDiC becomes associated with other technologies, such as dividing-wall column or reactive distillation (Yildirim, Kiss, and Kenig, 2011; Shah *et al.*, 2012).

REFERENCES

Aitken, W. (1998) Apparatus for combined heat and mass transfer, US Patent No. 5,722,258.

Aso, K., Matsuo, H., Noda, H. *et al.* (1998) Heat integrated distillation column, US Patent No. 5,783,047 A1.

Beggs, S. (2002) MEng Project Report, University of Edinburgh, UK.

Bruinsma, O.S.L., Krikken, T., Cot, J. *et al.* (2012) The structured heat integrated distillation column. *Chemical Engineering Research & Design*, *90*, 458–470.

Cabrera-Ruiz, J., Jimenez-Gutierrez, A., and Segovia-Hernandez, J.G. (2011) Assessment of the implementation of heat-integrated distillation columns for the separation of ternary mixtures. *Industrial & Engineering Chemistry Research*, *50*, 2176–2181.

Campbell, J.C., Wigala, K.R., van Brunt, V., and Kline, R.S. (2008) Comparison of energy usage for the vacuum separation of acetic acid(acetic anhydride using an internally heat integrated distillation column (HIDiC). *Separation Science and Technology*, *43* (9), 2269–2297.

Fitzmorris, R.E. and Mah, R.S.H. (1980) Improving distillation column design using thermodynamic availability analysis. *AIChE Journal*, *26*, 265–273.

Flores Landaeta, S.J. and Kiss, A.A. (2012) Selection of heat pump technologies for energy efficient distillation. *Computer Aided Chemical Engineering*, *30*, 267–271.

Flower, J.R. and Jackson, R. (1964) Energy requirements in the separation of mixtures by distillation. *Transactions of the Institution of Chemical Engineers*, *42*, 249–258.

Freshwater, D.C. (1951) Thermal economy in distillation. *Transactions of the Institution of Chemical Engineers*, *29*, 149–160.

Freshwater, D.C. (1961) The heat pump in multicomponent distillation. *British Chemical Engineering*, *6*, 388–391.

Fukushima, T., Kano, M., and Hasebe, S. (2006) Dynamics and control of heat integrated distillation column (HIDiC). *Journal of Chemical Engineering of Japan*, *39*, 1096–1103.

Gadalla, M.A. (2009) Internal heat integrated distillation columns (iHIDiCs) – New systematic design methodology. *Chemical Engineering Research & Design*, *87*, 1658–1666.

Gadalla, M., Jimenez, L.J., Olujić, Ž., and Jansens, P.J. (2007a) A thermo-hydraulic approach to conceptual design of an internally heat-integrated distillation column (i-HIDiC). *Computers & Chemical Engineering*, *31*, 1346–1354.

Gadalla, M., Olujić, Ž., de Rijke, A., and Jansens, P.J. (2006) Reducing CO2 emissions of internally heat-integrated distillation columns for separation of close boiling mixtures. *Energy*, *31*, 2409–2417.

Gadalla, M., Olujić, Ž., Esteller, L.J., and Guillen-Gosalbez, G. (2007b) A design method for internal heat integrated distillation columns (iHIDiCs). *Computer Aided Process Engineering*, *24*, 1041–1046.

Gadalla, M., Olujić, Ž., Sun, L. *et al.* (2005) Pinch analysis-based approach to conceptual design of internally heat-integrated distillation columns. *Chemical Engineering Research & Design*, *83*, 987–993.

Glenchur, T. and Govind, R. (1987) Study on a continuous heat integrated distillation column. *Separation Science and Technology*, *22*, 2323–2338.

Gokcen, N.A., and Reddy, R.G. (1996) Thermodynamics, 2nd ed., Plenum Press, New York, USA.

Govind, R. (1986) Distillation column and processes, US Patent No. 4,615,770.

Govind, R. (1987) Dual distillation columns, US Patent No. 4,681,661.

de Graauw, J., Steenbakker, M.J., de Rijke, A. *et al.* (2003) Distillation column with heat integration, Dutch Patent P56921NL00.

de Graauw, J., Steenbakker, M.J., de Rijke, A. *et al.* (2008) Distillation column with heat integration, US Patent No. 7,678,237 B2.

Haselden, G.G. (1958) An approach to minimum power consumption in low temperature gas separation. *Transactions of the Institution of Chemical Engineers*, *36*, 122–132.

Haselden, G.G. (1977) Distillation processes and apparatus, US Patent No. 4,025,398.

Huang, K., Iwakabe, K., Nakaiwa, M. *et al.* (2008a) Consider heat integration to improve separation performance. *Hydrocarbon Processing*, pp. 101–108.

Huang, K., Shan, L., Zhu, Q., and Qian, J. (2008b) A totally heat integrated distillation column (THIDiC) - The effect of feed pre-preheating by distillate. *Applied Thermal Energy*, *28*, 856–864.

Huang, K.J., Nakaiwa, M., Akiya, T. *et al.* (1996a) Dynamics of ideal heat integrated distillation columns. *Journal of Chemical Engineering of Japan*, *29*, 656–661.

Huang, K.J., Nakaiwa, M., Akiya, T. *et al.* (1996b) A numerical consideration on dynamic modeling and control of ideal heat integrated distillation columns. *Journal of Chemical Engineering of Japan*, *29*, 344–351.

Huang, K.J., Wang, S.J., Iwakabe, K. *et al.* (2007) Temperature control of an ideal heat-integrated distillation column (HIDiC). *Chemical Engineering Science*, *62*, 6486–6491.

Hugill, J.A. (2003) System for stripping and rectifying a fluid mixture, International Patent No. WO 03(011418 A1.

Humphrey, J.L., Seibert, A.F., and Koort, R.A. (1991) Separation Technologies: Advances and Priorities, Final report for US Department of Energy, Office of Industrial Technologies, Washington DC.

Iwakabe, K., Nakaiwa, M., Huang, K. *et al.* (2006) Energy saving in multicomponent separation using an internally heat-integrated distillation column (HIDiC). *Applied Thermal Engineering*, *26*, 1362–1368.

Iwakabe, K., Nakaiwa, M., Huang, K. *et al.* (2006) Performance of an internally heat-integrated distillation column (HIDiC) in separation of ternary mixtures. *Journal of Chemical Engineering of Japan*, *39*, 417–425.

Jansens, P.J., Fahkri, F., de Graauw, J., and Olujić, Z. (2001) Energy saving potential of a heat integrated distillation column, *AIChE 2001 Spring National Meeting April 22–26*, AIChE, New York, pp. 19–25.

Kaeser, M. and Pritchard, C.L. (2003) Heat transfer at the surface of sieve trays. Proceedings of the UK Heat Transfer Session.

Kansha, Y., Kishimoto, A., and Tsutsumi, A. (2011) Process design methodology for high-energy saving HIDiC based on self-heat recuperation. *Asia-Pacific Journal of Chemical Engineering*, *6*, 320–326.

Keller, G.E. and Humphrey, J.L. (1997) *Separation Process Technology*, McGraw-Hill, New York.

Kiran, B., Jana, A.K., and Samanta, A.N. (2012) A novel intensified heat integration in multicomponent distillation. *Energy*, *41*, 443–453.

Kiss, A.A., Flores Landaeta, S.J., and Infante Ferreira, C.A. (2012) Towards energy efficient distillation technologies – Making the right choice. *Energy*, *47*, 531–542.

de Koeijer, G.M., Kjelstrup, S., van der Kooij, H. *et al.* (2002) Positioning heat exchangers in binary tray distillation using isoforce operation. *Energy Conversion and Management*, *43*, 1571–1581.

Le Goff, P., Cachot, T., and Rivero, R. (1996) Exergy analysis of distillation processes. *Chemical Engineering and Technology*, *19*, 478–485.

Legatski, H. (1954) Fractionation distillation, US Patent No. 2,690,060.

Liu, X.G. and Qian, J.X. (2000) Modeling, control, and optimization of ideal internal thermally coupled distillation columns. *Chemical Engineering & Technology*, *23*, 235–241.

Mah, R.S., Nicholas, J.J., and Wodnik, R.B. (1977) Distillation with secondary reflux and vaporization - A comparative evaluation. *AIChE Journal*, *23*, 651–658.

Mane, A. and Jana, A.K. (2010) A new intensified heat integration in distillation column. *Industrial & Engineering Chemistry Research*, *49*, 9534–9541.

Matsuda, K., Iwakabe, K., and Nakaiwa, M. (2012) Recent advances in internally heat-integrated distillation columns (HIDiC) for sustainable development. *Journal of Chemical Engineering of Japan*, *45*, 363–372.

Matsuda, K., Kawazuishi, K., Kansha, Y. *et al.* (2011) Advanced energy saving in distillation process with self-heat recuperation technology. *Energy*, *36*, 4640–4645.

Naito, K., Nakaiwa, M., Huang, K. *et al.* (2000) Operation of a bench-scale ideal heat integrated distillation column (HIDiC): An experimental study. *Computers & Chemical Engineering*, *24*, 495–499.

Nakaiwa, M. (1988) Fundamental research on energy savings of distillation processes, Ph.D. Thesis, Kyoto University, Japan.

Nakaiwa, M., Huang, K., Owa, M. *et al.* (1997) Energy savings in heat-integrated distillation columns. *Energy*, *22*, 621–625.

Nakaiwa, M., Huang, K., Naito, K. *et al.* (2001) Parameter analysis and optimization of ideal heat-integrated distillation columns. *Computers & Chemical Engineering*, *25*, 737–744.

Nakaiwa, M., Huang, K., Endo, A. *et al.* (2003) Internally heat integrated distillation column: A review. *Transactions IChemE*, *81*, 162–177.

Noda, H., Takamatsu, T., Aso, K. *et al.* (2000) Development on a coaxial heat integrated distillation column (HIDiC). *Korean Journal of Chemical Engineering*, *17* (5), 593–596.

Null, H.R. (1976) Heat pumps in distillation. *Chemical Engineering Progress*, *78*, 58–64.

Olujić, Ž., Fakhri, F., de Rijke, A. *et al.* (2003) Internal heat integration - the key to an energy conserving distillation column. *Journal of Chemical Technology and Biotechnology*, *78*, 241–248.

Olujić, Ž., Sun, L., de Rijke, A., and Jansens, P.J. (2006) Conceptual design of an internally heat integrated propylene-propane splitter. *Energy*, *31*, 3083–3096.

Olujić, Ž., Sun, L., Gadalla, M. *et al.* (2008) Enhancing thermodynamic efficiency of energy intensive distillation columns via internal heat integration. *Chemical and Biochemical Engineering Quarterly*, *22*, 383–392.

Olujić, Ž., Jödecke, M., Shilkinb, A. *et al.* (2009) Equipment improvement trends in distillation. *Chemical Engineering and Processing*, *48*, 1089–1104.

Process Systems Enterprise Ltd . (2011) gPROMS Process Model Library documentation.

Pulido, J.L., Martinez, E.L., Maciel, M.R.W., and Maciel, R. (2011) Heat integrated reactive distillation column (r-HIDiC): Implementing a new technology distillation. *Chemical Engineering Transactions*, *24*, 1303–1308.

de Rijke, A. (2007) Development of a concentric internally Heat Integrated Distillation Column (HIDiC), PhD thesis, Delft University of Technology.

Rivero, R. (2001) Exergy simulation and optimization of adiabatic and diabatic binary distillation. *Energy*, *26*, 561–593.

Seader, J.D. (1980) Continuous distillation apparatus and method, US Patent No. 4,234,391 A1.

Seader, J.D. and Baer, S.C. (1984) Continuous distillation apparatus and method, Final report, University of Utah, Salt Lake City (US).

Schaller, M., Hoffmann, K.H., Rivero, R. *et al.* (2005) The influence of heat transfer irreversibilities on the optimal performance of diabatic distillation columns. *Journal of Non-Equilibrium Thermodynamics*, *27*, 257–269.

Schmal, J.P., van der Kooi, H.J., de Rijke, A. *et al.* (2006) Internal versus external heat integration - Operational and economic analysis. *Chemical Engineering Research & Design*, *84*, 374–380.

Sechrist, P. (2009) Heat Pump Distillation, US Patent Application No. 2009(0114524.

Shah, M., Kiss, A.A., Zondervan, E., and de Haan, A.B. (2012) A systematic framework for the feasibility and technical evaluation of reactive distillation processes. *Chemical Engineering and Processing*, *60*, 55–64.

Shenvi, A.A., Herron, D.M., and Agrawal, R. (2011) Energy efficiency limitations of the conventional heat integrated distillation column (HIDiC) configuration for binary distillation. *Industrial and Engineering Chemistry Research*, *50*, 119–130.

Stage H. (1974) Perforated tray column, US Patent No. 3,792,843.

Sunder, S., Houghton, P.A., and Kuznetsov, V.K. (2004) Plate-fin heat exchanger with textured surface, US Patent No. 6,838515 B2.

Suphanit, B. (2011) Optimal heat distribution in the internally heat-integrated distillation column (HIDiC). *Energy*, *36*, 4171–4181.

Takamatsu, T., Nakaiwa, M., and Nakanishi, T. (1996) The concept of an ideal heat-integrated distillation column (HIDiC) and its fundamental properties. *Kagaku Kogaku Ronbunshu*, *22*, 985–990.

Tondeur, D. and Kvaalen, E. (1987) Equipartition of entropy-production. An optimality criterion for transfer and separation processes. *Industrial & Engineering Chemistry Research*, *26*, 50–56.

Tung, H.H., Davis, J.F., and Mah, R.S.H. (1986) Fractionating condensation and evaporation in plate-fin device. *AIChE Journal*, *32* (7), 1116–1124.

Wang, J.C. and Henke, G.E. (1966) Tridiagonal matrix for distillation. *Hydrocarbon Processing*, *45*, 155–163.

Yildirim, O., Kiss, A.A., and Kenig, E.Y. (2011) Dividing wall columns in chemical process industry: A review on current activities. *Separation and Purification Technology*, *80*, 403–417.

9

Cyclic Distillation

9.1 INTRODUCTION

Process intensification in distillation (Olujic *et al.*, 2009) has received much attention in recent decades, following various integration routes, such as internal heat-integrated distillation columns (Harmsen, 2010), dividing-wall columns (Dejanović; Yildirim, Kiss, and Kenig, 2011), reactive distillation (Harmsen, 2007, 2010), and even reactive dividing-wall columns (Kiss, Pragt, and Strien, 2009; Kiss *et al.*, 2012a). Consequently, the design and control of such integrated systems is considered of utmost importance (Harmsen, 2010; Kiss and Bildea, 2011). Another route for process intensification in distillation is cyclic distillation, which presently receives revived attention. Cannon *et al.* (McWhirter and Cannon, 1961; Gaska and Cannon, 1961) proposed an innovative method of operating various types of existing distillation towers, namely, the controlled cycling distillation mode. Basically, a cyclic distillation column has an operating cycle consisting of two key operation parts: (i) a vapor flow period when vapor flows upwards through the column and liquid remains stationary on each plate and (ii) a liquid flow period when vapor flow is stopped, reflux and feed liquid are supplied, and liquid is dropped from each tray to the one below. Figure 9.1 illustrates these two operating periods (Maleta *et al.*, 2011). Remarkably, the throughput of such a column using the controlled cycle mode of operation is two or more times higher than that reachable with conventional operation (Schrodt *et al.*, 1967; Maleta *et al.*, 2011).

Although the Murphree vapor efficiency of a plate at any instant of time is assumed to be constant and equal to the local point efficiency, the

Advanced Distillation Technologies: Design, Control and Applications, First Edition.
Anton Alexandru Kiss.

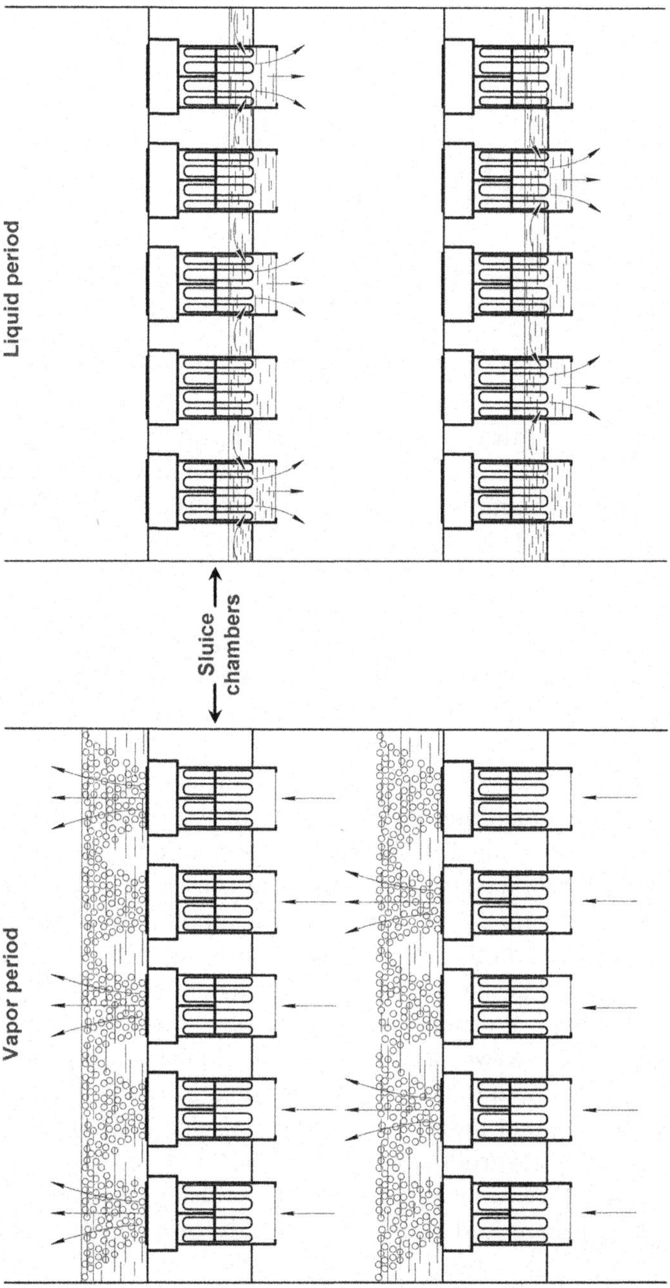

Figure 9.1 Schematics of the cyclic mode operation: vapor period followed by liquid period

effective stage efficiency based on the liquid-phase compositions is usually significantly greater than the point efficiency (Sommerfeld *et al.*, 1966). This leads to the improved separating ability achieved in controlled cyclic operation. A precise analogy exists between cyclic and conventional distillation with liquid-phase concentration gradients across the plates of the column. Basically, this analogy reduces to the substitution of time as the independent variable in the case of cyclic distillation, for distance in the case of conventional distillation. Note that the concept could be further extended to catalytic cyclic distillation (CCD), a novel setup that can add on top of the benefits of reactive distillation (RD) also those of the cyclic distillation operation mode—but this is a challenging topic that requires further research.

This chapter gives an overview of previous work on cyclic distillation and presents a novel model of a theoretical stage with perfect displacement, as well as basic design and control issues. Several relevant case studies illustrate the theoretical findings and practical applications of this work.

9.2 OVERVIEW OF CYCLIC DISTILLATION PROCESSES

The following literature review gives—in historical order—a good overview of the previously reported studies on the cyclic operation mode of various distillation systems. Controlled cycling distillation, in sieve and screen plate towers, was proposed by Cannon *et al.* (McWhirter and Cannon, 1961; Gaska and Cannon, 1961) as a new method of operating various types of existing equipment. This also permits new types of equipment to be designed for many conventional processes with key advantages such as: no downcomers needed on plates that are operated with controlled cycling and greater capacity than that attainable conventionally. Cannon *et al.* (1961) also reported the results of applying controlled cycling to several types of plates in distillation towers. Remarkably, a 48% increase in the total vapor load was possible at a fixed pressure drop. Their data illustrate an interesting fact, namely, that the maximum rate of the phase flow is not dictated by the physical dimensions of the equipment and the properties of the system only, but it is also a function of the method of operation employed. Thus, the authors (Gaska and Cannon, 1961) believed that the capacity of existing bubble-cap plate towers as well as other types can be increased by the use of controlled cycling.

Sommerfeld *et al.* (1966) studied controlled cyclic distillation and reported their computer simulations and an analogy with the conventional operation. At the same time, Chien *et al.* (1966) provided an analytical transient solution and reported the asymptotic plate efficiencies. These solutions were used then to establish asymptotic expressions for the compositions at the pseudo-steady-state condition of the cycling column. A method of numerical iterative solution for a nonlinear equilibrium relationship was demonstrated, and a simplified graphical method for calculating the number of stages required for a given separation was presented.

Robinson and Engel (1967) performed a theoretical analysis to demonstrate the advantages of cycled mass transfer operations, in which only one phase flows at any given time and where the phases use the same interstage flow passages during their respective flow periods.

Schrodt *et al.* (1967) carried out a plant-scale study of controlled cyclic distillation. A dual-purpose column (15 perforated trays, 305 mm diameter) for testing the controlled cycling concept was designed, installed, and operated successfully. The throughput for the column using the controlled cycle mode of operation was well over twice the throughput obtainable with conventional operation—at equivalent separation performance.

Gel'perin, Polotskii, and Potapov (1976) reported the similar operation of a bubble-cap fractionating column in a cyclic regime. Subsequently, Rivas (1977) showed that simple analytical equations can be used to calculate the ideal number of trays for cyclic countercurrent processes such as cyclic distillation. These equations represent an accurate solution of the set of ordinary differential equations that result from the material balances on each plate of a cyclic column. They simplify greatly the preliminary design of these columns since they are easy to use and similar to the well-known Kremser-Souders-Brown (KBS) equations that are used in conventional processes.

Furzer (1978) reported the discrete residence time distribution of a distillation column operated with microprocessor controlled periodic cycling. The fluid flow in a distillation column was studied by measurements of the discrete residence time distribution. The column was fitted with five sieve plates of 100 mm diameter, 14.8% free area, 6.1 mm hole diameter, and 635 mm plate spacing. The periodic control was obtained using a JOLT microprocessor system. An analysis of the discrete residence time distribution yielded the parameters in the (2S) model that describes the fluid flow. Modifications to the column internals were required to alter these parameters, if the maximum separation improvements were to be obtained.

Goss and Furzer (1980) then studied the mass transfer in periodically cycled plate columns containing multiple sieve plates. A five-plate distillation column (100 mm diameter) was used to study both the fluid mixing and the mass transfer separations of mixtures of methylcyclohexane and *n*-heptane, under periodically cycled conditions. Both the fluid mixing and mass transfer could be successfully analyzed using the (2S) model to describe the liquid movement in the column. However, only moderate improvements in separating ability could be obtained with sieve plates and packed sieve plate columns operated in the cycled mode.

Baron, Wajc, and Lavie (1980) investigated the stepwise periodic distillation under total reflux operation. They found a new periodic operating mode where the liquid flow was manipulated directly, and not indirectly through pulsations of the vapor flow rate (as in controlled cycling). They also presented the theory of stepwise periodic distillation, under assumptions allowing a fair comparison with known results for controlled cycling. They showed that the two processes have the same asymptotic efficiencies for large values of the number of trays, while periodic stepwise distillation is slightly more efficient than controlled cycling for a finite number of trays. In a follow-up study, Baron, Wajc, and Lavie (1981) then described the stepwise periodic operation of a distillation column in which a binary mixture was separated. A simple but realistic model and a simulation algorithm were proposed for both stepwise periodic and controlled cycling operations. The results of an extensive parametric study of stepwise periodic operation were presented along with those for ideal controlled cycling (i.e., no axial mixing). For reasonable tray efficiencies and difficult separations, stepwise periodic operation compares favorably with, or is superior to, ideal controlled cycling (Baron, Wajc, and Lavie, 1981).

Thompson and Furzer (1985) reported the hydrodynamic simulation and experimental confirmation of periodic cycled plate columns. Szonyi and Furzer (1985) investigated the periodic cycling of distillation columns using a new tray design. The simulations of a periodically cycled column predicted improvements in column performance of greater than 200% for systems with nonlinear equilibrium curves. Experiments in distilling various methanol–water mixtures validated the theoretical predictions for periodic cycling with a single plate. Moreover, a new tray design consisting of a sieve plate and inclined surfaces was introduced to obtain an effective liquid time delay. Remarkably, a wide range of operating conditions provided overall column efficiencies of over 140% (Szonyi and Furzer, 1985).

Matsubara, Watanabe, and Kurimoto (1985) investigated the binary periodic distillation scheme with enhanced energy conservation. They proposed a composite scheme in which the beneficial features of the Cannon (Gaska and Cannon, 1961) and Baron–Wajc (Baron, Wajc, and Lavie 1980, 1981) schemes were combined. The construction of the scheme and its superiority to conventional the steady-state operation scheme was ascertained first through numerical analyses (Matsubara, Watanabe, and Kurimoto, 1985). Afterwards, a laboratory-scale periodic distillation column using the idealized Cannon scheme was constructed and some basic experiments were carried out to examine its performance (Matsubara, Watanabe, and Kurimoto, 1985). The system employed was the binary mixture water–methanol in a five-stage distillation column. The average vapor rate, which is roughly proportional to the energy requirements rate, was 20–50% lower than that required to achieve the same separation by use of a conventional continuous column.

Toftegard and Jørgensen (1989) proposed an integration method for the dynamic simulation of cycled processes. The method is based on an efficient algorithm for dynamic simulation of nonsingular periodically cycled chemical processes. Application of the algorithm was successfully demonstrated on a dynamic simulation of controlled cycling distillation.

Other dynamic optimization methods can be applied to find the optimal profiles for all possible control variables. For an ideal ternary mixture separated into two or three fractions, it was demonstrated that appropriate control profiles can reduce the required energy supply compared to steady-state conditions (Bausa and Tsatsaronis, 2001).

More recently, Maleta *et al.* (2010, 2011) have shown the use of a theoretical stage model with perfect displacement in a tray column with separate movement of the vapor and liquid phases. Their work presents the complete theoretical model and the operating lines theory as well as two industrially relevant applications. MaletaCD have also published several patents related to cyclic distillation equipment (Maleta and Maleta, 2009, 2010). Moreover, the application of cyclic operation to batch distillation columns was recently reviewed by Flodman and Timm (2012), while Lita *et al.* (2012) explored the design and control of continuous cyclic distillation systems.

9.3 PROCESS DESCRIPTION

According to the work of Lewis (1936) the use of combinations of various hydrodynamic modes of liquid and vapor phases enhances the efficiency of component separation. The maximum effect is achieved upon perfect displacement of liquid and vapor, at single direction movement of the liquid on adjacent trays. Under such conditions, the Murphree efficiency of the tray may significantly exceed the local point efficiency and reach theoretical values of 200–300%.

Cannon *et al.* (1961) (McWhirter and Cannon, 1961; Gaska and Cannon, 1961) suggested the way of phase interaction: during the flow of vapor through the column the liquid does not overflow from tray to tray, and upon liquid overflow it does not mix on adjacent trays (cyclic process). In a series of works by Sommerfeld *et al.* (1966) and Thompson and Furzer (1985) it was shown that, upon comparison of stationary and cyclic processes, the cyclic process is similar to the stationary process upon single-direction movement of liquid on adjacent trays and perfect displacement of liquid and vapor.

The aim of obtaining separation efficiencies above the theoretical stage proposed by McCabe and Thiele (1925) promoted the development of constructive solutions—for a stationary mode towards longitudinal sectioning on the tray, and for the cyclic regime to enable the liquid overflow on the trays with minimum mixing of liquid on the adjacent trays.

The problem is that despite the many papers tackling this topic (see previous section) there is still no straightforward model describing the cyclic distillation process, and consequently there are no design methods for cyclic distillation columns. To solve this problem, Maleta *et al.* (2011) developed an adequate mass transfer model and a simple graphical mode for designing tray columns operating in a cyclic mode—thus providing valuable insight into the process intensification in a cyclic distillation process.

In the first attempts at industrial implementation, the mass transfer technology suggested by Cannon *et al.* (1961) used the ordinary overflow trays accompanied by changeable or cyclic supply of contacting phases to the column. The use of conventional trays in a non-conventional (cyclic) operation mode was unsuccessful. As shown in the literature review, the subsequent creation of special contact devices allowed a broad range of the required technical solutions to be obtained with various options of discrete and even uninterrupted flow inputs.

Therefore, it is suggested that the term *separate phase movement* (SPM) reflects the essence of this phenomenon in the most integral manner and it may be used as a synonym for the term *cyclic distillation process*. Basically, the cyclic process of mass transfer in tray columns reduces to satisfy the requirements of the liquid flow dynamics: (i) lack of liquid overflow on the trays during the vapor supply period and (ii) lack of mixing of liquids on adjacent trays during liquid overflow period - as shown in Figure 9.1 (Matela *et al.*, 2011). The constructive solution of such tasks involves paying attention to the enhancement of mass transfer efficiency and the reduction of energy requirement in such a distillation process. MaletaCD has obtained a series of patents (Maleta and Maleta, 2009, 2010a, 2010b) that allowed implementation of the advantages of cyclic distillation at an industrial scale.

The algorithm of engineering the design of such a special tray is relatively simple. The lack of liquid overflow on the trays during vapor supply is achieved by using perforated trays while the vapor velocity exceeds the column flooding. The lack of liquid mixing on adjacent trays during liquid overflow is realized by means of a sluice chamber located under the tray. This installed system works as follows. The liquid streams (e.g., feed supply, reflux) are continuously supplied to the corresponding tray. During the vapor supply period, the liquid does not overflow from tray to tray (in contrast to conventional operation) as the vapor velocity exceeds the column flooding. The vapor fed to the column controls the liquid overflow from tray to tray. When the vapor supply is cut for a few seconds it allows the liquid to overflow from the tray to the sluice chamber. Subsequent vapor supply opens the sluice chamber and the liquid flows by gravitation to the empty tray below. Such a sequence of actions takes place synchronically on all trays in the column. In practice, no complication related to increasing the number of trays was observed in terms of column operability.

A description of the liquid flow hydrodynamics can be clearly subdivided into two major stages: (i) vapor supply stage and (ii) liquid overflow stage. In the first case, during the mass transfer between vapor and liquid, the liquid is located on the tray only and it does not overflow from one tray to another. Hence, we can disregard the transfer of liquid to the tray below. The volume of liquid on the tray can be considered as a closed circuit and, hence, not a single droplet of liquid leaves this volume. As the liquid inside this volume is perfectly mixed under the influence of the vapor flowing through, there is no gradient of concentration and temperature at a certain point in time. Moreover, all the liquid on the tray has the same residence time, and there is equal speed of movement of the liquid volume—which in fact equals to zero.

Such conditions are typical of only one hydrodynamic mode, the perfect displacement mode. This can be achieved also in actual circumstances, and not only in a hypothetical piston flow. Therefore, during the transfer of vapor through the column (vapor period), the hydrodynamic model of the liquid flow on the tray in real circumstances is the perfect displacement model.

9.4 MATHEMATICAL AND HYDRODYNAMIC MODEL

9.4.1 Mathematical Model

The formalized description of a tray column with separate phase movement is provided by the mathematical model of the process that makes use of the following equations:

1. Mass balance of the light/volatile component (VC) on the contact stage, during the vapor supply:

$$\frac{dx_n}{d\tau_v} = -\frac{G}{H}(y_n - y_{n-1}) \tag{9.1}$$

   where n is the tray number, x_n is the molar fraction of VC in the liquid on the tray n, y_n is the mol fraction of VC in vapor leaving tray n, y_{n-1} is the molar fraction of VC in vapor coming from tray $n-1$, H is the amount of liquid on the tray (mol), G is the vapor flow rate (mol s^{-1}), and τ_{v} is the vapor supply time (s).

2. Liquid flow hydrodynamics during the liquid period (liquid overflow from tray to tray):

$$\begin{aligned} x_n(0) &= Fx_{n+1}(\tau_v) + (1-F)x_n(\tau_v) \\ F &= H_{\mathrm{l}}/H \quad \text{and} \quad 0 < F \leq 1 \end{aligned} \tag{9.2}$$

   where F is the multiplication factor of liquid transfer delay, H_{l} is the amount of liquid holdup flowing from the tray (mol), and τ_{v} is the vapor supply time (s).

3. Mass transfer kinetics, determined by the local point efficiency (E_{OG}):

$$E_{\mathrm{OG}} = \frac{y_n(\tau_k) - y_{n-1}(\tau_k)}{y_n^*(\tau_k) - y_{n-1}(\tau_k)} \tag{9.3}$$

   where τ_k is a random point in time: $\tau_k \in [0, \tau_v]$

4. Equilibrium line (vapor composition as function of liquid composition):

$$y^* = f(x) \tag{9.4}$$

This system of equations can be also solved analytically (Robinson and Engel, 1967). Consequently, the concentration profiles on each stage can be described as follows:

$$x_n(\tau) = e^{-\frac{Gm}{H}\tau} \sum_{i=1}^{n} C_i \frac{\left(\frac{Gm}{H}\tau\right)^{n-1}}{(n-i)!}, i = \overline{1,n} \tag{9.5}$$

with the boundary condition:

$$x_i(0) = C_i \tag{9.6}$$

Note that for tray columns the most widespread model is the theoretical stage introduced by McCabe and Thiele (1925). The model is based on the column material balance by distributed components and it is postulated as follows: (i) The concentration of the volatile component on the contact stage is constant and (ii) the vapor leaving the stage is in equilibrium with the liquid leaving the same stage. Therefore, the hydrodynamic models of liquid and vapor flows are expressed as *perfectly mixed models*. The process operating line determines the concentrations of the distributed component in the liquid/vapor on each tray. Moreover, it is discrete and the coordinates of the operating line points (trays) in the Y–X coordinates are given by $A(x_n, y_{n-1})$.

Here, we use a similar technique to obtain the cyclic process operating line, as previously suggested by Maleta, Taran, and Dubovik (1986). On knowing the interim concentration profiles in the liquid and vapor on all trays, at a random point in time $\tau_k \in [0, \tau_v]$, one can determine the operating line point coordinates as: $A[x_n(\tau_k), y_{n-1}(\tau_k)]$.

Figure 9.2 illustrates the methodology of constructing the perfect displacement mode operating line (Maleta *et al.*, 2011): temporary concentration profiles in liquid (Figure 9.2a) and the perfect displacement operating line in coordinates (Figure 9.2b).

By changing the time points—$\tau_k \in [0, \tau_v]$—one can find graphically any number of points referring to the operating line (line 3 in Figure 9.2b). Therefore, the operating line of the cyclic process is built on the basis of differential equations of the material balance by distributed component on the contact stages. It is drawn by definition as a line of points where the vapor rising from the tray below meets the liquid located on the tray above. The line is uninterrupted and straight for $y^* = mx$. The tangent of the perfect displacement operating line inclination is less than the values of L/G. Using this line one can determine the values of the concentration of the distributed component in vapor and liquid for each tray and any point in time.

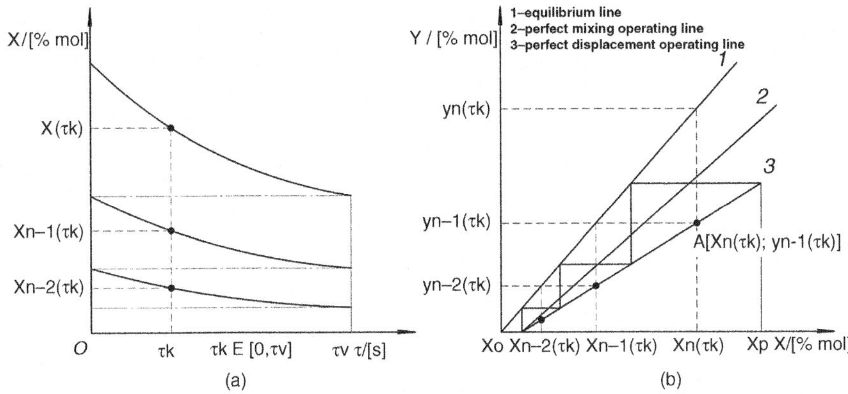

Figure 9.2 Methodology of construction of the perfect displacement mode operating line: temporary concentration profiles in liquid (a) and depiction of the perfect displacement operating line in coordinates (b)

The contact stage built on this line may be called a *theoretical contact stage of perfect displacement,* since its operation conditions are the same as those of the hydrodynamic mode of perfect displacement of liquid and vapor in a single-direction liquid movement on adjacent trays suggested by Lewis (1936). The difference between the cyclic and the stationary process consists in changing the coordinate system: for the cyclic process during vapor supply $dx_n/d\tau$, and for the stationary distillation process dx_n/dl, along the liquid movement route on the tray.

9.4.2 Hydrodynamic Model

Upon constructing the theoretical stage of perfect displacement we considered previously the case of a lack of liquid mixing on adjacent trays, $F = 1$. According to the theoretical stage model with perfect displacement, the liquid flow hydrodynamics upon overflow result in the distribution of components on contact stages during vapor supply, as given by Eq. (9.2). In other words, this is equivalent to the determination of concentration in Eq. (9.4). Therefore, the task of describing the imperfect flows is significantly simplified. One can use the simplest—historically the first—cell model, in which the only parameter is the number of cells. To determine the actual stage of liquid mixing, one needs to study the distribution of separate liquid volumes according to their residence time on the tray—that is, to determine what share of the liquid flow resides on the tray at a certain point in time. Data on the residence time distribution of separate liquid volumes on the tray are obtained by plotting the system disturbance source (tracer agent input) and analysis of the dependence curves of the tracer agent concentration on time $C(\tau)$ in the liquid outlet.

For analysis of the mixing degree, one can use three values: (i) average residence time of the liquid on the tray, (ii) dispersion of the residence time of the liquid on the tray, and (iii) the number of perfect mixing cells. Modeling of the liquid flow hydrodynamics, in the range of $0 < F \leq 1$, allows one to obtain the following dependencies:

Average residence time of the liquid on the tray:

$$\bar{\tau} = \frac{\tau_v}{F} \tag{9.7}$$

Degree of dispersion around the mean value:

$$\sigma^2 = 1 - F \tag{9.8}$$

Hydrodynamic cell model:

$$N = 1/(1 - F) \tag{9.9}$$

The validity of the obtained dependency of the cell model on the liquid flow hydrodynamics can be determined by studying their behavior upon extreme values of F. When $F = 1$ the average residence time $\bar{\tau}$ is equal to the vapor supply time (τ_v), the dispersion is $\sigma^2 = 0$, and the number of cells is $N \to \infty$, so these are the *perfect displacement* conditions. Similarly, when $F = 0$, the average residence time is $\bar{\tau} \to \infty$, the dispersion is $\sigma^2 = 1$, and the number of cells is $N = 1$; these are in fact the *perfect mixing* conditions.

When the number of cells is sufficiently high ($N \geq 10$), it is considered that the cell model approximates very well the perfect displacement. In this case, the condition is met in the range $0.9 < F \leq 1$. Therefore, the liquid flow hydrodynamics determines the theoretical stage—from the theoretical stage with perfect mixing to the theoretical stage with perfect displacement. For the cyclic operation mode, the physical interpretation of transfer of the theoretical stage with perfect displacement into the theoretical stage model results in an increase of the amount of liquid on the tray up to infinity, $N \to \infty$, and in this case the concentration of the volatile component on the tray during the vapor supply time will not change.

9.4.3 Sensitivity Analysis

Sensitivity analysis can be used as a powerful tool to explore the effect of several key parameters of the cyclic distillation model described earlier. The cyclic distillation model parameters include the values of the multiplication factor of liquid delay transfer, $0 < F \leq 1$, and the local point efficiency of contact $0 < E_{OG} \leq 1$. The mathematical model includes also the system of nonlinear differential equations that can be solved only numerically. Any continuous phenomenon may be considered as a limit to which the discrete phenomenon tends to, if the interval between component events tends to zero. In this case, instead of describing the object with differential equations one can use the system of finite differences. The continuous signal (concentration of volatile component in vapor and liquid phases) can be presented as a sequence of discrete signals quantified simultaneously by level and time. This is the so-called *double quantification* that presents the signal in the form of a sequence of numbers and their delivery at discrete moments of time. Using the above technique, a calculation algorithm can be created and implemented based on iterative methods. The solution of the mathematical description represents a limit of a certain sequence of approximations. The calculation results in the form of the number of trays, tray efficiency according to Murphree, distribution of concentration of volatile component in liquid and vapor on the trays, and the operating lines in coordinates $Y - X$.

The influence of the equilibrium line shape on the position and shape of the operating line was applied to an equimolar binary mixture ethanol–water. Figure 9.3 presents the modeling results (Maleta *et al.*, 2011). The perfect displacement operating line (line 3, $F = 1$, $E_{OG} = 1$) is not a straight but a concave continuous line that is in fact a mirror reflection of the equilibrium line.

The liquid flow dynamics, expressed through the value of the multiplication factor of liquid delay transfer (F), influence the mass transfer process as follows. At $F < 1$ (line 4 plotted in Figure 9.4), the operating line has a discrete nature, while the solid parts are determined by the dependency expressed in Eq. (9.2) (Maleta *et al.*, 2011). When $F \to 0$, the average residence time of liquid on the tray tends to infinity ($\bar{\tau} \to \infty$), the mixture separation efficiency reduces, and the operating parts converge into a single point. In this case the process operating line coincides with the operating line of perfect mixing with the inclination tangent L/G (line 2, Figure 9.4). For linear equilibrium dependency, $y^* = mx$, the above position is proven analytically if the VC concentration on the tray is not changed. According to the suggested calculation technique, this

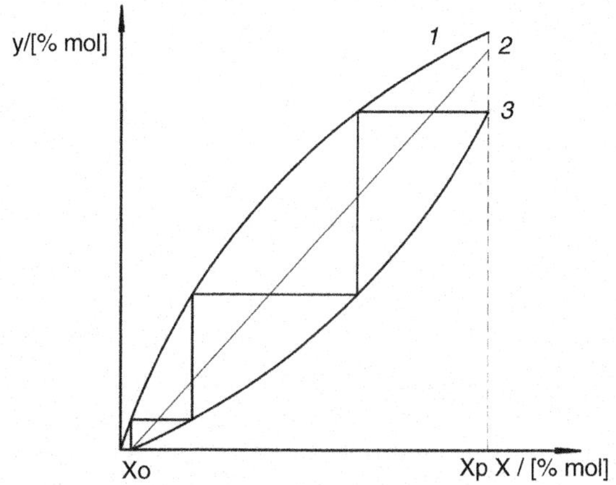

Figure 9.3 Influence of model parameters on the position and shape of the operating line: (1) equilibrium line, (2) operating line (L/G), and (3) perfect displacement operating line ($F=1$; $E_{OG}=1$)

condition is met when the amount of liquid on the contact stage is $H \rightarrow \infty$, $F \rightarrow 0$, and $\bar{\tau} \rightarrow \infty$. Using Eq. (9.4) one can create an equation for the operating line:

$$y = Ax + b \tag{9.10}$$

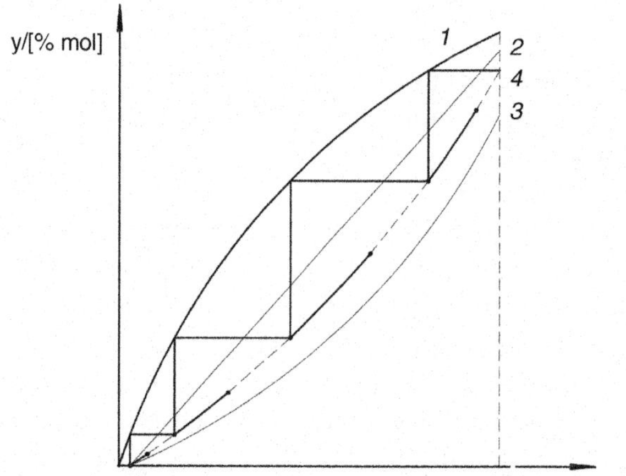

Figure 9.4 Influence of model parameters on the position and shape of operating line: (1) equilibrium line, (2) operating line (L/G), (3) perfect displacement operating line ($F=1$; $E_{OG}=1$), and (4) perfect displacement operating line ($F \neq 1$, $E_{OG}=1$)

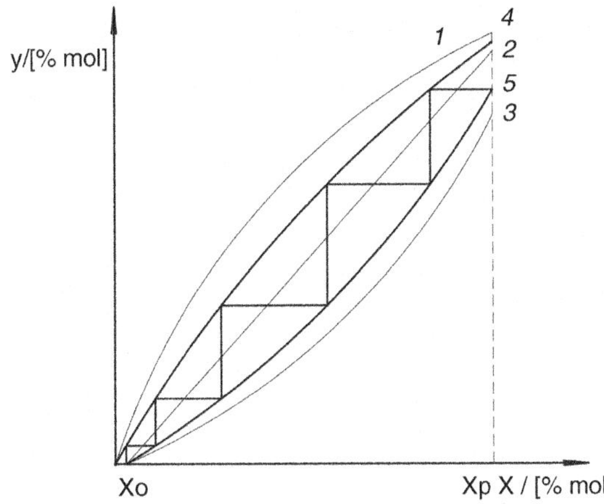

Figure 9.5 Influence of model parameters on position and shape of the operating line: (1) equilibrium line, (2) operating line (L/G), (3) perfect displacement operating line ($F=1$, $E_{OG}=1$), (4) kinetic curve, and (5) perfect displacement operating line ($F=1$; $E_{OG}\neq 1$)

where:

$$A=\frac{(x_n-x_{n-1})m}{x_{n+1}-x_n}\ \text{and}\ b=\frac{m(x_{n-1}x_n-x_n^2)}{x_{n+1}-x_n}\tag{9.11}$$

By means of serial transformations, it is clear that when $F\to 0$ the tangent of this line is equal to L/G at any values of τ_k in the interval $t_k\in[0,t_v]$. The limit is given by:

$$\lim_{F\to 0}A=\lim_{F\to 0}\frac{mF}{a\tau_v}=\frac{mFL}{mGF}=\frac{L}{G},\tau_k\in[0,\tau_v]\tag{9.12}$$

The leading role in the separation process is played by the local point efficiency value (E_{OG}). According to Figure 9.5 it becomes clear that E_{OG} influences not only the degree of equilibrium—this known fact is used in the calculation of stationary process columns—but also the position of the operating line (Maleta *et al.*, 2011). In view of E_{OG} reduction, the kinetic curve 4 and the operating line 5 come close to each other, thus reducing the separation efficiency. The ultimate position of both lines at $E_{OG}\to 0$ will be the operating line L/G.

Figure 9.6 (Maleta *et al.*, 2011) presents the quantitative characteristics of the cyclic process in the form of the dependency of the Murphree efficiency (E_{MV}) of the tray on the local point efficiency (E_{OG}), the liquid transfer delay factor (F), and the diffusion potential factor ($\lambda=mG/L$).

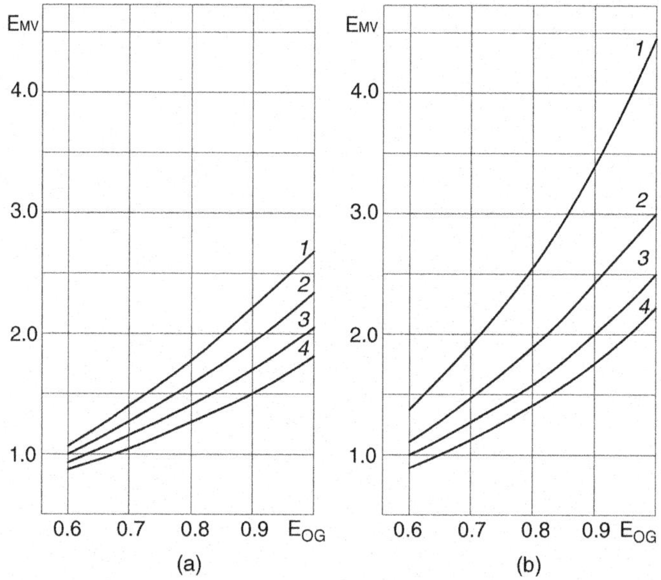

Figure 9.6 Dependency of efficiency of stage according to the Murphree efficiency E_{MV} from a point efficiency of E_{OG}, for various stripping factor values (1: $\lambda = 2.2$, 2: $\lambda = 1.9$, 3: $\lambda = 1.6$, and 4: $\lambda = 1.3$) at $F = 0.7$ (a) and $F = 1$ (b)

The value of E_{MV} continuously increases, subject to the increase of E_{OG}, F, and λ. When $F \to 0$, E_{MV} tends to the local point efficiency ($E_{MV} \to E_{OG}$). The obtained results are in line with the theoretical stage model when the point efficiency and the Murphree efficiency of the tray have the same value. Significant differences between the efficiency of the stationary and the cyclic distillation processes can be observed at higher values of $F \to 1$, and $E_{OG} \to 1$. This conclusion points in the right direction for engineering developments of contact devices for cyclic distillation processes. The influence of model parameters on efficiency is clearly illustrated by analysis of the mass transfer rate equation:

$$n_A = k_c A \Delta c_A \tag{9.13}$$

where A is the effective mass transfer area, k_c is the mass transfer coefficient, and Δc_A is the driving force concentration difference.

The local point efficiency (E_{OG}) determines the product $k_c A$, while the value of λ along with the liquid transfer delay factor F determines the process driving force Δc_A. Obviously, an increase of each term of the right-hand side of the mass transfer rate equation results in an increase in the amount of mass transferred from one stage to another.

9.5 MODELING AND DESIGN OF CYCLIC DISTILLATION

Although the advantages of a cyclic operation have been demonstrated experimentally, information about the modeling and design of cyclic distillation columns is rather limited. Nonetheless, the literature reveals that many experimental and theoretical studies have been carried out so far, including several models for simulating cyclic distillation (Maleta *et al.*, 2011). When the assumption of linear equilibrium is employed, it is indeed possible to derive analytical solutions of the model equations. However, although these analytical expressions could be used to design cyclic operation columns, the accuracy of the results is limited due to the linear equilibrium assumption. Moreover, these models can be used for rating studies only and no design method has been described so far for the general case of nonlinear equilibrium. However, the recent study of Lita, Bildea, and Kiss (2012) fills this gap by presenting a novel approach for the design of cyclic distillation systems and an insightful comparison with conventional distillation. This section provides the complete model of a cyclic distillation system, an innovative graphical design method, and some basic control considerations of cyclic distillation.

As already mentioned, the cyclic operating mode consists of two parts: (i) a vapor-flow period, when vapor flows upward through the column, while liquid remains stationary on each plate and (ii) a liquid-flow period, when vapor flow is stopped and reflux and feed are supplied to the column, while the liquid holdup is dropped from each tray to the tray below. This mode of operation can be easily achieved by using perforated trays, without downcomers, combined with sluice chambers located under each tray. If the vapor velocity exceeds the flooding limit, the liquid does not overflow from tray to tray during the vapor-flow period. When the vapor supply is interrupted, the liquid drops down by gravitation to the sluice chamber. When the vapor supply is started again, the sluice chambers open, and the liquid is transferred to the tray below. The principle of cyclic operation of a cyclic distillation column is schematically illustrated in Figure 9.7 (Lita, Bildea, and Kiss, 2012).

The notations used hereafter are as follows: B = bottoms, D = distillate, L = liquid reflux, V = vapor flow rate and F = feed (all expressed in kmol per cycle), M = holdup (kmol), NF = feed stage, NT = total number of stages, t = time, t_{vap} = duration of the vapor-flow period, x = mole fraction (liquid phase), y = mole fraction (vapor phase). The subscript k means the stage number (1 – condenser, NT – reboiler), while superscripts (V) or (L) refer to the end of the vapor/liquid flow, respectively.

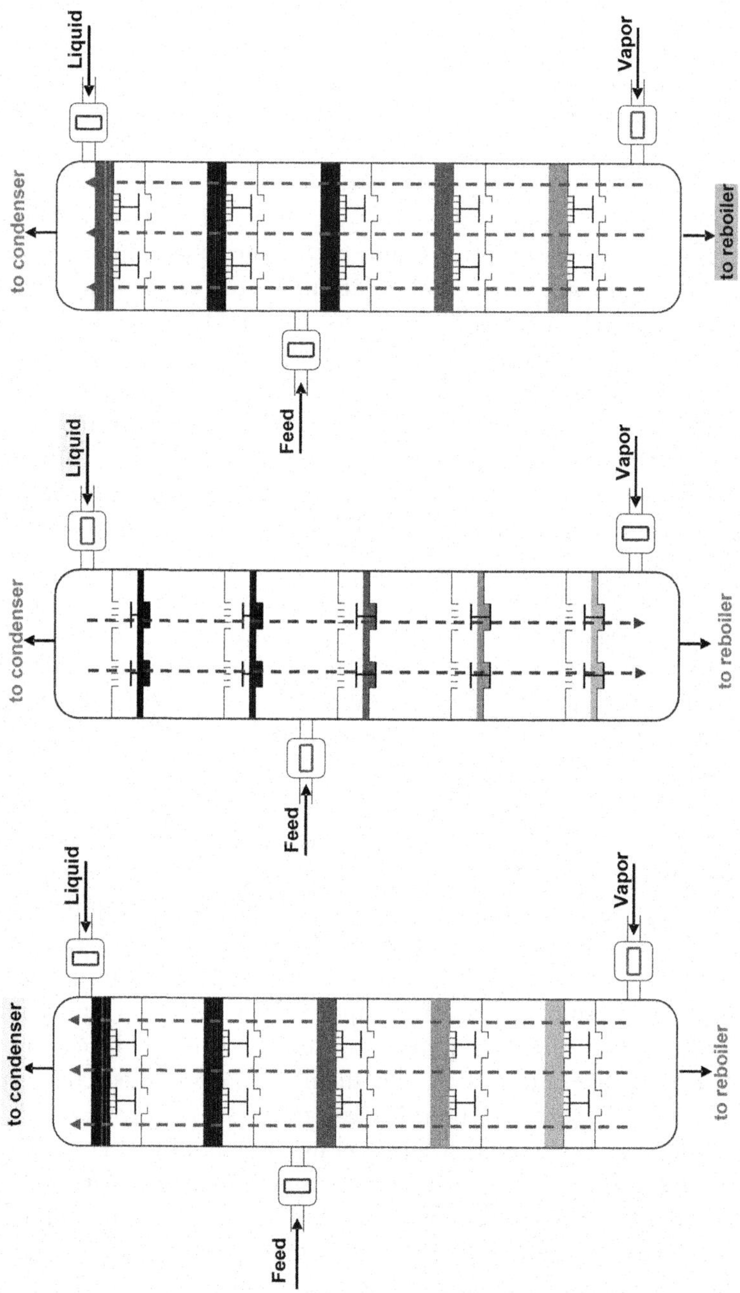

Figure 9.7 Schematic illustration of the working principle of cyclic distillation

9.5.1 Modeling Approach

The model of the cyclic distillation column is derived under the following assumptions:

- Binary (mixture) distillation,
- ideal stages (vapor–liquid equilibrium is reached),
- equal heat of vaporization (this implies constant molar holdup and vapor flow rate),
- perfect mixing on each stage,
- negligible vapor holdup,
- saturated liquid feed

Note that in contrast to previously cited papers, no assumptions are made here with regard to the linearity of vapor–liquid equilibrium, infinite reboiler holdup, or zero condenser holdup.

In terms of *operational constraints,* from the condenser/reboiler mass balance, written for one operating cycle:

$$V \cdot t_{\text{vap}} = D + L \tag{9.14}$$

$$L + F = V \cdot t_{\text{vap}} + B \tag{9.15}$$

the following feasibility condition follows:

$$L < V \cdot t_{\text{vap}} < L + F \tag{9.16}$$

A *model of the vapor-flow period* involves the following equations used to describe the evolution in time of stage holdup and composition, during the vapor-flow period (n.b. extendable to multi-component systems):

$$\text{Condenser:} \quad \frac{\mathrm{d}M_{1\,V\cdot t_{\text{vap}}=D+L}}{\mathrm{d}t} = V;\, M_1\frac{\mathrm{d}x_1}{\mathrm{d}t} = V(y_2 - x_1) \tag{9.17}$$

$$\text{Trays:} \quad \frac{\mathrm{d}M_k}{\mathrm{d}t} = 0;\, M_k\frac{\mathrm{d}x_k}{\mathrm{d}t} = V\left(y_{k+1} - y_k\right) \tag{9.18}$$

$$\text{Reboiler:} \quad \frac{\mathrm{d}M_{\text{NT}}}{\mathrm{d}t} = -V;\, M_{\text{NT}}\frac{\mathrm{d}x_{\text{NT}}}{\mathrm{d}t} = V(x_{\text{NT}} - y_{\text{NT}}) \tag{9.19}$$

$$\text{Initial conditions: at}\, t = 0, (M, x) = \left(M^{(\text{L})}, x^{(\text{L})}\right) \tag{9.20}$$

Integration from $t = 0$ to $t = t_{vap}$ gives the state of the system at the end of the vapor flow period, $\left(M^{(V)}, x^{(V)}\right)$ (Lita, Bildea, and Kiss, 2012).

A *model of the liquid-flow period* involves the following equations, which give the stage holdup and composition at the end of the liquid-phase period (n.b. also extendable to multi-component systems):

$$\text{Condenser: } M_1^{(L)} = M_1^{(V)} - D - L;\ x_1^{(L)} = x_1^{(V)} \tag{9.21}$$

$$\text{Trays, rectifying section: } M_k^{(L)} = L;\ x_k^{(L)} = x_{k-1}^{(V)} \tag{9.22}$$

$$\text{Feed tray: } M_{\text{NF}+1}^{(L)} = L + F;\ x_{\text{NF}+1}^{(L)} = \frac{Lx_{\text{NF}}^{(V)} + Fx_{\text{F}}}{(L+F)} \tag{9.23}$$

$$\text{Trays, stripping section: } M_k^{(L)} = L + F;\ x_k^{(L)} = x_{k-1}^{(V)} \tag{9.24}$$

$$\text{Reboiler: } M_{\text{NT}}^{(L)} = M_{\text{NT}}^{(V)} - B + M_{\text{NT}-1}^{(V)};$$

$$x_{\text{NT}}^{(L)} = \frac{\left(M_{\text{NT}}^{(V)} - B\right)x_{\text{NT}}^{(V)} + M_{\text{NT}-1}^{(V)}x_{\text{NT}-1}^{(V)}}{M_{\text{NT}}^{(L)}} \tag{9.25}$$

In the *solution method*, the previous equations are written in the following condensed form, where $\Phi^{(V)}$ and $\Phi^{(L)}$ are mappings relating the state at the start and the state at the end of the vapor- and liquid-flow periods, respectively:

$$\left(M^{(V)}, x^{(V)}\right) = \Phi^{(V)}(M, x) \tag{9.26}$$

$$\left(M^{(L)}, x^{(L)}\right) = \Phi^{(L)}(M, x) \tag{9.27}$$

Note that the periodicity condition requires:

$$\left(M^{(L)}, x^{(L)}\right) = \Phi^{(L)} \circ \Phi^{(V)}\left(M^{(L)}, x^{(L)}\right) \tag{9.28}$$

A straightforward solution of this equation can be obtained by considering an initial state and applying relationships for $\left(M^{(V)}, x^{(V)}\right)$ and $\left(M^{(L)}, x^{(L)}\right)$ until the difference between two iterations becomes small. However, the convergence can be accelerated by applying algebraic equation numerical methods (e.g., Newton or Broyden). The model

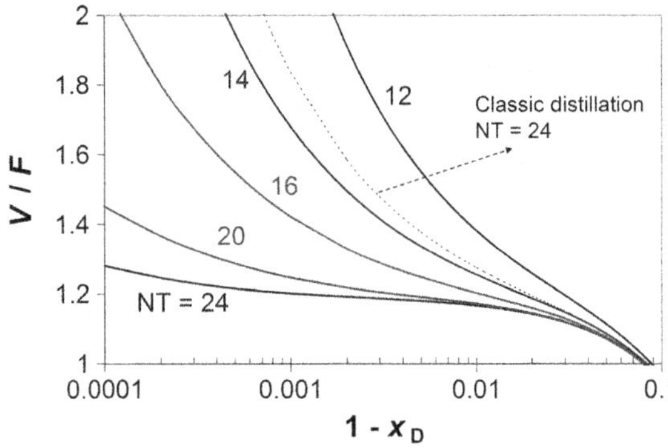

Figure 9.8 Comparison of energy requirements in cyclic distillation versus conventional distillation

can be conveniently solved in MathWorks MATLAB® or other similar software (Lita, Bildea, and Kiss, 2012).

9.5.2 Comparison with Classic Distillation

The binary separation of an equimolar mixture of benzene-toluene was considered for this comparison. Figure 9.8 compares the vapor/feed ratio versus the product purity ($x_D = 1 - x_B$) in conventional distillation and cyclic distillation employing different numbers of trays (NT): 24 trays for conventional distillation and 12–24 trays for the cyclic distillation (Lita, Bildea, and Kiss, 2012). It can be observed that the energy requirements—which are directly proportional to the vapor flow rate—are greatly reduced, especially for high purity products. Moreover, the number of trays required by cyclic distillation is almost half that needed for conventional distillation, when the same purity is obtained with the same vapor flow rate.

9.5.3 Design Methodology

- Initial data: given the feed (F, x_F) and the required performance (x_D, x_B) solve the mass balance over one operating cycle, to find the product flow rates (D, B):

$$F = D + B \tag{9.29}$$

$$Fx_F = Dx_D + Bx_B \tag{9.30}$$

Hydrodynamics: specify the vapor flow rate V and the duration of the vapor-flow period, t_{vap}, and calculate the amount of liquid transferred from condenser to the top tray:

$$L = V \cdot t_{vap} - D \quad (9.31)$$

The tray holdups are given by: $M_k = L$ (rectifying section) and $M_k = L + F$ (stripping section). Afterwards, check the hydrodynamics (column diameter, vapor velocity, pressure drop).

- The state of the reboiler at the end of the vapor-flow period, is specified as follows:
  - Holdup, $M_{NT}(t_{vap})$. The result of the design procedure is independent of this value, whose specification is nevertheless necessary.
  - Composition, $x_{NT}(t_{vap}) = x_B$. At the end of vapor-flow period, the reboiler is richer in the heavy component and, therefore, this is the moment to take out the bottom product.
- Integrate the reboiler equations, from $t = t_{vap}$ to $t = 0$, to find the holdup and reboiler composition at the beginning of the vapor-flow period, $M_{NT}(0)$, $x_{NT}(0)$.
- Find the state of the last tray (stage NT − 1) at the end of the vapor-flow period, using the mass balance for the liquid-flow period:

$$M_{NT-1}(t_{vap}) = L + F \quad (9.32)$$

$$x_{NT-1}(t_{vap}) = \frac{M_{NT}(0) \cdot x_{NT}(0) - [M_{NT}(t_{vap}) - B] \cdot x_{NT}(t_{vap})}{L + F} \quad (9.33)$$

Find the state of the reboiler and the last tray at the beginning of the vapor flow period by integrating the equations for the trays and reboiler, from $t = t_{vap}$ to $t = 0$.

- Add one more tray, whose state at the end of the vapor-flow period is given by:

$$M_{NT-2}(t_{vap}) = L + F \quad (9.34)$$

$$x_{NT-2}(t_{vap}) = x_{NT-1}(0) \quad (9.35)$$

and integrate the resulting set of equations from $t = t_{vap}$ to $t = 0$. Repeat until the feed composition is reached for the tray NF + 1.

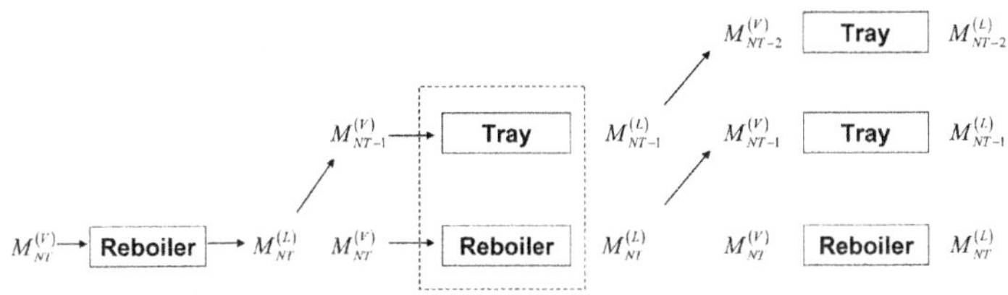

Figure 9.9 First three steps of the cyclic distillation design procedure

- Find the state of the feed tray at the end of the vapor-flow period:

$$M_{\mathrm{NF}}(t_{\mathrm{vap}}) = L \tag{9.36}$$

$$x_{\mathrm{NF}}(t_{\mathrm{vap}}) = \frac{M_{\mathrm{NF+1}} \cdot x_{\mathrm{NF+1}}(0) - F \cdot x_{\mathrm{F}}}{M_{\mathrm{NF}}} \tag{9.37}$$

  and integrate the resulting set of equations from $t = t_{\mathrm{vap}}$ to $t = 0$.
- Similarly, repeat the addition of one tray, finding its state at the end of vapor-flow period and integration of the resulting equations until the distillate composition is reached.

Figure 9.9 readily illustrates the first three steps of the cyclic distillation design procedure (Lita, Bildea, and Kiss, 2012).

9.5.4 Demonstration of the Design Procedure

The following model system is used to demonstrate the design procedure: an ideal binary mixture, with a relative volatility $\alpha = 3.5$ (n.b. the large volatility is chosen for a clear graphical illustration of the results). The following flow rates and compositions are specified:

- Feed: $F = 0.375$ kmol per cycle, $x_{\mathrm{F}} = 0.5$;
- distillate: $D = 0.1875$ kmol per cycle, $x_{\mathrm{D}} = 0.995$;
- bottoms: $B = 0.1875$ kmol per cycle, $x_{\mathrm{B}} = 0.005$;
- vapor flow: $V = 2.21$ kmol min^{-1}, $t_{\mathrm{vap}} = 0.2$ min (0.442 kmol per cycle).

Figure 9.10 shows the evolution of the liquid phase composition over four vapor flow–liquid flow cycles, for the model system (Lita, Bildea, and Kiss, 2012). The tray numbers are indicated on the right-hand side (condenser not shown). The lines indicate the flow of the liquid from one tray to tray (the liquid-flow period). The feed is on tray NT-4, as the

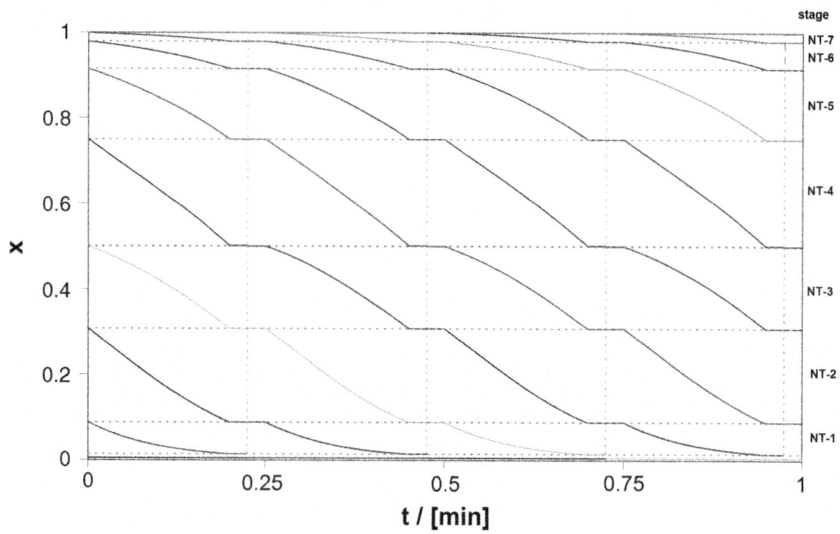

Figure 9.10 Evolution of liquid-phase mole fraction over four operating cycles

composition at the end of vapor-flow period/beginning of the liquid-flow period matches the feed composition, $x_F = 0.5$. If the vapor flow rate is set to 2.21 kmol min^{-1}, nine separation stages are needed to perform the required separation (reboiler, seven trays, condenser). The bottom mole fraction is $x_B = 0.005$, while the distillate purity $x_D = 0.9993$ exceeds the specification.

It is useful to present the results in a plot similar to the McCabe–Thiele diagram (Figure 9.11) (Lita, Bildea, and Kiss, 2012). The diagram contains the equilibrium line and the operating line. The coordinates of the points on the operating line are the liquid mole fraction on tray k, $x_k(t)$, and the vapor mole fraction on the tray below, $y_{k+1}(t)$. As the time varies between 0 and t_{vap}, one segment corresponds to each tray k. Moreover, these segments are connected because at the end of the vapor-flow period the liquid from tray k is moved to tray $k+1$, therefore:

$$x_k(0) = x_{k+1}(t_{vap}) \tag{9.38}$$

and:

$$y_{k+1}(0) = y_k(t_{vap}) \tag{9.39}$$

The dotted lines represent the operating lines for the classic distillation corresponding to the same reflux and vapor flow. Clearly, fewer trays are

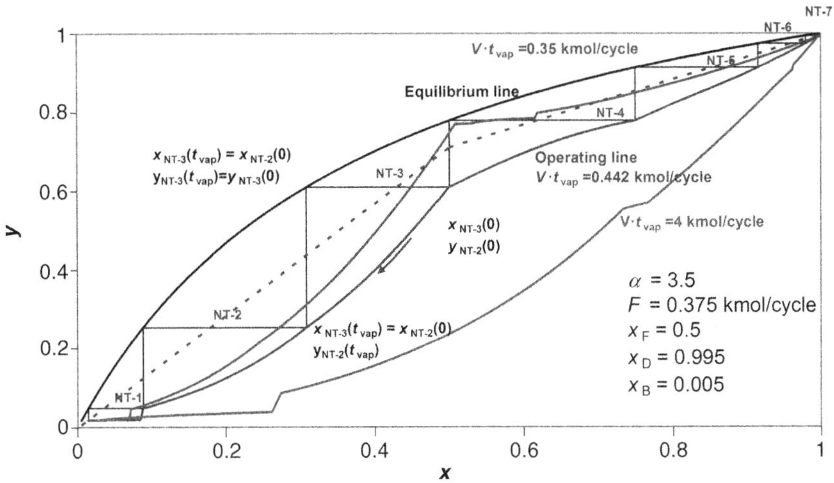

Figure 9.11 Representation of the cyclic operation mode using equilibrium and operating lines

required in cyclic operation. Moreover, Figure 9.11 illustrates the existence of a minimum vapor flow rate—corresponding to an infinite number of trays—and also a minimum number of trays, corresponding to an infinite vapor flow rate.

After the number of trays is found, the cyclic operation mode is simulated and the vapor flow rate is adjusted such that the purity specifications are exactly matched. For the test case, the new vapor flow rate is $V = 2.12\,\text{kmol min}^{-1}$.

Figure 9.12 shows the evolution of the liquid mole fraction in condenser, reboiler, and selected trays over several operating cycles (Lita, Bildea, and Kiss, 2012). Note that a stationary periodic state is reached. Moreover, at the end of vapor-flow period, the reboiler and condenser mole fractions are equal to the bottoms and distillate specifications.

9.6 CONTROL OF CYCLIC DISTILLATION

Whenever process intensification methods are introduced, more challenging and complex control might be required. However, from the standpoint of practical control of the process, cyclic distillation is relatively simple and robust—similar to conventional distillation. Nonetheless, additional optimal control (dynamic optimization) problems could be conveniently solved by employing model based control

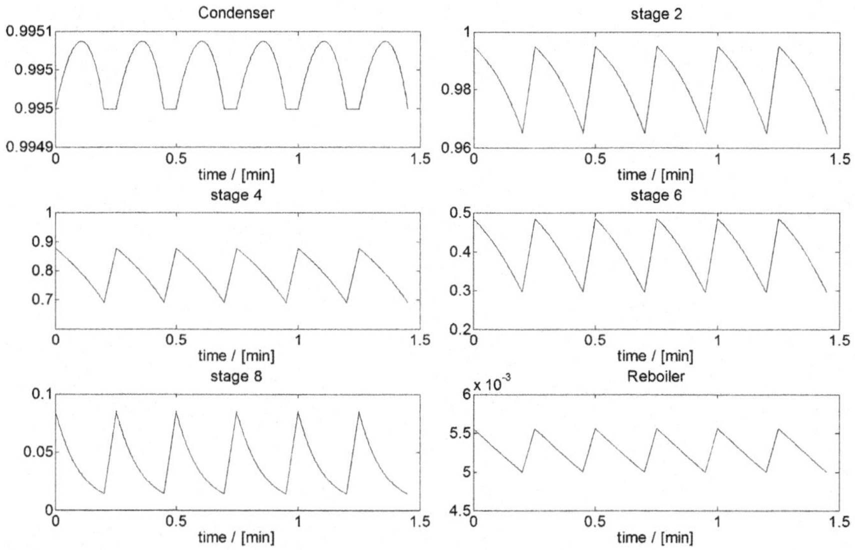

Figure 9.12 Evolution of the liquid composition in the condenser, reboiler, and selected trays, over several operation cycles

techniques such as model predictive control (MPC) (Camacho and Bordons, 2004; Nagy *et al.*, 2007).

To illustrate the good controllability of cyclic distillation, the design procedure described in the previous section was applied here for the separation of an equimolar mixture of benzene and toluene. The feed flow rate was set to 0.375 kmol per cycle, while the distillate and bottoms composition were set to 0.995 and 0.005, respectively. The cyclic distillation column has 14 stages in total, with the feed located on stage 7. The required vapor and liquid flows were set to $V = 0.496$ kmol per cycle and $L = 0.312$ kmol per cycle, respectively.

The control strategy measures the reboiler and condenser compositions at the end of the vapor-flow period and uses the discrete PI-algorithm to adjust the values of the vapor and reflux, according to:

$$u_{k+1} = u_k + \alpha\varepsilon_{k+1} + \beta\varepsilon_k \tag{9.40}$$

where u and ε are the manipulated and control errors, respectively, while α and β are the control tuning parameters.

In addition, the bottoms and distillate amounts are used to control the reboiler and condenser holdups (e.g., level control). The performance of the control system was tested for a 10% increase of the feed flow rate, as

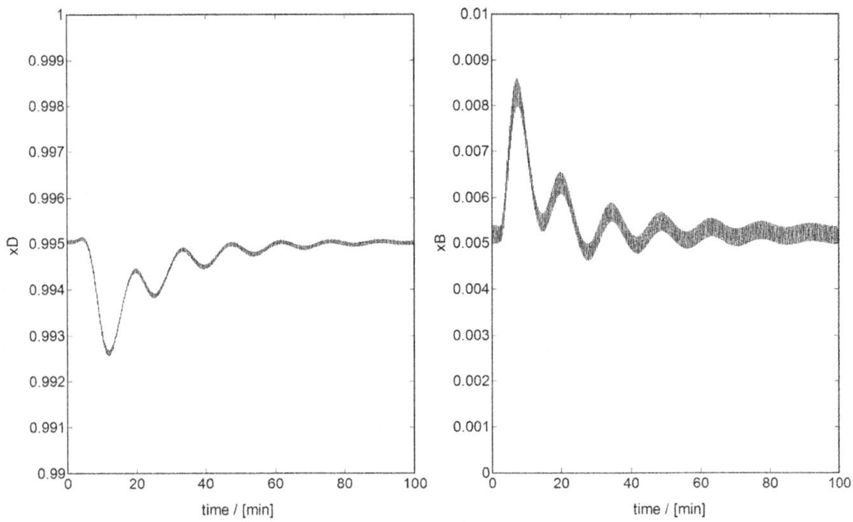

Figure 9.13 Performance of the control system for a 10% increase in the feed flow

well as for a change of the feed composition from $x_F = 0.5$ to 0.6. The results presented in Figures 9.13 and 9.14 show that the product purities can be indeed kept to their set points (Lita, Bildea, and Kiss, 2012). Remarkably, the disturbances in the feed flow rate and composition are

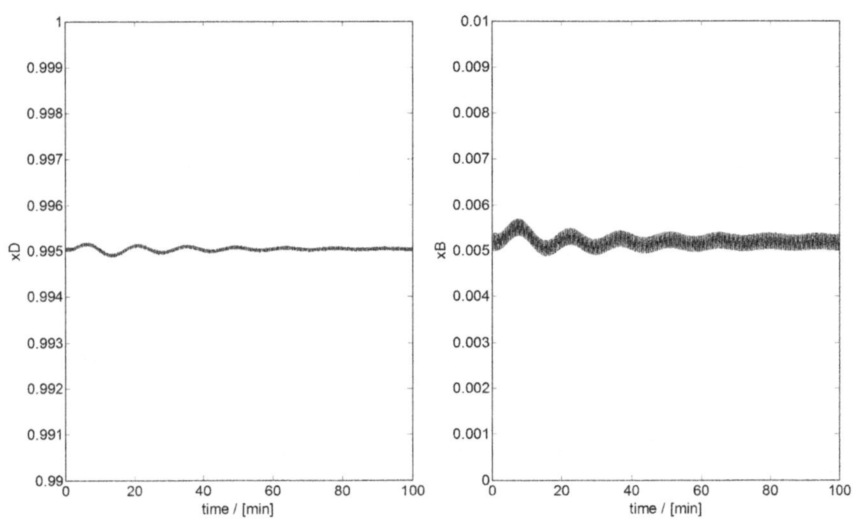

Figure 9.14 Performance of the control system for a change in feed composition from $x_F = 0.5$ to 0.6

rejected successfully with short settling times and low overshooting—proving indeed the good controllability of cyclic distillation.

9.7 CYCLIC DISTILLATION CASE STUDIES

Two industrially relevant case studies are presented below: ethanol–water stripping and concentration (from the food industry) and methanol–water separation (from the chemical process industry).

9.7.1 Ethanol–Water Stripping and Concentration

Here we describe briefly the modeling of a cyclic distillation column implemented at industrial scale in the food industry (Lipnitsky alcohol plant, Ukraine). The cyclic distillation column described hereafter acts mainly as a stripper that increases the alcohol concentration to a higher grade. In the beer production process, a mixture of mainly ethanol–water is obtained by fermentation. We model the stripping (beer) column for the real plant production of 20 000 l per day of food-grade ethanol. There are also additional components (~30 impurities) but these constitute less than 0.2 mol.% in the feed stream. The concentration of alcohol in the feed stream is 10 vol.%—equivalent to 3.29 mol.%. The feed stream is supplied on the top tray of the cyclic distillation column (stripping column), while direct steam injection is used in the bottom of the column. The specification for ethanol concentration in the top distillate varies between 13 and 24 mol.%, with a typical value of ~18.25 mol.%, while the concentration of ethanol in the bottom product must not exceed 0.004 mol.%. For this system a simple modeling of the cyclic distillation column was performed, based on the theoretical background presented earlier (Maleta *et al.*, 2011).

Table 9.1 illustrates the influence of the diffusion potential factor ($\overline{\lambda} = mG/L$), local point efficiency (E_{OG}), and the factor of liquid delay transfer (F) on the Murphree efficiency ($\overline{E_{MV}}$) and the number of trays (N) in a cyclic distillation column for ethanol concentration (Maleta *et al.*, 2011). Consequently, fewer trays and a higher efficiency are possible when the perfect displacement mode of operation is achieved.

The column for concentrating impurities is also part of an industrial plant producing food grade alcohol (Lipnitsky alcohol plant, Ukraine). Figure 9.15 shows the flowsheet of the ethanol concentration section, with the cyclic distillation column as the key operating unit (Maleta *et al.*,

Table 9.1 Influence of the diffusion potential factor ($\overline{\lambda} = mG/L$), local point efficiency (E_{OG}), and liquid transfer delay factor (F) on the Murphree efficiency ($\overline{E_{MV}}$) and the number of trays (N) in a cyclic distillation column for ethanol concentration

L (kmol h^{-1}) (feed)	x_F (mol.%) (feed)	G (kmol h^{-1}) (steam)	y_F (mol.%)	L/G	λ (= mG/L)	E_{OG} (point efficiency)	F (= H_l/H)	x_B (mol.%) (bottom)	x_D (mol.%) (distillate)	E_{MV} (Murphree efficiency)	N (trays)
617	3.29	152.78	0	4.04	**2.2**	1	1	0.003	13.269	**4.91**	3
617	3.29	138.89	0	4.44	**2**	1	1	0.001	14.61	**4.29**	4
617	3.29	125	0	4.94	**1.8**	1	1	0.003	16.22	**3.64**	4
617	3.29	111.11	0	5.55	**1.6**	1	1	0.002	18.25	**3.19**	5
617	3.29	97.22	0	6.35	**1.4**	1	1	0.002	20.866	**2.77**	7
617	3.29	83.33	0	7.4	**1.2**	1	1	0.003	24.345	**2.35**	10
617	3.29	111.11	0	5.55	1.6	**1**	1	0.002	18.25	**3.19**	5
617	3.29	111.11	0	5.55	1.6	**0.9**	1	0.002	18.254	**2.68**	6
617	3.29	111.11	0	5.55	1.6	**0.8**	1	0.002	18.255	**1.82**	7
617	3.29	111.11	0	5.55	1.6	**0.7**	1	0.003	18.252	**1.48**	8
617	3.29	111.11	0	5.55	1.6	**0.6**	1	0.003	18.257	**1.12**	10
617	3.29	111.11	0	5.55	1.6	**0.5**	1	0.002	18.267	**0.79**	13
617	3.29	111.11	0	5.55	1.6	**0.1**	1	0.003	18.898	**0.1**	84
617	3.29	111.11	0	5.55	1.6	1	**1**	0.002	18.25	**3.19**	5
617	3.29	111.11	0	5.55	1.6	1	**0.9**	0.001	18.258	**2.81**	6
617	3.29	111.11	0	5.55	1.6	1	**0.8**	0.002	18.253	**2.45**	6
617	3.29	111.11	0	5.55	1.6	1	**0.7**	0.004	18.246	**2.13**	6
617	3.29	111.11	0	5.55	1.6	1	**0.6**	0.002	18.225	**1.89**	7
617	3.29	111.11	0	5.55	1.6	1	**0.5**	0.004	18.249	**1.67**	7
617	3.29	111.11	0	5.55	1.6	1	**0.1**	0.001	18.388	**1.05**	12

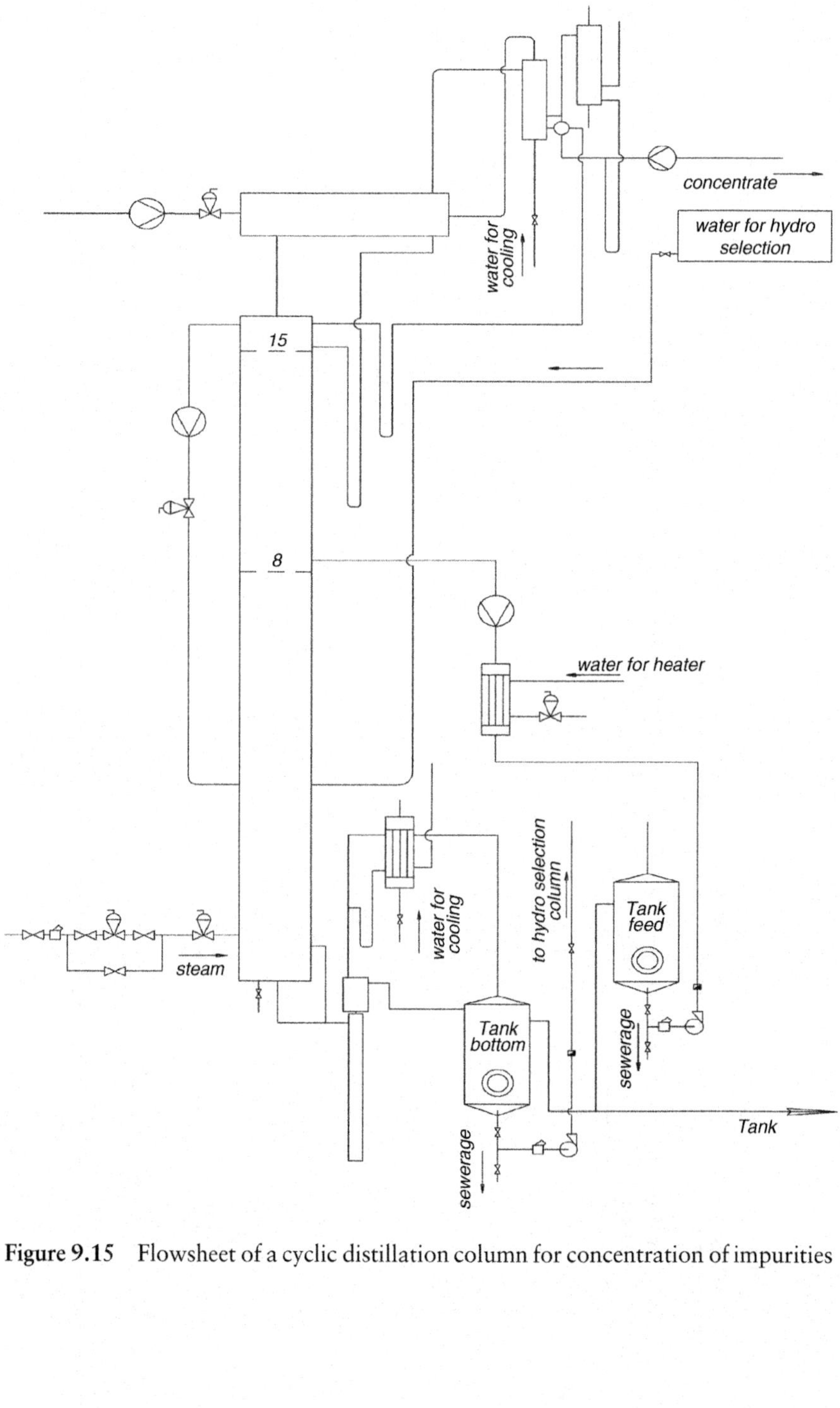

Figure 9.15 Flowsheet of a cyclic distillation column for concentration of impurities

2011). The wastes from this column represent 5% of the plant capacity. The column concentrating the impurities recycles waste back to the plant. Remarkably, the wastes of this column are only 0.5% of the plant capacity, so an additional alcohol output (4.5% of the total plant capacity) is possible.

Note that the hydro-selection column and the column for concentrating impurities are the same type of column, performing the same task of removing the light components, while sharing identical impurities and working under the same conditions. The main difference between them lies in the fact that the concentration of impurities in the cyclic distillation column is much higher than in the hydro-selection column. This makes it possible to increase the yield of the desired product. The efficiency of both columns is compared under conditions close to the ones determined by Fenske. For both columns, the reflux ratio is about 50 mol mol^{-1}. Table 9.2 presents the concentration of impurities (ppm) in the liquid stream for the hydro-selection column and also for the column concentrating the impurities (Maleta *et al.*, 2011).

For example, a simple comparison of the key components listed in Table 9.2 shows that in the conventional hydro selection column (56 trays) the concentration of acetic aldehyde in the bottom is 0.461 ppm and in the top is 660.38 ppm. For the cyclic distillation column (15 trays) the bottom concentration of acetic aldehyde is 0.36 ppm while the top concentration reaches 17730 ppm. A similar trend is observed for the other components in the hydro selection column: methyl acetate (conventional: bottom 0 ppm and top 103.72 ppm, cyclic distillation: bottom 0 ppm and top 2394 ppm), acetone (conventional: bottom 0 ppm and top 7.3 ppm, cyclic distillation: bottom 0 ppm and top 192.2 ppm), and ethyl acetate (conventional: bottom 0 ppm and top 3806 ppm, cyclic mode: bottom 0 ppm and top 94527 ppm). According to the reported industrial data (Lipnitsky alcohol plant, Ukraine) it can be concluded that the separation efficiency in the cyclic distillation column (15 trays) is much higher than that of the conventional bubble cap trays column (56 trays)—exceeding 200% efficiency.

9.7.2 Methanol–Water Separation

This case study describes briefly the modeling of a cyclic distillation column used for the separation of a methanol–water mixture. The model is based on detailed equations for the differential mass balance of the stage during vapor supply, the liquid flow hydrodynamics, and the mass

Table 9.2 Impurity content (ppm) in feed and products for a capacity of 20 000 l per day

Component	Group	Hydro selection column $D=1000$ mm, 56 trays, $W=550$ kW			Column concentrating the impurities, cyclic mode $D=400$ mm, 15 trays, $W=50$ kW		
		Feed	Top	Bottom	Feed	Bottom	Top
Ethylic ether	Ether	—	0.223	—	0.223	—	1.404
Acetic aldehyde	Aldehyde	88.29	660.38	0.461	660.382	0.36	17730
Methyl acetate	Ester	6.597	103.72	—	103.727	—	2394
Acetone	Ketone	0.727	7.3	—	7.3	—	192.2
Ethyl acetate	Ester	261.5	3806	—	3806.69	—	94527
Methanol	Alcohol	0.038	0.15	0.005	0.150	0.009	1.94
2-Butanone	Ketone	—	5.38		5.383	—	115.37
2-Propanol	Alcohol	5.77	75.17	1.339	75.178	2.93	1796
Isobutyl acetate	Ether	2.00	57.85	0.393	57.856	0.17	1465
1-Propanol	Alcohol	13762	83.76	2289	83.763	80.04	1192
Croton aldehyde	Aldehyde	0.655	5.22	0.075	5.226	0.201	143.6
Isobutyl alcohol	Alcohol	23476	84.64	2661	84.642	139.86	57.6
Isoamyl acetate	Ester	—	57.17	—	57.175	—	1272
1-Butanol	Alcohol	130.4	—	18.72	—	0.744	6.62
Isoamyl alcohol	Alcohol	42982	3.43	4295	3.431	205.64	78.49
1-Pentanol	Alcohol	11.82	—	1.193	—	—	3.64
1-Hexanol	Alcohol	20.20	—	1.936	—	0.142	0.26
Benzaldehyde	Aromatic aldehyde	6.42	—	0.809	—	0.309	—
2-Phenyl ethanol	Aromatic alcohol	518.5	—	42.43	—	1.088	—

transfer kinetics using local point efficiency. Aspen Custom Modeler (ACM) was chosen for model implementation due to its rich features as a computer-aided process engineering tool. As ACM has built-in connectivity for the thermodynamics and flowsheet calculations, the model allows a straightforward performance evaluation for any binary separation process, filling the gap between design and application of cyclic distillation systems.

Considering a generic tray i (Figure 9.1) during the vapor period, a vapor flow V_{i+1} with a composition y^a_{i+1} enters the plate, opening the liquid valve of the previous sluice chamber. Then, a liquid flow L_i with composition x^a_i enters the current tray, forming the tray hold-up H_i. The desired mass transfer is only carried out during the vapor period, conferring a new composition y^a_i to the vapor flow V_i leaving the plate (Robinson and Engel, 1967). Note that in the feed tray, V_{i+1} and L_i must accordingly include the feed flow rate F, feed composition z_a, and quality q (Rivas, 1977). Thus, the global mole balance is expressed as:

$$\frac{dH_i}{dt} = V_{i+1} + L_i - V_i - L_{i-s} \tag{9.41}$$

The mole balance of the more volatile component describes the liquid hold up for that component (H^a_i) and the liquid composition x^a_{i+1}, as follows (Skogestad, 1997):

$$\frac{dH^a_i}{dt} = L_i \cdot x^a_i + V_{i+1} \cdot y^a_{i+1} - L_{i-s} \cdot x^a_{i+1} - V_i \cdot y^a_i \tag{9.42}$$

Additionally, the relative volatilities calculated with the Aspen Properties interface define the liquid–vapor equilibrium. Furthermore, the local point efficiency E_{OG}, an analog of the Murphree efficiency, defines the mass transfer for the present model by (Maleta *et al.*, 2011):

$$E_{OG} = \frac{y^a_i - y^a_{i+1}}{y^a_{eq} - y^a_{i+1}} \tag{9.43}$$

The liquid period begins when the vapor flow stops. During this time, the liquid runs from the tray to the sluice chamber with a flow $L_{i\text{-}s}$ to build up the sluice holdup H_s. Afterwards, the vapor flow starts again and opens the sluice valves, allowing the liquid flow L_{i+1} to reach the tray $i+1$ below. During the liquid period, the liquid composition x^a_{i+1} remains unchanged from the vapor period. Thus, the global balance for the sluice chamber is given by:

$$\frac{\mathrm{d}H_s^i}{\mathrm{d}x} = L_{i\text{-s}} - L_{i+1} \tag{9.44}$$

The definition of a liquid flow L_j that is falling by gravity from the plates or sluices is given by hydrodynamic relations. These equations are associated with the respective molar holdup H_j, but they are heavily dependent on the tray design. Thus, the adjustable parameter k_{hydro} is introduced here to obtain a model as generic as possible. The parameter k_{hydro} also allows matching the average residence time for liquid in the trays as previously suggested by Maleta *et al.* (2011). Therefore, the liquid flows are defined as:

$$L_j = -k_{\text{hydro}} \cdot H_j \tag{9.45}$$

The proposed model was implemented in ACM, and solved using the Newton solver and the variable-step-implicit-Euler integrator. ACM also has the advantage of easy integration with other AspenTech software. A separate property model is not required in this case, since a convenient Aspen Properties interface can be used instead. Consequently, a wide variety of processes can be evaluated with the same model by simply selecting the required components from the Aspen Plus database. Moreover, the so-called "*tasks*" routines of ACM allow not only modeling of the regular pseudo-steady-state operation of cyclic distillation, but also evaluation of its operability when significant inlet changes occur. SPM was implemented by setting directly the vapor flow values to simulate the vapor period, while the parameter k_{hydro} is adjusted accordingly, defining indirectly the liquid flow during the liquid periods. In addition, ACM allows connection of the cyclic distillation model with any process flowsheet to evaluate the overall process dynamics in a simple manner.

The model performance is evaluated for the separation of an equimolar methanol–water mixture. A vapor flow rate of 8 mol s^{-1} is fed to the tray during the vapor period, while a liquid flow rate of 4 mol s^{-1} is introduced during the remaining period using a modified sluice model. For this study, the tray local point efficiency was considered at ideal conditions (100%). Figure 9.16 shows the overall performance regarding the hydrodynamics and the separation task for the cyclic distillation tray model (Kiss *et al.*, 2012). The SPM conditions were satisfied by basing the model on the newest hardware design (Maleta and Maleta, 2009, 2010).

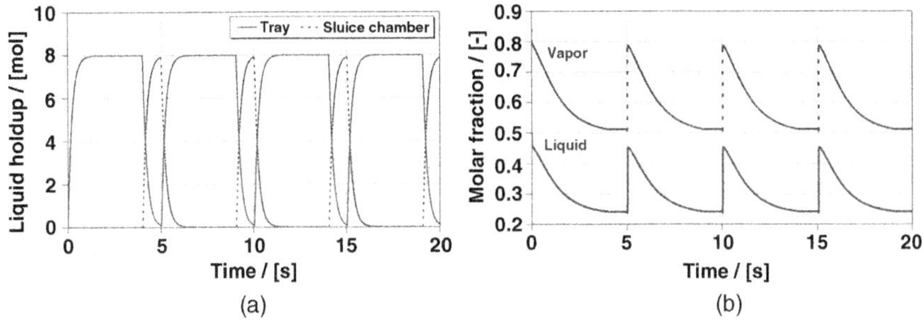

Figure 9.16 Overall performance of the hydrodynamics (a) and the separation task (b) of the proposed cyclic distillation model

Based on the liquid and vapor flows, the total cycle is set to 5 s, starting with a 4 s vapor period. Figure 9.16a shows the typical increase in the tray hold up at the start of the vapor period, reproducing the opening of the previous sluice valves by the upcoming vapor flow. The vapor flow also prevents liquid drainage to the sluice chambers, thus stabilizing the liquid holdup on the tray. The molar composition plot (Figure 9.16b) shows how the vapor becomes richer in the most volatile compound, while the liquid becomes poorer during the vapor period. The mass transfer may continue until the inlet and outlet vapor compositions are equal, indicating that the system has reached equilibrium. For the current model, the mass transfer is disrupted at 4 s, when the vapor feed stops. Then, the liquid starts draining from the plate and filling the sluice chamber according to the hydrodynamic conditions, as seen in Figure 9.16a (Kiss *et al.*, 2012). Note that there is no mass transfer during the liquid period—Figure 9.16b shows unchanged liquid and vapor compositions. When all the liquid from the tray is stored in the sluice chamber, the cycle resumes its vapor period once more, repeating the process.

Figure 9.17 illustrates the influence of local point efficiency (E_{OG}) on the concentration profile of methanol in the liquid phase (Kiss *et al.*, 2012). Lower efficiencies diminish the speed at which the methanol concentration is changing. Similarly, the time required to reach mass transfer equilibrium becomes higher when constant vapor flow is considered. The relative difference in concentrations at the end of the vapor period increases up to 16% (at a local point efficiency of 50%) as compared to the ideal case ($E_{OG} = 100\%$). Moreover, Figure 9.17b shows the hydrodynamic results for disturbances in the liquid inlet flow rate, which was increased from 8 to 10 mol s^{-1} at $t = 4$ s, and then decreased to 5 mol s^{-1} at $t = 14$ s. Remarkably, cyclic distillation

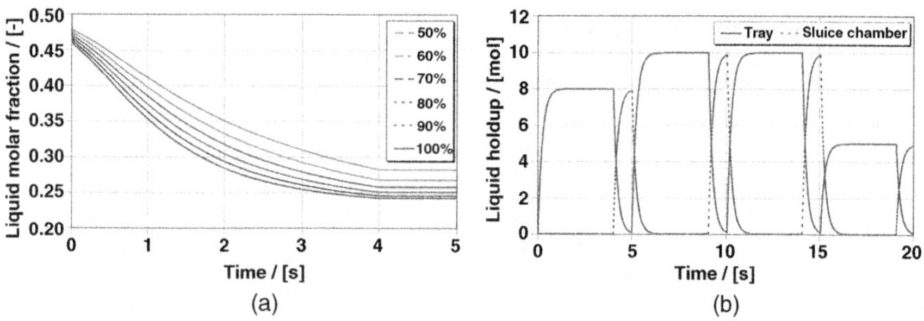

Figure 9.17 (a) System behavior for changes in local point efficiency (top line corresponds to 50% and the bottom to 100%) and (b) liquid inlet flow

can easily account for the capacity changes in straightforward fashion, as long as there are no hardware restrictions. In practice the liquid fed from an upstream plant is sent to buffer tanks (reflux drums also included). From there, the liquid is periodically pumped to the cyclic distillation column during the liquid period. Thus, the inlet liquid flow is increased or reduced accordingly to account for production capacity changes.

Experimental studies and pilot-scale tests of cyclic distillation were conducted several decades ago, as reported in the literature (McWhirter and Cannon, 1961; Maleta *et al.*, 2011). Consequently, the model proposed here was validated against experimental data to check its accuracy. The details are shown in Table 9.3 and Figure 9.18 for the methanol–water separation (Kiss *et al.*, 2012). The experimental data were chosen mainly because a relatively complex column design was used to achieve separate phase movement (Matsubara, Watanabe, and Kurimoto, 1985). Remarkably, the generic model developed here predicts the final liquid composition in the vapor period within less than 1% relative error. Additionally, the liquid composition trend achieved with the current model matches the experimental claims reported in the literature (Matsubara, Watanabe, and Kurimoto, 1985).

Table 9.3 Model parameters of the simulation and experiments

Parameter (unit)	Symbol	Value
Local point efficiency (–)	E_{OG}	0.846
Molar liquid holdup (mol)	H_i	21.8
Mol fraction vapor (–)	y^a_{i+1}	0.652
Vapor flow rate (mol s^{-1})	V_{i+1}	0.0324
Temperature (°C)	T	75
Mol fraction liquid (experimental) (–)	$x^a_{i+1,\mathrm{exp}}$	0.445
Mol fraction liquid (simulation) (–)	$x^a_{i+1,\mathrm{sim}}$	0.442
Relative error (%)	—	−0.6

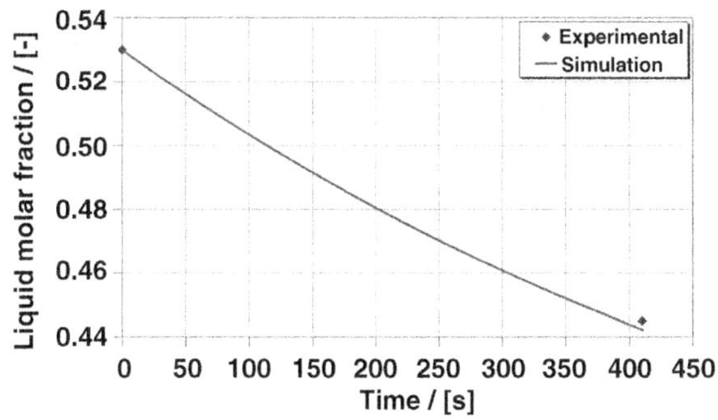

Figure 9.18 Model validation: simulation results versus experimental data

The industrial case studies included here demonstrate that cyclic distillation could now be considered as proven technology. Note that, practically, cyclic distillation could be used in batch or continuous distillation, and absorption and reactive distillation. However, some restrictions are imposed on the velocity of the gas (steam) in the column ($0.3–2\,m\,s^{-1}$) and the load of liquid ($1–30\,m^{-3}\,m^{-2}\,h^{-1}$).

9.8 CONCLUDING REMARKS

Cyclic operation can be achieved by controlled cycling, stepwise periodic operation, a combination of these two, or by stage switching. Controlled cycling appears to be the simplest scheme and it is therefore preferred. The cyclic operation was demonstrated on columns equipped with various types of internals: plates (brass, mesh-screen, bubble cap, sieve, or packed-plate) and trays with sluice chambers. Cyclic distillation can bring new life to old distillation columns, providing key benefits such as high column throughput, low energy requirements, and high separation performance. Moreover, the mass transfer technology with separate phase movement (SPM) has more degrees of freedom that may contribute to better process control. The review of the available literature indicates several benefits of cyclic operation as compared to conventional continuous operation:

- Higher tray efficiencies, meaning that at the same vapor flow rate a required purity can be obtained using fewer trays—thus also using lower capital expenditure (CapEx);
- reduced energy requirements, meaning that a required purity can be achieved with lower vapor flow rates, with the same number of trays—thus high savings in operating costs (OpEx);

- higher throughput and equipment productivity than conventional distillation;
- increased quality of the products, due to higher separation efficiency;
- smaller pressure drops on trays as compared to steady-state operation;
- flexible operation with maximum capacity or maximum efficiency.

Nevertheless, one must also pay attention to the limitations of cyclic operation. For example, the application of cyclic operation to vacuum distillation seems rather difficult. Moreover, the performance enhancement critically depends on the complete separation between the liquid and vapor flow periods. However, with simple trays, both theoretical models and experiments indicate that this is difficult to achieve when the column has more than ten trays—installing a pressure equalizing manifold has been suggested as a solution. The more recently proposed sluice-chamber trays also seem to avoid the limitations related to the number of trays (Maleta *et al.*, 2011).

The model of a theoretical stage with perfect displacement and the operating lines theory described in this chapter provide significant insight into understanding process intensification in cyclic distillation, and makes available adequate tools for designing tray distillation columns operated in a controlled periodic mode. The process operating line is a continuous concave line that reflects the shape of the equilibrium line. The mixing of liquid during liquid overflow from tray to tray enhances the separation efficiency. The operating line has a discrete nature and at its limit, upon perfect mixing, can be approximated as the operating line L/G. The mass transfer kinetics expressed through the local point efficiency (E_{OG}) influence not only the position of the kinetic curve but also the position and the shape of the operating line. The ultimate position of both lines at $E_{OG} \to 0$ is the operating line L/G. The material differences between the efficiency of the stationary and the cyclic processes are observed at high values of $F \to 1$, and $E_{OG} \to 1$. The Murphree efficiency of a tray continuously increases, subject to the increase of the local point efficiency, stripping factor, and hydrodynamic conditions of liquid overflow on the trays.

In addition, the novel approaches for the modeling and design of cyclic distillation systems allow an insightful comparison with conventional distillation (Lita, Bildea and Kiss, 2012). The key findings are:

- The energy requirements for cyclic distillation are greatly reduced, especially for high purity products, while the number of trays required is almost half that needed for conventional distillation—when the same purity is obtained with the same vapor flow rate.

- A minimum vapor-flow rate exists in cyclic distillation, corresponding to an infinite number of trays—and a minimum number of trays that corresponds to an infinite vapor flow rate.
- Cyclic distillation columns can be easily controlled by adjusting the reflux and the vapor flow rate, to keep the required product purities. Disturbances in the feed flow rate and composition are successfully rejected with low settling times and low overshooting.

REFERENCES

Baron, G., Wajc, S., and Lavie, R. (1980) Stepwise periodic distillation - I: Total reflux operation. *Chemical Engineering Science*, *35*, 859–865.

Baron, G., Wajc, S., and Lavie, R. (1981) Stepwise periodic distillation - II: Separation of a binary mixture. *Chemical Engineering Science*, *36*, 1819–1827.

Bausa, J. and Tsatsaronis, G. (2001) Reducing the energy demand of continuous distillation processes by optimal controlled forced periodic operation. *Computers & Chemical Engineering*, *25*, 359–370.

Camacho, E.F. and Bordons, C. (2004) *Model Predictive Control*, 2nd edn, Springer, London.

Chien, H.H., Sommerfeld, J.T., Schrodt, V.N., and Parisot, P.E. (1966) Studies of controlled cyclic distillation: II. Analytical transient solution and asymptotic plate efficiencies. *Separation Science and Technology*, *1*, 281–317.

Dejanović, I., Matijasevic, Lj., and Olujic, Z. (2010) Dividing wall column - A breakthrough towards sustainable distilling. *Chemical Engineering and Processing: Process Intensification*, *49*, 559–580.

Flodman, H.R. and Timm, D.C. (2012) Batch distillation employing cyclic rectification and stripping operations. *ISA Transactions*, *51*, 454–460.

Furzer, I.A. (1978) Discrete residence time distribution of a distillation column operated with microprocessor controlled periodic cycling. *Canadian Journal of Chemical Engineering*, *56*, 747–750.

Gaska, R.A. and Cannon, M.R. (1961) Controlled cycling distillation in sieve and screen plate towers. *Industrial & Engineering Chemistry*, *53*, 630–631.

Gel'perin, N.I., Polotskii, L.M., and Potapov, T.G. (1976) Operation of a bubble-cap fractionating column in a cyclic regime. *Chemical and Petroleum Engineering*, *11*, 707–709.

Goss, D.W. and Furzer, I.A. (1980) Mass transfer in periodically cycled plate columns containing multiple sieve plates. *AIChE Journal*, *26*, 663–669.

Harmsen, J. (2007) Reactive distillation: The front-runner of industrial process intensification. A full review of commercial applications, research, scale-up, design and operation. *Chemical Engineering and Processing*, *46*, 774–780.

Harmsen, J. (2010) Process intensification in the petrochemicals industry: Drivers and hurdles for commercial implementation. *Chemical Engineering and Processing*, *49*, 70–73.

Kiss, A.A., Pragt, J.J., and Strien, C.J.G. (2009) Reactive dividing-wall columns - How to get more with less resources? *Chemical Engineering Communications*, *196*, 1366–1374.

Kiss, A.A. and Bildea, C.S. (2011) A control perspective on process intensification in dividing-wall columns. *Chemical Engineering and Processing: Process Intensification*, *50*, 281–292.

Kiss, A.A., Segovia-Hernandez, J.G., Bildea, C.S. *et al.* (2012a) Reactive DWC leading the way to FAME and fortune. *Fuel*, *95*, 352–359.

Kiss, A.A., Flores Landaeta, S.J., and Zondervan, E. (2012b) Cyclic distillation – Towards energy efficient binary distillation. *Computer Aided Chemical Engineering*, *30*, 697–701.

Lewis, W.K. (1936) Rectification of binary mixtures. *Industrial & Engineering Chemistry*, *28*, 399–402.

Lita, I., Bildea, C.S., and Kiss, A.A. (2012) Modeling, design and control of cyclic distillation systems. *Procedia Engineering*, *42*, 1311–1322.

Maleta, V.N., Taran, V.M., and Dubovik, V.A. (1986) Comparison of cyclic and steady-state rectification processes. *Proceedings of the Universities of USSR, Food Technology*, *6*, 52–55.

Maleta, V.N., Taran, V.M., and Maleta, B.V. (2010) Use of a theoretical stage model with perfect displacement with a trayed column with separate movement of the vapour and liquid phases. Presented at the 19th International Conference of Chemical and Process Engineering, CHISA 2010, Paper number 0285.

Maleta, B.V. and Maleta, O.V. (2009) Mass exchange contact device, US Patent Application No., 20090145738.

Maleta, B.V. and Maleta, O.V. (2010) Mass exchange contact device - mass-exchange contact device is proposed comprising an upper tray contact element including a barbotage unit configured, US Patent Application No., 20100221156.

Maleta, B.V. and Maleta, O.V. (2010) Mass-exchange contact device – is proposed comprising upper and lower flat ring-shaped trays attached to a column, US Patent Application No., 20100219060.

Maleta, V.N., Kiss, A.A., Taran, V.M., and Maleta, B.V. (2011) Understanding process intensification in cyclic distillation systems. *Chemical Engineering and Processing: Process Intensification*, *50*, 655–664.

Matsubara, M., Watanabe, N., and Kurimoto, H. (1985) Binary periodic distillation scheme with enhanced energy conservation - I: Principle and computer simulation. *Chemical Engineering Science*, *40*, 715–721.

Matsubara, M., Watanabe, N., and Kurimoto, H. (1985) Binary periodic distillation scheme with enhanced energy conservation - II: Experiment. *Chemical Engineering Science*, *40*, 755–758.

McCabe, W.L. and Thiele, E.W. (1925) Graphical design of fractionating columns. *Industrial & Engineering Chemistry*, *17*, 605–611.

McWhirter, J.R. and Cannon, M.R. (1961) Controlled cycling distillation. *Industrial & Engineering Chemistry*, *53*, 632–634.

Nagy, Z.K., Mahn, B., Franke, R., and Allgöwer, F. (2007) Evaluation study of an efficient output feedback nonlinear model predictive control for temperature tracking in an industrial batch reactor. *Control Engineering Practice*, *15*, 839–850.

Olujic, Z., Jodecke, M., Shilkin, A. *et al.* (2009) Equipment improvement trends in distillation. *Chemical Engineering and Processing*, *48*, 1089–1104.

Rivas, O.R. (1977) An analytical solution of cyclic mass transfer operations. *Industrial & Engineering Chemistry – Process Design and Development*, *16*, 400–405.

Robinson, R.G. and Engel, A.J. (1967) Analysis of controlled cycling mass transfer operations. *Industrial & Engineering Chemistry*, *59*, 22–29.

Sommerfeld, J.T., Schrodt, V.N., Parisot, P.E., and Chien, H.H. (1966) Studies of controlled cyclic distillation: I. Computer simulations and the analogy with conventional operation. *Separation Science and Technology*, *1*, 245–279.

Schrodt, V.N., Sommerfeld, J.T., Martin, O.R. *et al.* (1967) Plant-scale study of controlled cyclic distillation. *Chemical Engineering Science*, *22*, 759–767.

Skogestad, S. (1997) Dynamics and control of distillation columns - A tutorial introduction. *Trans IChemE*, *75*, 539–562.

Szonyi, L. and Furzer, I.A. (1985) Periodic cycling of distillation-columns using a new tray design. *AIChE Journal*, *31*, 1707–1713.

Thompson, M.F. and Furzer, I.A. (1985) Hydrodynamic simulation of periodic cycled plate columns. *AIChE Journal*, *31*, 1275–1287.

Toftegard, B. and Jørgensen, S.B. (1989) An integration method for dynamic simulation of cycled processes. *Computers & Chemical Engineering*, *13*, 927–930.

Yildirim, O., Kiss, A.A., and Kenig, E.Y. (2011) Dividing wall columns in chemical process industry: A review on current activities. *Separation and Purification Technology*, *80*, 403–417.

10

Reactive Distillation

10.1 INTRODUCTION

Most chemical processes involve reaction and separation operations that are typically carried out in different sections of the plant and use different equipment types—such as continuous stirred-tank reactor (CSTR), plug-flow reactor (PFR), batch reactor, distillation column—operated under a wide variety of conditions (Luyben and Yu, 2008). Recycle streams between these operating units are often used to improve conversion and selectivity, to minimize the production of undesired by-products, to reduce the energy requirements, and to improve the process controllability. Recent economic and environmental considerations have encouraged the chemical process industry to focus on technologies based on process intensification. Reactive distillation (RD) is an integrated operation that conveniently combines reaction and separation in a single unit, allowing the simultaneous production and removal of products, thus improving the productivity and selectivity, reducing the energy use, eliminating the need for solvents, and leading to intensified, high-efficiency systems with green engineering attributes (Malone and Doherty, 2000;; Malone, Huss, and Doherty, 2003). Some of these benefits are realized by using the reaction to improve separation (e.g., overcoming azeotropes, reacting away contaminants, avoiding or eliminating difficult separations) while others are realized by using separation to improve reactions (e.g., enhancing overall rates, overcoming reaction equilibrium limitations, improving selectivity, removing catalyst poisons)—with the maximum effect being achieved when both aspects are important (Malone and Doherty, 2000; Harmsen, 2007, 2010). However, the

Advanced Distillation Technologies: Design, Control and Applications, First Edition.
Anton Alexandru Kiss.

application of RD is somewhat limited by specific constraints, such as common operation range for distillation and reaction (similar temperature and pressure), proper boiling point sequence (product should be the lightest or heaviest component, with side or by-products as the mid boiling ones) and difficulties in providing proper residence time characteristics (Krishna, 2002; Noeres, Kenig, and Gorak, 2003).

A small number of industrial applications of RD have been around for many decades (Luyben and Yu, 2008), but even today the RD crown is held by the Eastman process that reportedly replaced a methyl acetate production plant (Figure 10.1) with a single RD column using 80% less energy at only 20% of the investment costs (Agreda, Partin, and Heise, 1990; Siirola, 1996). Nowadays, the application with the largest number of installations is methyl tertiary butyl ether (MTBE), which is used in gasoline blending. Other esters are also produced now by RD, such as ethyl *tert*-butyl ether (ETBE), *tert*-amyl methyl ether (TAME) or fatty acid methyl esters (FAME) (Harmsen, 2007; Kiss and Bildea, 2012).

Considering the large extent of the topic, this chapter gives only a brief overview of RD. For more detailed information, the reader it directed to several books and reviews published in the last decade covering a large range of subjects about RD, namely, process synthesis, conceptual design, control, optimization, operation, as well as industrial applications (Sundmacher and Kienle, 2003; Sundmacher, Kienle, and Seidel-Morgenstern, 2005; Luyben and Yu, 2008; Malone and Doherty, 2000; Harmsen, 2007, 2010; Tuchlenski *et al.*, 2001; Schoenmakers and Bessling, 2003; Stankiewicz, 2003; Hiwale *et al.*, 2004).

10.2 PRINCIPLES OF REACTIVE DISTILLATION

RD is especially attractive in systems where certain chemical and phase equilibrium conditions co-exist. The reaction and distillation take place in the same zone of a distillation column, the reactants being converted with the simultaneous separation of the products and recycle of unused reactants. As the products must be separated from reactants by distillation, this implies that the products should be lighter and/or heavier than the reactants. The ideal case is when one product is the lightest and the other product is the heaviest, with the reactants being the intermediate boiling components (Luyben and Yu, 2008). Moreover, as both operations occur simultaneously in the same unit, there must be a proper match between the temperatures and pressures required for reaction and separation (Noeres, Kenig, and Gorak, 2003; Thery, Meyer, and Joulia,

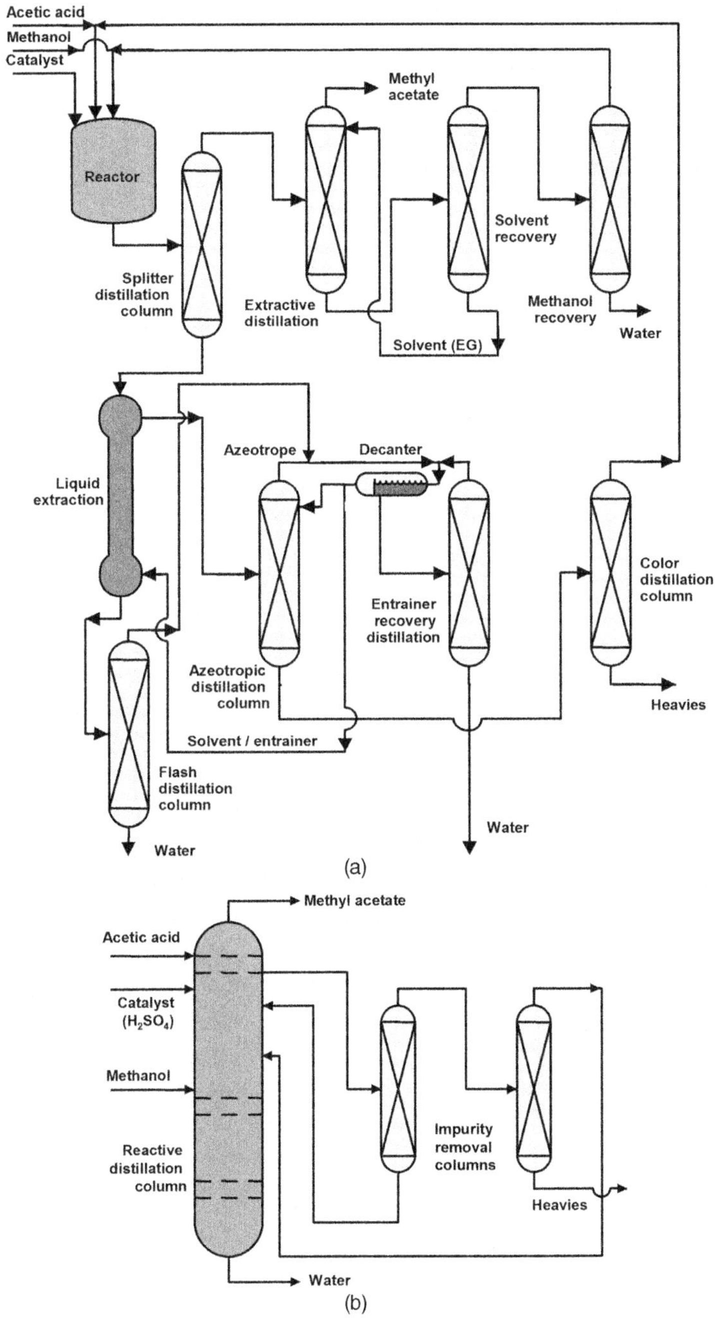

Figure 10.1 Methyl acetate production: conventional process (a) versus reactive distillation (b)

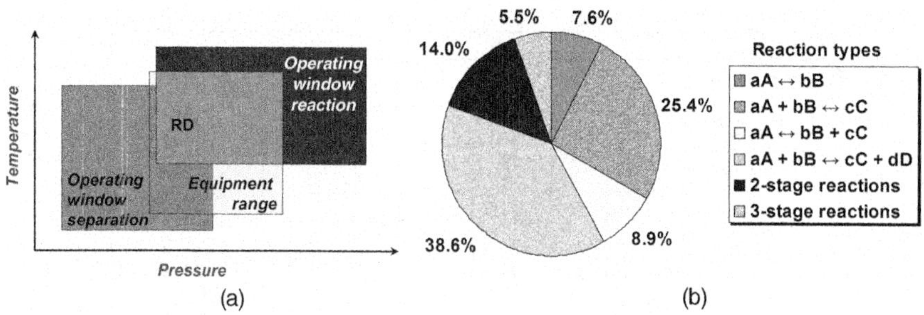

Figure 10.2 Overlapping of operating windows for reaction, separation, and equipment (a); reaction types used in reactive distillation (b)

2005)—as clearly illustrated by Figure 10.2a. If there is no significant overlap of the operating conditions of reaction and separation, then the combination of reaction and distillation is not possible (e.g., a high-pressure reaction cannot be combined with a vacuum distillation). Furthermore, one must also consider that working in the limited overlapping window of operating conditions is not always the optimal solution, but merely a trade-off (Thery, Meyer, and Joulia, 2005). Conversely, in a conventional multiunit flow sheet, the reactors can be operated at their optimum parameters that are most favorable for the chemical kinetics while the distillation columns can be operated at their optimal pressures and temperatures that are most favorable for the vapor–liquid equilibrium (VLE) properties (Luyben and Yu, 2008). In this respect, the residue curve maps (RCMs) are an invaluable tool for the initial screening and flowsheet development of RD.

The pressure and temperature effects are much more pronounced in RD than in conventional distillation as these parameters affect both the phase equilibrium and chemical kinetics (Stichlmair and Fair, 1998; Doherty and Malone, 2001; Schmidt-Traub and Gorak, 2006; Luyben and Yu, 2008). A low temperature that gives high relative volatilities may give small reaction rates that require large amounts of catalyst or liquid holdups to achieve the required conversion. In contrast, a high temperature may promote undesirable side reactions or give a low equilibrium constant that makes it difficult to drive the reaction to completion (Luyben and Yu, 2008; Thery, Meyer, and Joulia, 2005).

RD is typically applied to equilibrium reactions, such as esterification, etherification, hydrolysis, and alkylation. Remarkably, over 1100 articles and 800 US patents on RD were published during the past 40 years, covering in total over 235 reaction systems (Luyben and Yu, 2008). Figure 10.2b provides a convenient overview of these systems classified into various reaction types. Most of the reactions types belong to the $aA + bB \leftrightarrow cC + dD$ or $aA + bB \leftrightarrow cC$ or $aA \leftrightarrow bB + cC$ class, with the rest of them falling into other categories of two- or three-stage reaction (Luyben and Yu, 2008).

10.3 DESIGN, CONTROL AND APPLICATIONS

As compared to conventional distillation, RD sets specifications on both product compositions and reaction conversion. Consequently, the degrees of freedom (DoF) in a RD column must be adjusted to accomplish these specifications while optimizing an objective function such as the total annual cost (TAC). These design DoF include pressure–temperature, number of reactive trays, holdup of the reactive trays, location of reactant feed streams, number of stripping and rectifying trays, reflux ratio and reboiler duty (Sundmacher and Kienle, 2003; Nagy *et al.*, 2007; Luyben and Yu, 2008).

Note that in conventional distillation the typical design specifications are the concentration of heavy key component in the distillate and the concentration of light key component in the bottom product. The holdup has no effects whatsoever on the steady-state design of a distillation column, only on the dynamic behavior. The column diameter is easily determined from the maximum vapor-loading correlations, after calculating the vapor rates required to achieve the desired separation. However, the holdup is very important in RD as the reaction rates depend directly on the liquid holdup and the amount of catalyst available on each tray. Accordingly, RD requires an iterative design procedure since the liquid holdup must be known before the column can be designed. This means that a tray holdup is assumed and the column is designed to achieve the desired conversion and product purities. Then, the column diameter is calculated as well as the required height of liquid on the reactive trays corresponding to the assumed holdup. Note that liquid heights of over 10–15 cm are not recommended due to the hydraulic pressure-drop limitations. If the calculated liquid height is too large, then a smaller holdup is assumed and the calculations are repeated (Luyben and Yu, 2008).

For a long time it was assumed that reaction and distillation can be favorably combined in a column enhanced with special internals or additional exterior volume. However, distillation columns are an appropriate solution only for reactions that are sufficiently fast in order to reach high conversions within the residence time range of such columns. Schoenmakers and Bessling (2003) showed that two operating conditions can be distinguished: (i) distillation controlled range, where conversion is

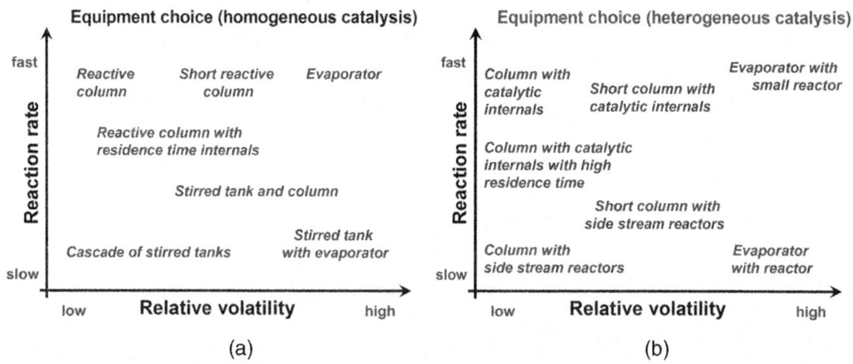

Figure 10.3 Equipment selection in the case of homogeneous (a) and heterogeneous (b) catalysis

influenced by the concentration of the components to be separated, and (ii) kinetics controlled range, in which the conversion is influenced mainly by the residence time and the reaction constant. The industrial design of RD should aim at operating conditions within these two ranges: just sufficient residence time and only the necessary expenditure for distillation. Figure 10.3 presents the resulting principles for the choice of equipment for homo- and heterogeneous catalysis (Schoenmakers and Bessling, 2003). Note that in the case of heterogeneous catalysis additional, separate reaction volumes are necessary to retain the catalyst inside the column. These volumes can be arranged either within the equipment or in a side reactor unit coupled by recycle streams.

Figure 10.4 shows the most common RD configuration alternatives, from conventional RD column (Serafimov, Pisarenko, and Kulov, 1999; Subawalla and Fair, 1999; Tuchlenski *et al.*, 2001; Daza *et al.*, 2003; Thery, Meyer, and Joulia, 2005; Damartzis and Seferlis, 2010), to reactive dividing-wall column (Mueller and Kenig, 2007; Kiss, Pragt, and van Strien, 2009; Hernandez *et al.*, 2009), and RD column combined with pre-reactor and/or side reactors (Klöker *et al.*, 2003; Baur and Krishna, 2004; Kaymak and Luyben, 2007).

RD processes can be divided into homogeneous (auto-catalyzed or homogeneously catalyzed) and heterogeneous processes catalyzed by a solid catalyst (catalytic distillation). Note that the rate of autocatalytic reactions can only be influenced by the temperature or pressure of the RD equipment. Homogeneous catalysis allows the reaction rate to be influenced also by changing the concentration of catalyst and, thus, the reaction rate can be adapted over a wide range to the needs of the RD equipment. While homogeneous catalysis is more flexible, it requires

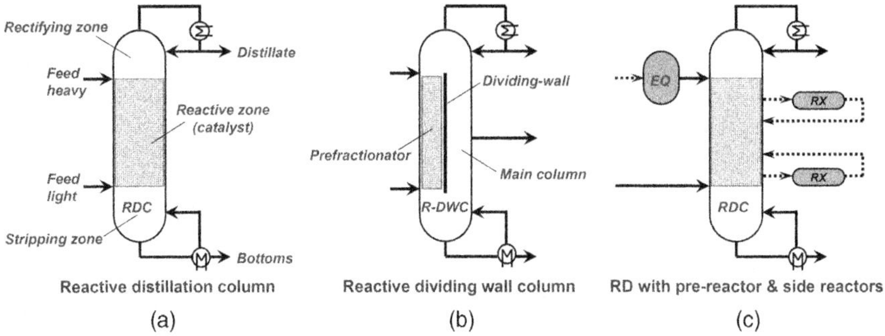

Figure 10.4 Conventional reactive distillation (a), reactive dividing-wall column (R-DWC) (b), and RD column with pre-reactor and/or side-reactors (c)

expensive separation steps for catalyst recovery. Heterogeneous catalysis requires a special construction—catalyst packed in "tea bags" on trays, or sandwiched in structured packing such as Sulzer Katapak®—to fix the catalytic particles in the reactive zone, thus limiting the catalyst concentration that can be achieved. Figure 10.5 conveniently illustrates the main

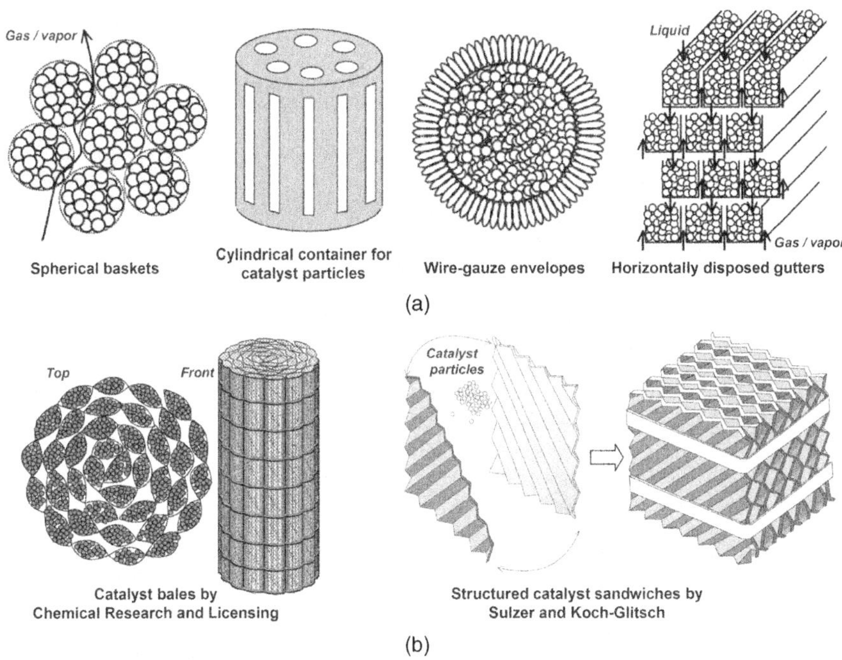

Figure 10.5 Examples of internals for catalytic distillation: various "tea-bag" configurations for trays, gutters (a), and structured catalyst-sandwiches for packed columns (c)

types of internals used in catalytic distillation (Taylor and Krishna, 2000; Krishna, 2002). Note that the reaction rate can be enhanced only to the limit set by the attainable concentration range. Heterogeneous catalysis is simpler in principle, but it needs more equipment volume and it suffers from increased operating temperature and limited catalyst life time (Krishna, 2002; Schoenmakers and Bessling, 2003).

Table 10.1 (Kiss, 2012) lists the most important applications of RD to date: (trans-)esterification, etherification, hydrolysis, (de-)hydration,

Table 10.1 Main industrial applications of reactive distillation

Reaction type	Catalyst/internals
Alkylation	
Alkyl benzene from ethylene/propylene and benzene	Zeolite β, molecular sieves
Amination	
Amines from ammonia and alcohols	H_2 and hydrogenation catalyst
Carbonylation	
Acetic acid from CO and methanol/dimethyl ether	Homogeneous
Condensation	
Diacetone alcohol from acetone	Heterogeneous
Bisphenol-A from phenol and acetone	N/A
Trioxane from formaldehyde	Strong acid catalyst, zeolite ZSM-5
Esterification	
Methyl acetate from methanol and acetic acid	H_2SO_4, Dowex 50, Amberlyst-15
Ethyl acetate from ethanol and acetic acid	N/A
2-Methylpropyl acetate from 2-methyl propanol and acid	Katapak-S
Butyl acetate from butanol and acetic acid	Cation-exchange resin
Fatty acid methyl esters from fatty acids and methanol	H_2SO_4, Amberlyst-15, Metal oxides
Fatty acid alkyl esters from fatty acids and alkyl alcohols	H_2SO_4, Amberlyst-15, metal oxides
Cyclohexyl carboxylate from cyclohexene and acids	Ion-exchange resin bags
Etherification	
MTBE from isobutene and methanol	Amberlyst-15
ETBE from isobutene and ethanol	Amberlyst-15/pellets, structured
TAME from isoamylene and methanol	Ion-exchange resin
DIPE from isopropanol and propylene	ZSM 12, Amberlyst-36, zeolite
Hydration/dehydration	
Mono ethylene glycol from ethylene oxide and water	Homogeneous

Hydrogenation/dehydrogenation	
Cyclohexane from benzene	Alumina-supported Ni catalyst
MIBK from benzene	Cation-exchange resin with Pd/Ni
Hydrolysis	
Acetic acid and methanol from methyl acetate and water	Ion-exchange resin bags
Acrylamide from acrylonitrile	Cation exchanger, copper oxide
Isomerization	
Iso-paraffins from *n*-paraffins	Chlorinated alumina and H_2
Nitration	
4-Nitrochlorobenzene from chlorobenzene and nitric acid	Azeotropic removal of water
Transesterification	
Ethyl acetate from ethanol and butyl acetate	Homogeneous
Diethyl carbonate from ethanol and dimethyl carbonate	Heterogeneous
Vinyl acetate from vinyl stearate and acetic acid	N/A
Unclassified reactions	
Monosilane from trichlorosilane	Heterogeneous
Methanol from syngas	$Cu/Zn/Al_2O_3$ and inert solvent
DEA from monoethanolamine and ethylene oxide	N/A

alkylation, isomerization, (de-)hydrogenation, amination, condensation, chlorination, nitration, and so on (Noeres, Kenig, and Gorak, 2003; Sundmacher and Kienle, 2003). Presently, the application with the largest number of installations is methyl *tert*-butyl ether (MTBE), used as a gasoline additive. CDTECH—the major commercial provider of RD technology—has licensed to date over 200 commercial-scale processes operated worldwide at capacities of 100–3000 ktpy for the production of ethers (MTBE, TAME, ETBE), hydrogenation of aromatics and light sulfur hydro-desulfurization (HDS), and ethylbenzene and isobutylene production (Harmsen, 2007). Sulzer ChemTech has also reported several industrial-scale applications, such as synthesis of ethyl, butyl, and methyl acetates, hydrolysis of methyl acetate, synthesis of methylal (dimethoxymethane), methanol removal from formaldehyde, and production of fatty acid esters (Harmsen, 2007).

RD provides an excellent example of the ever-present interaction between design and control. Both steady-state and dynamic aspects of a chemical process must be considered at all stages of the development and commercialization of a chemical process: laboratory, pilot plant, and production (Luyben and Yu, 2008; Sharma and Singh, 2010). Typically, the controllability of a RD column is improved by adding more reactive trays while taking into account the conflict between the steady-state design and dynamic controllability (Luyben and Yu, 2008). In neat operation mode, the reactants are fed according to the stoichiometric ratio. Consequently, the control system must be able to detect any imbalance that would inevitably result in a gradual accumulation of one of the reactants and a loss of conversion and product purities (Kiss, 2011). An alternative is the operation with an excess of one reactant. This makes control of the RD column easier, but it requires the recovery and recycle of the reactant that is in excess (Luyben and Yu, 2008). The case studies that follow hereafter provide some practical examples of classic and advanced control implementation.

10.4 MODELING REACTIVE DISTILLATION

RD is a process of multicomponent nature and, hence, it is qualitatively more complex than similar binary processes. Thermodynamic and diffusional coupling in the phases and at the interface are accompanied by complex chemical reactions (Noeres, Kenig, and Gorak, 2003). Consequently, the RD models must take into consideration the column hydrodynamics, mass transfer resistances, and reaction kinetics. The complexity of a RD model depends on its application and the problem to be solved: plant design, model-based control, or real time optimization. A process model consists of sub-models for mass transfer, reaction, and hydrodynamics of various complexities. The mass transfer between the vapor and liquid phase can be described by the rigorous rate-based approach (Maxwell–Stefan diffusion equations) or it can be accounted for by the simple equilibrium (EQ) stage model assuming thermodynamic equilibrium between the vapor and liquid phases (Noeres, Kenig, and Gorak, 2003; Taylor and Krishna, 2000).

Detailed modeling of RD has been covered by several studies: Grosser, Doherty, and Malone, 1987; Taylor and Krishna, 1993, 2000; Lee and Dudukovic, 1998; Kreul, Górak, and Barton, 1999; Baur *et al.*, 2000; Noeres, Kenig, and Gorak, 2003; Klöker *et al.*, 2003, 2005; Kenig *et al.*, 2004; Egorov *et al.*, 2005; Mueller *et al.*, 2007; Mueller and Kenig, 2007. Figure 10.6 illustrates both the equilibrium and non-equilibrium stage models (Taylor and Krishna, 2000). The equilibrium stage model assumes that each vapor stream leaving a tray or a packing segment is in thermodynamic equilibrium with the correspondent liquid stream leaving the same tray or segment. In the case of RD, the chemical reaction

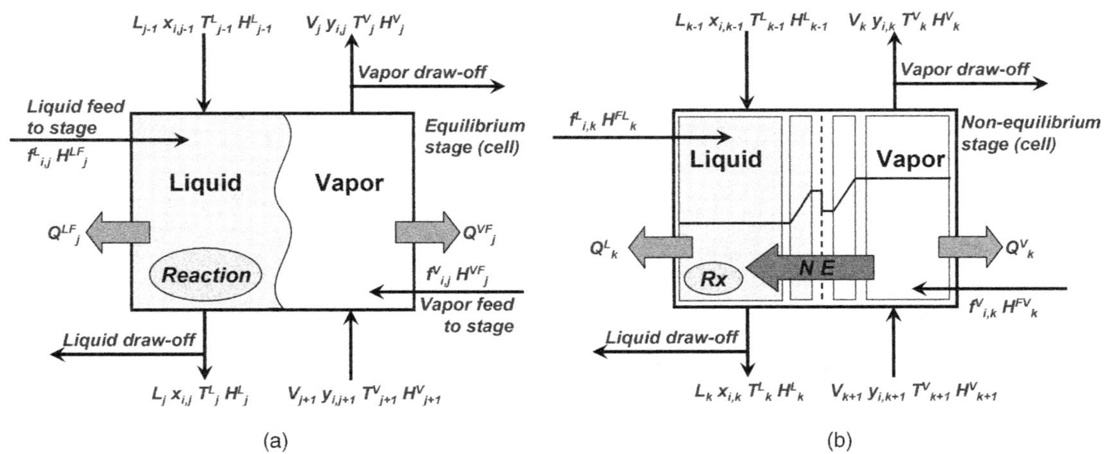

Figure 10.6 Modeling reactive distillation: equilibrium (a) and non-equilibrium (b) stage

is additionally considered via reaction equilibrium equations or via rate expressions integrated into the mass and energy balances (Taylor and Krishna, 2000; Noeres, Kenig, and Gorak, 2003). The use of the Hatta number (Ha) representing the reaction rate in reference to that of the mass transfer helps to discriminate between very fast, fast, average, and slow chemical reactions. If a fast reaction system is considered, RD can be satisfactorily described assuming reaction equilibrium. A more physically consistent way to describe a column stage is the rate-based approach that takes directly into account the actual rates of multicomponent mass and heat transfer and chemical reactions (Taylor and Krishna, 1993, 2000; Noeres, Kenig, and Gorak, 2003; Kenig *et al.*, 2004; Mueller *et al.*, 2007; Mueller and Kenig, 2007). Both approaches can be conveniently simulated in commercial process simulators, such as Aspen Plus® (e.g., RADFRAC distillation unit with RateSep model), Aspen Custom Modeler® (AspenTech, 2009a–c), ChemCAD, PRO/II, or *g*PROMS.

It is also important to note that the introduction of an *in situ* separation function within the reaction zone leads to complex interactions between VLE, vapor–liquid mass transfer, catalyst diffusion, and chemical kinetics. Such interactions were shown to lead to the phenomenon of multiple steady-states and complex dynamics, which have been verified in experimental laboratory and pilot plant units (Güttinger and Morari, 1997; Venimadhavan, Malone, and Doherty, 1999; Taylor and Krishna, 2000).

10.5 FEASIBILITY AND TECHNICAL EVALUATION

10.5.1 Feasibility Evaluation

The development of a realistic process model is the key step in evaluating the feasibility of a reactive distillation process. Moreover, a successful modeling effort requires a complete understanding of both equilibrium and kinetic limits of the process. Thery, Meyer, and Joulia (2005) gave an overview of the state-of-the-art methods available in the literature for the feasibility analysis and design of RD processes. A recent approach for the economical evaluation of RD was proposed by Kiss *et al.* (2011). This is based on fundamental explanations as well as industrial applications of RD reported in the scientific literature. Figure 10.7 presents the proposed framework for checking whether a RD process is economically attractive (Kiss, Singh, and van Strien, 2011; Kiss and Shah, 2012). To perform a feasibility evaluation, some basic information on the chemical process is

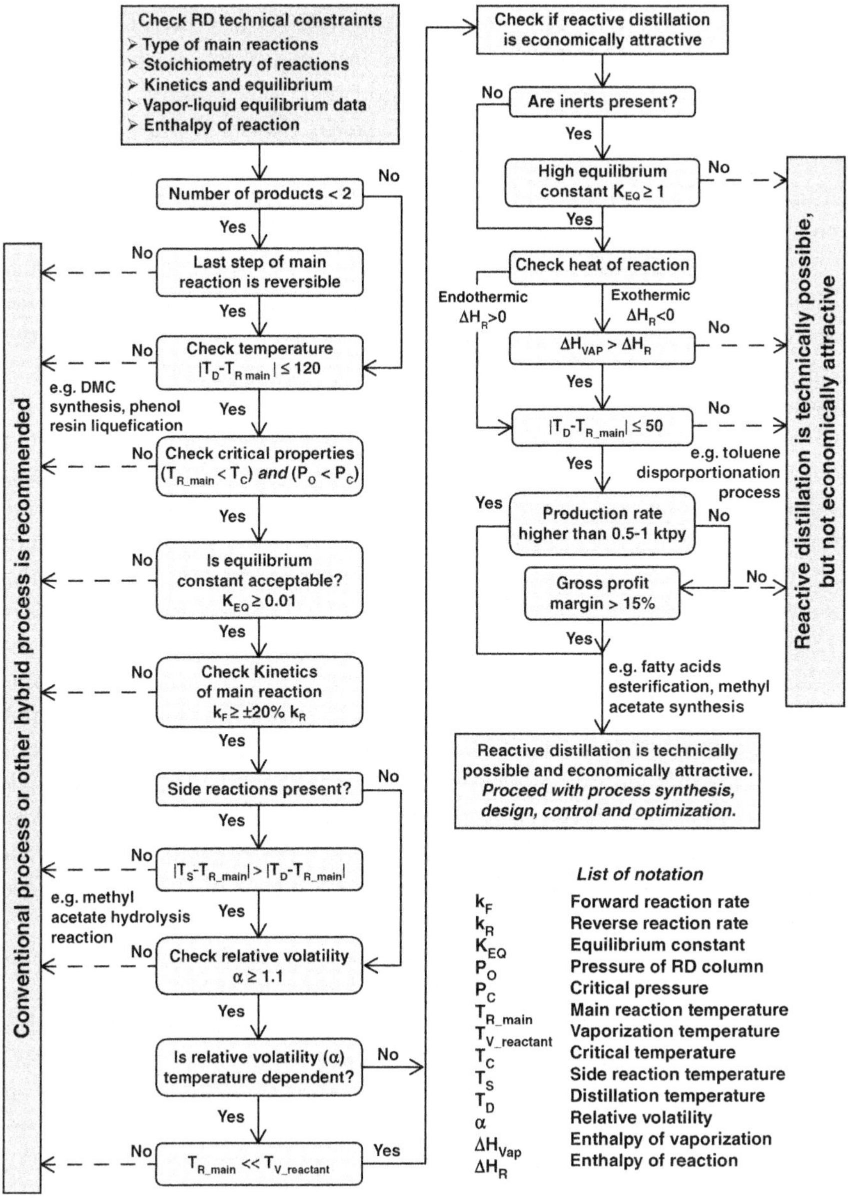

Figure 10.7 Framework for checking if reactive distillation is an attractive option or not

required, such as vapor–liquid equilibrium (VLE data), stoichiometry of reactions, kinetics, and enthalpy of reactions. Note that the boundary values mentioned in this framework should only be taken as guideline figures and not interpreted as very strict limits of the RD process.

The first step is to check the number of products and the reaction type. If there is only one product present, the last step of the main reaction is irreversible, and when there are no side reactions present there is no advantage in using RD over a simple reactor (Kenig *et al.*, 1999). One must not forget that the main advantages of RD rely on overcoming equilibrium limitations and enhancing selectivity towards the desired product (Taylor and Krishna, 2000; Harmsen, 2007).

As both operations occur simultaneously in the same unit, there must be a proper and significant match between the temperatures required for reaction and separation (Taylor and Krishna, 2000), otherwise the combination of reaction and distillation is not possible (e.g., a high-pressure reaction cannot be combined with a vacuum distillation). Moreover, one must also keep in mind that working in the limited overlapping window of operating conditions is not always the optimal solution, but merely a trade-off. For a feasible RD process the temperature difference between separation and main reaction should be less than about 50 °C. For a conventional hydro-dealkylation (HDA) process the temperature difference between the reaction and the separation process is 120 °C. Consequently, although RD was found to be technically applicable it was not sufficiently attractive from an economical viewpoint (Kiss, Singh, and van Strien, 2011). Moreover, the operating pressure and temperature should not be close to the critical region of the key components, since that would potentially lead to a supercritical phase (Kaymak and Luyben, 2004). If the column operates at the critical pressure of key components, these will be present in the vapor—while in the vast majority of RD processes the reaction takes place in the liquid phase (Harmsen, 2007).

The temperature dependence of the vapor pressure of individual components can results in decreased relative volatility as temperature increases in multicomponent systems. This can create a mismatch of favorable temperature for kinetics and relative volatilities that can make the RD process unattractive (Luyben and Yu, 2008). A relative volatility of minimum 1.1 was chosen here, as this is the typical minimum value for distillation process (Luyben and Yu, 2008; Sundmacher and Kienle, 2003). For instance, in the hydrolysis of methyl acetate the reactant itself is the lightest component and it is difficult to keep it in the reactive zone.

The chemical equilibrium constant influences the reactant flow and residence time in a RD column. A low equilibrium constant would require a significant excess of reactants and also a higher number of reactive trays for the RD column (Kaymak and Luyben, 2003), thus resulting in an increased cost of the RD process. Consequently, a value of

$K_{EQ} > 0.01$ (equivalent to an equilibrium conversion of at least 10%) is recommended, in order to be economically attractive. Otherwise, the cost of recycling the excess reactant would outweigh the benefits of integrating reaction and separation.

For various reasons, inerts are sometime present in many chemical processes as well as in some RD commercial systems where the lighter reactants are fed together with an inert. However, the presence of inerts reduces the concentration of reactants and results in lower reaction rates as well as reduced K_{EQ}. Nevertheless, a certain amount of inerts can be beneficial for optimum conversion—for example, in MTBE production, *n*-butene serves as a coolant for the reactive zone, thereby keeping the temperature of the reactive zone at a level where the equilibrium is favorable for MTBE conversion (Higler *et al.*, 1999). Note that in the RD processes, the specific reaction rate for the main reaction cannot be too low as that would require large liquid holdups, huge amounts of catalyst on each reactive tray, and eventually a larger column (Kaymak and Luyben, 2003). Therefore, the reaction rate for the main reaction should be higher than 10^{-5} kmol kg-cat^{-1} s^{-1}—for example, methyl acetate hydrolysis (Lin *et al.*, 2008). In RD processes, the desired column temperature should be selected such that the secondary reactions are minimized while the productivity is still sufficiently high. Moreover, the heat of reaction should be lower than the heat of vaporization of key components. A higher heat of reaction results in drying out of trays and reduced conversion.

The last criterion is checking the production rate—if it is above 0.5–1 ktonne yr^{-1} then the proposed RD process becomes economically attractive. A gross profit higher than 15% for small-scale production is also suitable for RD process application (e.g., pharmaceutical industry). Ultimately, only if all conditions are fulfilled can the RD process be considered economically attractive—for example, fatty acids esterification (Dimian *et al.*, 2009; Kiss, 2010, 2011)—and the procedure continue with the detailed technical evaluation. Just in case that one or two checks are not satisfied, it may be that there are technical alternatives able to overcome the problem (Kiss and Shah, 2012).

10.5.2 Technical Evaluation

The model of a RD process must account for the phase equilibrium, the rate of reaction, and mass transfer in the bulk as well as in the film. The equilibrium behavior of most RD systems varies between two boundaries, corresponding to phase equilibrium and chemical equilibrium control. A residue curve map (RCM) derived from the chemical and phase equilibrium for a reactive and non-reactive section of the RD column represents variations of the stationary points—corresponding to the pure components and azeotropes—between equilibrium boundaries for given temperature and pressure. Buzad and Doherty (1994), Venimadhavan *et al.* (1994), and Okasinski and Doherty (1998) have identified that the variations in these singular points can be represented by the Damköhler number (Da). The Damköhler number is defined as,

$$\mathrm{Da} = \frac{H_0 k_f}{V} \tag{10.1}$$

where, H_0 is the liquid holdup (mol), k_f is a pseudo-first-order rate constant (s^{-1}), and V is the vapor rate ($mol\ s^{-1}$). The Damköhler number is the ratio of characteristic residence time (H_0/V) to characteristic reaction time ($1/k_f$). For low values of Da ($\mathrm{Da} \leq 0.1$) the reaction rate on each stage is relatively slow as compared to the residence time available on each stage, and the system is dominated by phase equilibrium. For large values of Da ($\mathrm{Da} > 10$) the reaction rate is fast and chemical equilibrium is approached on the reactive stages. If the Damköhler number does not lie between these values, then neither the phase equilibrium nor the chemical equilibrium is controlling—hence the process is in fact kinetically controlled (Venimadhavan *et al.*, 1994).

Moreover, the combinations of the Damköhler number and the chemical equilibrium constant can be used to perform a preliminary screening of the suitability of the process for an RD application. Reactive distillation is significantly beneficial when the process demonstrates a combination of low Da ($\mathrm{Da} \leq 0.1$) and high K_{EQ} ($K_{EQ} > 1$) or high Da ($\mathrm{Da} > 0.1$) and low K_{EQ} ($K_{EQ} \leq 1$), as proposed by Venimadhavan *et al.* (1994). The combination of low Da and high K_{EQ} represents a slow forward reaction but a slower reverse reaction leads to high product formation. Reactive distillation offers benefits as long as the required holdup is not too large. The combination of high Da and low K_{EQ} represents a fast product formation but a fast reverse reaction also leads to little product formation. RD is beneficial because the product can be removed quickly from the reactive zone and allows the equilibrium shifts to the product side. Since the product removal rate solely depends on the rate of mass transfer between phases, the process must not be mass transfer controlled.

The combination of high Da and high K_{EQ} represents a fast forward reaction and a slower reverse reaction, leading to the situation of an instantaneous irreversible reaction; therefore, a simple reactor is sufficient to carry out the process. For this class of processes, reactive distillation can only be useful compared to a simple reactor when the higher selectivity of the main reaction over the side reaction is noticeable. The reactive stage can be assumed to be at chemical equilibrium and such a reactive stage can be modeled as a vapor–liquid equilibrium stirred tank reactor. Reactive distillation is not beneficial when the process demonstrates a combination of low Da and low K_{EQ}. Such a combination represents a slow forward reaction and a fast reverse reaction, which essentially leads to no product formation. This class of processes requires an optimally designed reactor with a large holdup. This technical feasibility analysis of RD process and process limitations based on Da and the chemical equilibrium is represented in Figure 10.8 (Shah *et al.*, 2012).

The working regime of the process must be identified to confirm whether the process is mass transfer or kinetically controlled, whether the reaction takes place only in the bulk or also in the film. This allows one to establish the requirements of internals for the RD column and the modeling approach that needs to be applied to design the RD process. Identification of the working regime can be performed by using the Hatta number (Ha), pseudo-first-order rate constant (k_f), and the product of mass transfer coefficient and interfacial area ($k_L a$). The Hatta number is the ratio of the maximum possible conversion in the film to the maximum diffusion transport through the film. For higher order reactions of two components, the Hatta number is defined as (Roizard and Wild, 2002):

$$\mathrm{Ha} = \sqrt{\frac{\frac{2}{n+1} k_f C_A^{n-1} C_B D_A}{k_L^2}} \tag{10.2}$$

where n is the order of reaction (–), k_f is the forward reaction rate constant (s^{-1}), C is the concentration ($mol\,m^{-3}$), D is the diffusivity ($m^2\,s^{-1}$), and k_L is the mass transfer coefficient ($m\,s^{-1}$). A Hatta number of less than one represents a slow RD process where the reactions take place in the bulk, while a Hatta number greater than one represents a fast RD process with the reaction taking place in the bulk as well as in the film. This identification of the working regime based on the Hatta number is illustrated in Figure 10.8. The pseudo-first-order rate constant (k_f) and the product of mass transfer coefficient and interfacial area ($k_L a$) in particular determine whether the process operates under the slow kinetic regime ($k_f < k_L a$) or slow diffusion regime ($k_f > k_L a$) or slow mixed regime ($k_f = k_L a$). Depending on the working regime the proper selection of internals can be readily performed (Kiss and Shah, 2012).

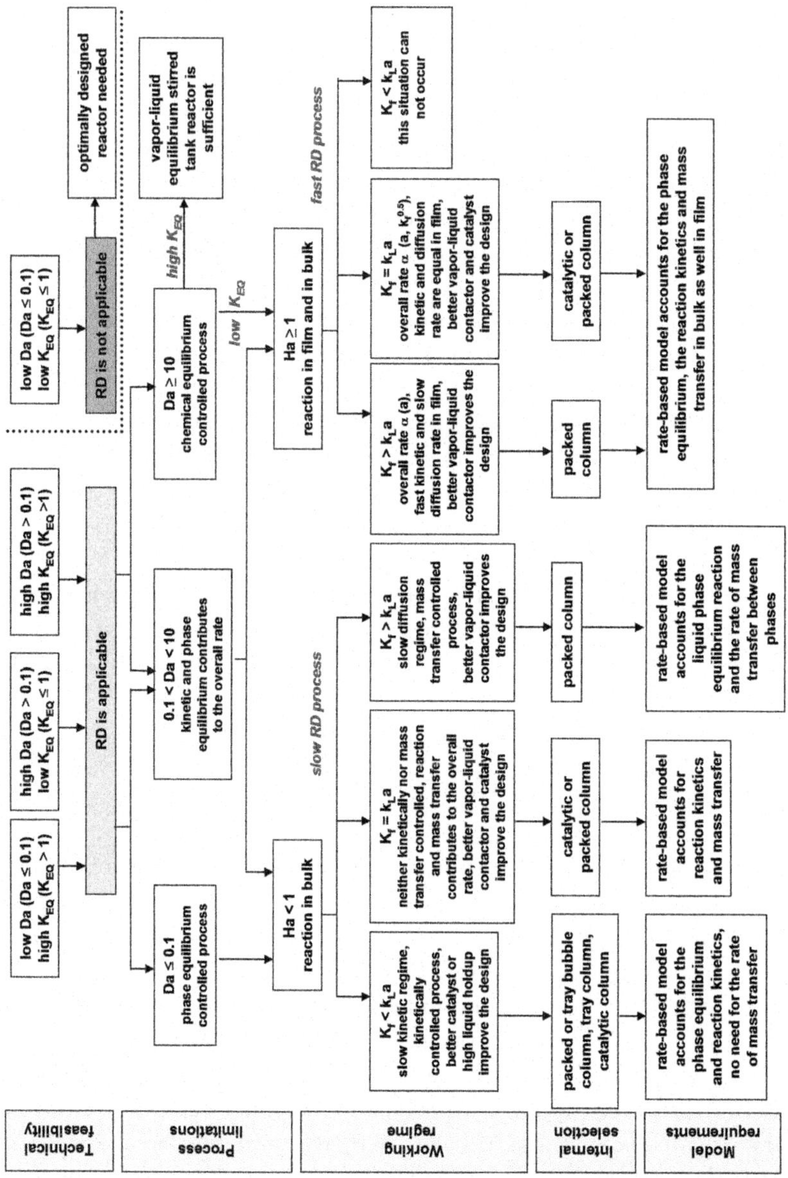

Figure 10.8 Systematic framework for the technical evaluation of reactive distillation processes

Figure 10.8 also summarizes the internals and model requirements for different working regimes. As clearly shown here, the technical evaluation of any RD process can be quickly but systematically performed based only on the Damköhler number, chemical equilibrium constant, Hatta number, pseudo-first-order rate constant and product of mass transfer coefficient and the interfacial area (Kiss and Shah, 2012).

10.6 CASE STUDY: ADVANCED CONTROL OF A REACTIVE DISTILLATION COLUMN

Despite the complex nonlinear dynamics of the RD systems, caused by the coupling between the reaction and separation processes, the main research focused mainly on the steady-state design of RD systems (Taylor and Krishna, 2000; Malone and Doherty, 2000). Several control approaches were proposed based on linear (Sneesby *et al.*, 1999; Luyben and Yu, 2008) or nonlinear control approaches (Kumar and Daoutidis, 1999; Nagy *et al.*, 2007; Sharma and Singh, 2010). The objective of this case study is to evaluate the feasibility of different control techniques from both linear and nonlinear controllers by considering the application of nonlinear model predictive control (NMPC), linear model predictive control (LMPC), and coupled PI control to a sample reactive distillation process. In addition to the simulation of the different control strategies, a systematic analysis of the control performances is carried out based on a detailed open-loop analysis of the dynamic behavior of the system at low and high purity operating conditions (Nagy *et al.*, 2007).

This case study is based on a general mathematical representation of a RD column (Figure 10.4a), where the following chemical reaction (e.g., esterification) takes place in the liquid phase:

$$\mathrm{A} + \mathrm{B} \overset{r}{\longleftrightarrow} \mathrm{C} + \mathrm{D} \tag{10.3}$$

10.6.1 Mathematical Model

The RD column is considered to have 41 theoretical trays, with 1 being the reboiler and 41 the condenser. The feed of component A and B enter the column on tray numbers 10 and 30, respectively. The products C and D are removed as top and bottom products, respectively. The process control objective is to maximize conversion of the reactants and separation of the products. As manipulated variables the reboiler vapor flow and the liquid reflux in the condenser are used. The distillate flow and the bottom liquid flow are used to control the total liquid holdup in the reboiler and the condenser, respectively. The reaction is assumed to take place only in the liquid phase. The tray-to-tray model equations for all the column trays are given by:

$$\frac{\mathrm{d}M_i^{\mathrm{l}}}{\mathrm{d}t} = F_i + L_{i+1} + V_{i-1} - L_i - V_i + \mathrm{Vol}_i \sum_{j=0}^{\mathrm{NC}} \mu_j r_i \tag{10.4}$$

$$\frac{\mathrm{d}(M_i^{\mathrm{l}} x_{i,j})}{\mathrm{d}t} = F_i x_{\mathrm{f}} + L_{i+1} x_{i+1,j} + V_{i-1} y_{i-1,j} - L_i x_{i,j} - V_i y_{i,j} + \mathrm{Vol}_i \mu_j r_i \tag{10.5}$$

for $i = \{2, \ldots, 40\}$, $j = \{1, \ldots, 4\}$. Here, $x_{i,j}$ and $y_{i,j}$ are the mol fractions of component j in the liquid and in the vapor of tray i. M_i^{l} is the total molar liquid hold-up of a tray i. The model equations for the condenser and the reboiler are as follows:

$$\frac{\mathrm{d}M_{\mathrm{b}}^{\mathrm{l}}}{\mathrm{d}t} = L_2 - Bf - V_{\mathrm{b}} + \mathrm{Vol}_{\mathrm{b}} \sum_{j=1}^{\mathrm{NC}} \mu_j r_{\mathrm{b}} \tag{10.6}$$

$$\frac{\mathrm{d}(M_{\mathrm{c}}^{\mathrm{l}})}{\mathrm{d}t} = V_{40} - L_{\mathrm{c}} - Df + \mathrm{Vol}_{\mathrm{c}} \sum_{j=1}^{\mathrm{NC}} \mu_j r_{\mathrm{c}} \tag{10.7}$$

$$\frac{\mathrm{d}(M_{\mathrm{b}}^{\mathrm{l}} x_{i,j})}{\mathrm{d}t} = L_2 x_{2,j} - Bf x_{\mathrm{b},j} - V_{\mathrm{b}} y_{\mathrm{b},j} + \mathrm{Vol}_{\mathrm{b}} \mu_j r_{\mathrm{b}} \tag{10.8}$$

$$\frac{\mathrm{d}(M_{\mathrm{c}}^{\mathrm{l}} x_{\mathrm{c},j})}{\mathrm{d}t} = V_{40} y_{40,j} - L_{\mathrm{c}} x_{\mathrm{c},j} - Df x_{\mathrm{c},j} + \mathrm{Vol}_{\mathrm{c}} \mu_j r_{\mathrm{c}} \tag{10.9}$$

where Bf and Df are the product flow from the reboiler and the condenser, respectively. The reaction rate is given by:

$$r_i = k_{\mathrm{or}} \left(x_{i,\mathrm{A}} x_{i,\mathrm{B}} - \frac{1}{k_{\mathrm{eq}}} x_{i,\mathrm{C}} x_{i,\mathrm{D}} \right) \tag{10.10}$$

The vapor mol fraction is calculated based on constant volatilities by:

$$y_{i,j} = \frac{\alpha_j \cdot x_{i,j}}{1 + \sum_{j=1}^{\mathrm{NC}} \left(\alpha_j - 1\right) \cdot x_{i,j}} \tag{10.11}$$

Finally, the change in component composition is given by:

$$\frac{\mathrm{d}x_{i,j}}{\mathrm{d}t} = \frac{1}{M_i} \left[\frac{\mathrm{d}\left(M_i^{\mathrm{l}} x_{i,j}\right)}{\mathrm{d}t} - x_{i,j} \frac{\mathrm{d}M_i^{\mathrm{l}}}{\mathrm{d}t} \right] \tag{10.12}$$

The reactive distillation column model was implemented in the MATLAB/Simulink environment. The column is represented by a Simulink S-Function, which is written in C-language and compiled with the Watcom compiler to a .dll file for fast simulation. Figure 10.9 illustrates the efficient multistage solution of the nonlinear MPC optimization using OptCon (Nagy *et al.*, 2007). The implementation was performed using MathWorks MATLAB and HQP (huge quadratic programming) based package OptCon. This approach has several key benefits: parameterized control $[u(t) = u(uk),\ k = 0,\ 1, \ldots,\ N-1]$, conversion of a dynamic optimization problem to a discrete time optimal control problem, solves as large-scale NLP, sparse NLP, horizon divided in stages, equations integrated independently, states prediction not consistent until convergence, and additional computational advantages (Nagy *et al.*, 2007).

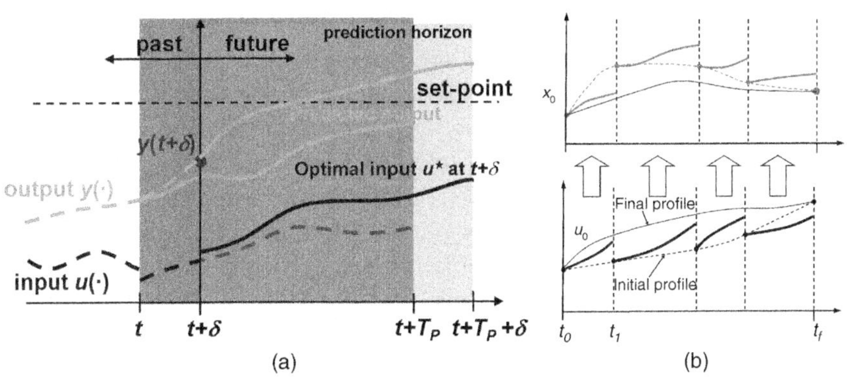

Figure 10.9 Efficient multistage solution of the NMPC (a); optimization using OptCon (b)

10.6.2 Open-Loop Dynamic Analysis

RD is an extremely nonlinear process with highly interacting dynamics. The characteristics of the process change under different operating conditions and, thus, linear models are only good approximations of the real process for very small deviations of the steady state. Prior to designing a controller, several simulations were carried out to characterize the sensitivity of the control variables with respect to small input changes and disturbances. Steady state deviations have been analyzed starting from two different steady states. The vapor flow V and the condenser reflux L were varied with $\pm 10\%$ of the respective steady stated values. Figure 10.10a,b shows the new steady states resulting from the input variation (Nagy *et al.*, 2007). The trend is different for small regions of the steady state for operation in a low or high purity region, but the overall trend for larger input changes are similar. A high composition for both C_{41} in the condenser and D_1 in the reboiler is achievable only in a narrow regime; thus, disturbances are expected to have high influences in the composition of the outflows. Figure 10.10c,d presents the step responses of the column with regard to very small steps (at $t = 1000$ min) in the vapor and liquid flow. The responses are completely different in the two operating regimes. Note the sign reversal in the gain, and also the longer response time for the high purity regime. This already suggests that a linear controller would have to be tuned as quite *sluggish* if it is to be used in the whole operating region.

10.6.3 Closed-Loop Performance

Different controller and control structures have been tested, such as decentralized PID, coupled PID, and linear and nonlinear model predictive control. The decentralized PID could not cope with large set point changes and any disturbances in the high purity region. The PID-like control structure was based on open-loop analysis, RGA calculations, and trial and error simulations (intuition and experience) with the nonlinear model. The controller parameter found by considerations of a linear model (Taylor approximation around a steady state) failed to stabilize the system over the whole operating regime. Since the column should be operated under different conditions, two different steady states were analyzed: one at low product purity (0.8, 0.8) and one at high product purity (0.99, 0.99). Linear models were derived by first-order power series expansion and controllability analysis performed using

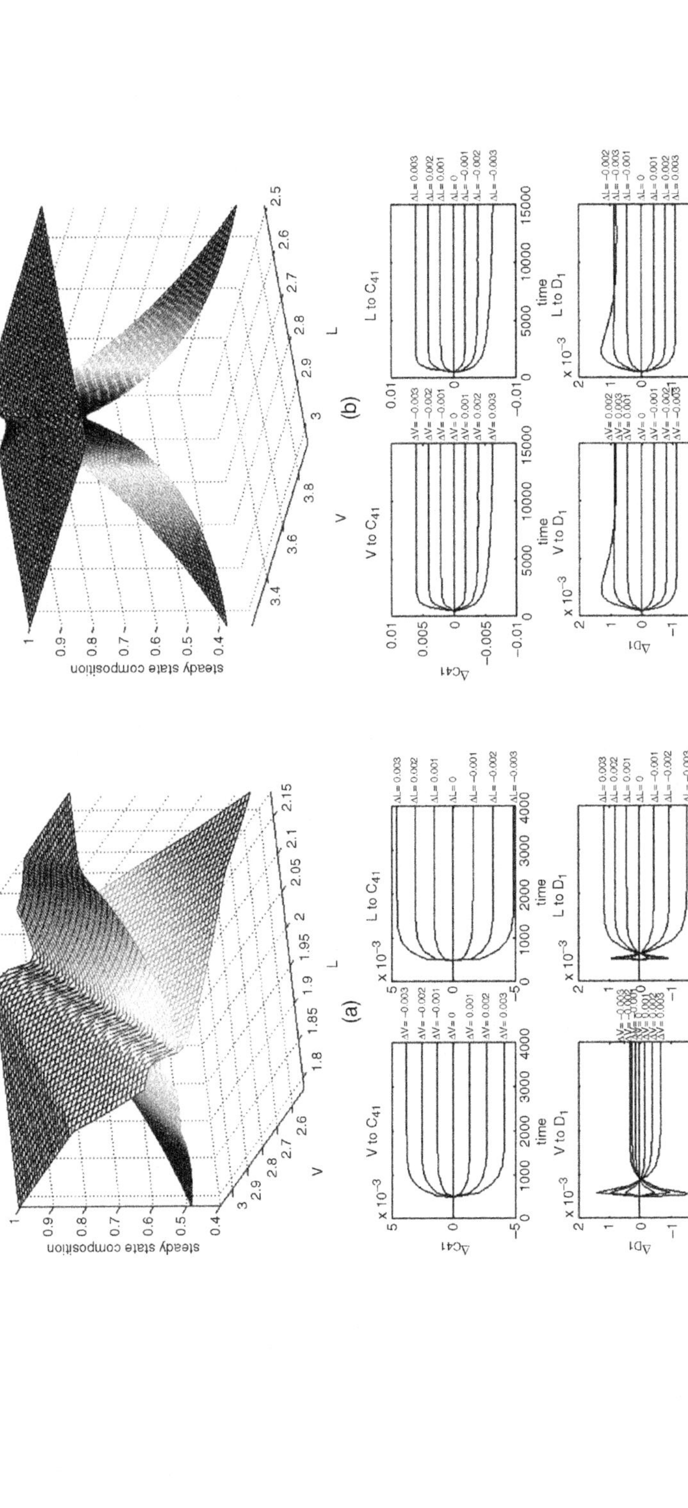

Figure 10.10 Steady state analysis within the low purity (0.79, 0.81) and high purity (0.99, 0.99) operating regime in open loop simulations (a, b). Step response of the RD column for very small steps in the inputs in the low and high purity region (c, d)

RGA. The values of the entries in the RGA were quite large and negative at frequency zero (steady state), indicating a difficult control problem. In addition, we observed sign reversal and inverse response. For systems with negative entries in the RGA it is not recommended to choose diagonal PID, since the system might be unstable if a loop becomes open (e.g., due to saturation). Interaction is taken into account using a coupled PID structure. Problems due to saturation were handled by considering an anti-windup structure and less aggressive tuning. If the controller is tuned more aggressively, problems with saturations are more pronounced, and for different cases (e.g., large disturbances or set point changes) the controller fails.

Figure 10.11a shows the control performance of the best PI structure and tuning, coupled with anti-windup (Nagy *et al.*, 2007). The closed-loop response is very slow, but a faster controller would lead to a highly oscillating response and even instability of the closed system. Linear MPC is regarded as a more advanced control technique than simple PIDs (Bemporad, Morari, and Ricker, 2009); however, for such a highly nonlinear process with a wide operating regime and large disturbances, the LMPC is not able to stabilize the process at the desired set point using a single linear model. Multiple linear models, identified under different operating conditions, or a model of the measured disturbances can be employed to enhance the performance. But since the effects are expected to vary over the operating regime, this control technique has been abandoned in favor of the nonlinear MPC, where the aforementioned problems are inherently considered in the controller (Morari and Lee, 1999; Maciejowski, 2002; Agachi *et al.*, 2006). The NMPC controller uses a multiple shooting optimization solver called HQP and the optimization package OptCon. Owing to the multiple shooting approach a much larger optimization problem arises that is, however, highly structured. Therefore, sparse techniques can be applied, such that the optimization problem is solved very rapidly. The excellent control performance achieved by the NMPC controller is shown in Figure 10.11b. The results prove that in comparison with a well-tuned linear controller the nonlinear MPC shows a superior performance, by keeping a tight product composition, with respect to set-point changes and disturbances.

The application of several advanced control approaches was clearly demonstrated here in the case of a reactive distillation column. The dynamic and steady state analysis of the open loop system illustrates the difficulty of control that depends on the operating regime. Linear MPC was not able to control the system, whereas a coupled PID gave acceptable results, with a very slow response time in the high purity operating

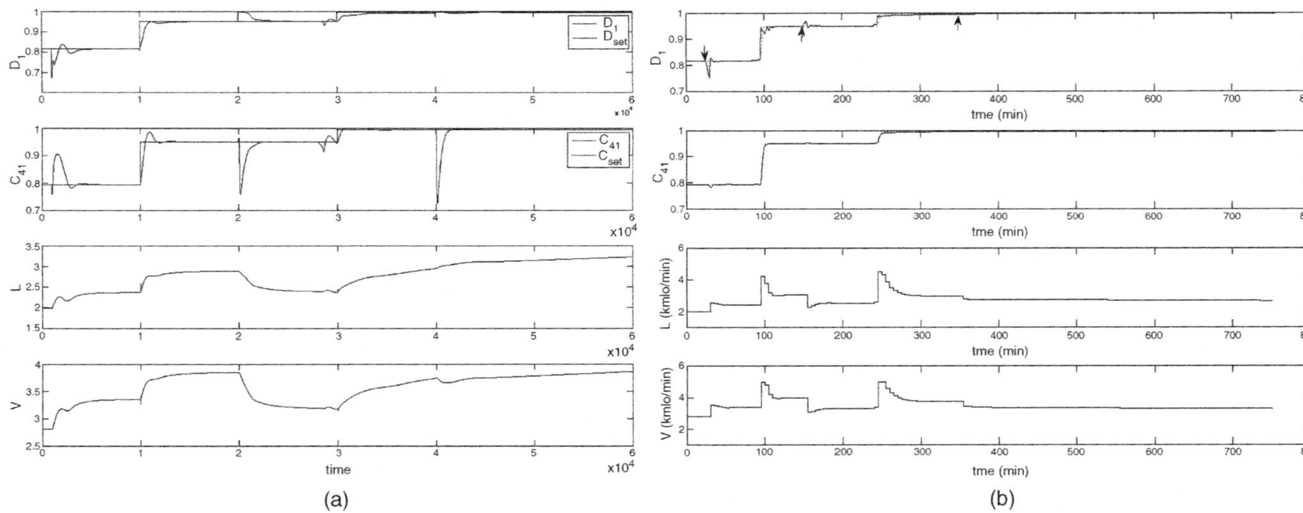

Figure 10.11 Performance of the coupled PI control for set point changes and various disturbances using coupled PID control (a)—disturbances in feeds A and B occurred at times 1000, 20 000, and 40 000 min (+20%, nominal and −20% disturbance in the nominal flow). NMPC performance for set point changes from low purity to high purity performed in two steps (b)—arrows indicate unmeasured disturbances in the flow rate for feeds A and B: +20%, nominal, and −20%

zone. The implemented nonlinear MPC based on efficient multiple shooting algorithms and the software package OptCon showed excellent control performance.

10.7 CASE STUDY: BIODIESEL PRODUCTION BY HEAT-INTEGRATED RD

Biodiesel is an alternative, renewable, and biodegradable fuel produced mainly from green sources such as vegetable oils, animal fat, or even waste cooking-oils from the food industry. However, waste raw materials can contain a substantial amount of free fatty acids (FFAs), up to 100%. Accordingly, the development of an efficient continuous biodiesel process is required, in which the use of a solid catalyst is especially wanted to suppress the costly processing steps and waste treatment. Note that the common problem of all conventional processes is the use of liquid catalysts that require neutralization and an expensive multistep separation that generates salt waste streams.

To solve these problems, solid acids can be applied in an esterification process based on catalytic reactive distillation. Such an integrated process is able to shift the chemical equilibrium to completion and preserve the solid catalyst activity by continuously removing the products, while also leading to lower investment and operating costs (Kiss *et al.*, 2006a,b; Kiss, Dimian, and Rothenberg, 2008; Kiss, 2010, 2011).

The integrated reactive distillation process was designed according to previously reported process synthesis methods for reactive separations (Schembecker and Tlatlik, 2003; Noeres, Kenig, and Gorak, 2003). Rigorous simulations embedding experimental results were performed using Aspen Plus (AspenTech, 2009). The RD column was simulated using the rigorous RADFRAC unit with the RateSep (rate-based) model enabled, and explicitly considering three phase balances.

The physical properties required for the simulation and the binary interaction parameters for the methanol–water and acid–ester pairs were available in the Aspen Plus database of pure components, while the other interaction parameters were estimated using the UNIFAC-DMD group contribution method. The fatty components were conveniently lumped into one fatty acid and its fatty acid methyl ester (FAME)—according to the reaction: R-COOH + $CH_3OH \leftrightarrow$ R-COO-CH_3 + H_2O. Dodecanoic (lauric) acid/ester was selected as lumped component due to the availability of experimental results, kinetics, and VLLE parameters for this system (Kiss *et al.*, 2006; Dimian *et al.*, 2009). The lumping of components is very reasonable since fatty acids and their corresponding esters have similar properties. This approach was already reported to be successfully used to simulate other fatty ester production processes (Dimian *et al.*, 2009, Kiss, Dimian, and Rothenberg, 2008; Kiss, 2009).

In this case study, sulfated zirconia is considered as solid acid catalyst due to available kinetics (Kiss, Dimian, and Rothenberg, 2008; Dimian *et al.*, 2009). The esterification reaction is a second-order reversible reaction and, hence, the reaction rate accounts for both direct and reverse reactions:

$$r = (k_1 W_{cat}) C_{Acid} C_{Alcohol} - (k_2 W_{cat}) C_{Ester} C_{Water} \tag{10.13}$$

where k_1 and k_2 are the kinetic constants for the direct (esterification) and reverse (hydrolysis) reactions, W_{cat} is the amount of catalyst, and $C_{component}$ is the molar concentration of the components present in the system. As water is continuously removed from the system, the reverse hydrolysis reaction is extremely slow—hence the second term of the reaction rate can be practically neglected.

The conceptual design of the process is based on a reactive distillation column (RDC) that integrates the reaction and separation steps into a single operating unit able to shift the reaction equilibrium towards product formation by continuous removal of reaction products, instead of using an excess of reactant. An additional flash and a decanter are used to guarantee the high purity of the products. RDC consists of a core reactive zone completed by rectifying and stripping separation sections, whose extent depends on the separation behavior of the reaction mixture. Since methanol and water are much more volatile than the fatty ester and acid, these will separate easily in the top. Figure 10.12a presents the flow sheet of this biodiesel process based on conventional reactive distillation, as reported by Kiss, Dimian, and Rothenberg, (2008) and Kiss (2011).

The reference flow sheet presented in Figure 10.12a is relatively simple, with just a few operating units, two cold streams that need to be preheated (fatty acid and alcohol) and two hot streams that have to be cooled down (top water and bottom fatty esters). Therefore, heat-integration was performed by applying previously reported heuristic rules (Chen and Yu, 2003; Dimian and Bildea, 2008). Consequently, a feed-effluent heat exchanger (FEHE) should replace each of the two heat exchangers HEX1 and HEX2. Figure 10.12b illustrates the improved process design including heat-integration around the RD

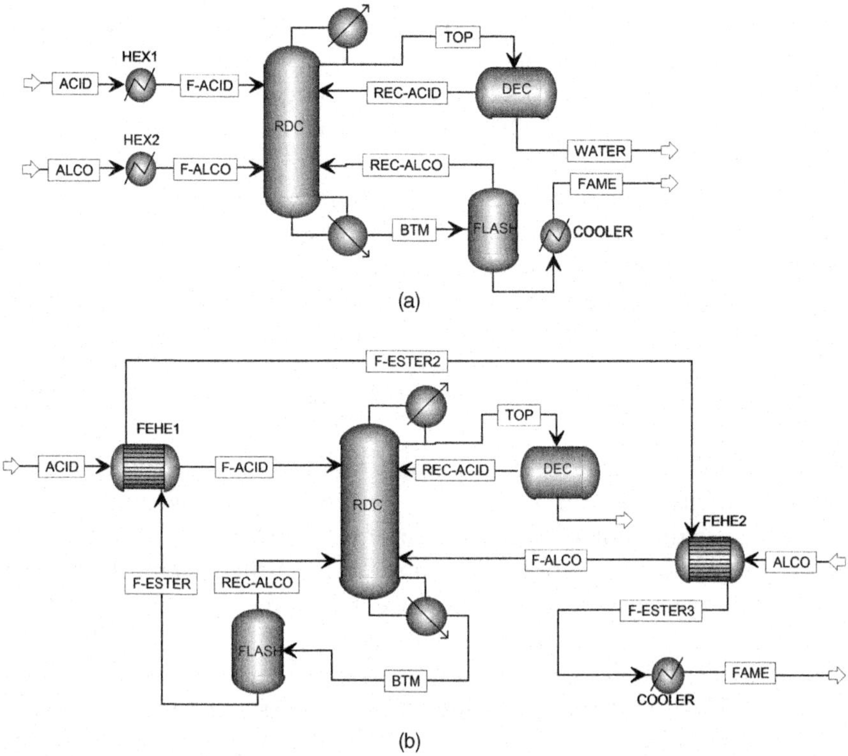

Figure 10.12 Synthesis of fatty esters by reactive distillation: base case flow sheet (a) and heat-integrated RD flow sheet (b)

column (Kiss, 2011). The hot bottom product of the column (FAME) is used to preheat both reactants: the fatty acid and alcohol feed streams. Notably, there is no longer need for an external hot utility to preheat the reactant feed streams and no additional heat exchanger is required by this heat-integrated setup.

The main design parameters—such as column size, catalyst loading, and feed condition—are listed in Table 10.2 (Kiss, 2011). High conversion of the reactants is achieved, with the productivity of the RD unit exceeding 20 kg fatty ester/kg catalyst/h and purity specifications of over 99.9 wt% for the final biodiesel product (FAME stream). Figure 10.13a–c shows the liquid and vapor composition, as well as the reaction rate and temperature profiles, along the reactive distillation column (Kiss, 2011). The RD column is operated in the temperature range 70–210 °C, at ambient pressure. As the reaction takes place mainly in the reactive zone, the reaction rate exhibits a maximum in the middle of the column. The concentration of water

Table 10.2 Design parameters for simulating the heat-integrated reactive distillation column

Parameter (units)	Value
Total number of theoretical stages	15 (reactive from 3 to 12)
Column diameter (m)	0.4
HETP (m)	0.5
Valid phases (–)	VLL
Volume liquid holdup per stage (l)	18
Mass catalyst per stage (kg)	6.1
Catalyst bulk density (kg m^{-3})	1050
Fatty acid conversion (%)	> 99.99
Fatty acid feed, on stage 3 (liquid, at 145 °C) (kg h^{-1})	1167
Methanol feed, on stage 10 (liquid, at 65.4 °C) (kg h^{-1})	188
Reboiler duty (kW)	136
Condenser duty (kW)	−72
Reflux ratio (mass ratio R/D) (kg kg^{-1})	0.10
Boil-up ratio (mass ratio V/B) (kg kg^{-1})	0.12
Production of biodiesel (FAME) (kg h^{-1})	1250
RD column productivity (kg FAME kg-cat^{-1} h^{-1})	20.4
Specific energy requirements (kW·h per ton ester)	108.8

increases from the bottom to the top of the column, while the concentration of fatty ester increases from the top to bottom. Therefore, in the top of the reactive separation column there is mainly water with negligible amounts of fatty acids, while in the bottom there is liquid fatty esters product (biodiesel) with a very limited amount of methanol.

Heat-integrated reactive distillation indeed offers major advantages such as reduced capital investment and operating costs, as well as no catalyst-related waste streams and no soap formation. However, the controllability of the process is just as important as the savings in capital and operating costs. In processes based on reactive separations, feeding the reactants according to their stoichiometric ratio is essential to achieve high products purity (Dimian *et al.*, 2009; Bildea and Kiss, 2011). This constraint must be fulfilled not only during normal operation but also during the transitory regimes arising due to planned production rate changes or unexpected disturbances. Despite the high degree of integration the heat-integrated RD process is well controllable—a key result being an efficient control structure that can ensure the reactants ratio required for the total conversion of fatty acids and for the prevention of difficult separations (for details see Kiss, 2011).

Figure 10.13d,e depicts the dynamic simulation results (Kiss, 2011). The simulation starts from the steady state. At time $t = 1$ h, the acid flow

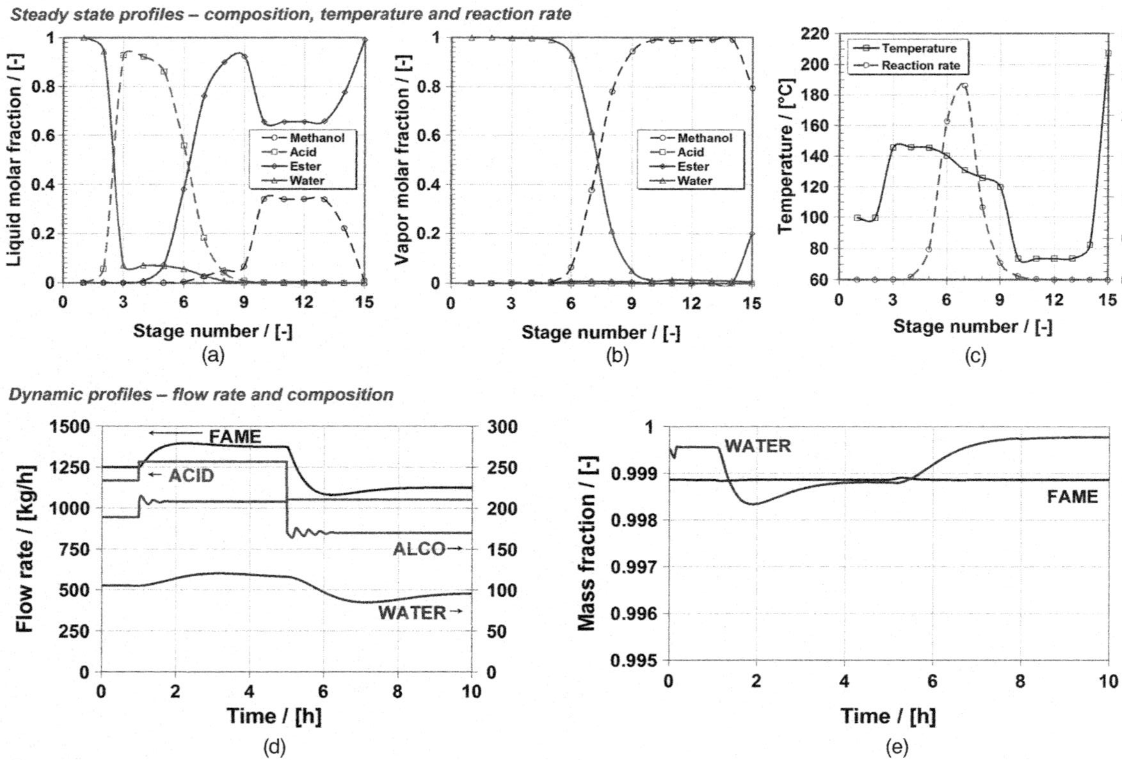

Figure 10.13 Steady state profiles: liquid (a) and vapor (b) molar composition; temperature and reaction rate profiles along the reactive column (c). Dynamic simulation results (d, e): acid flow rate disturbance of +10% at 1 h, and −10% at 5 h

rate is increased by 10%, from 1168 kg h^{-1} to 1284.4 kg h^{-1}. Then, at time $t = 5$ h, the acid flow rate is decreased to 1051.2 kg h^{-1}, representing a 10% decrease with respect to the nominal value. The new production rate is achieved in about 2 h. The purity of FAME remains practically constant throughout the dynamic regime, the main impurity being methanol. Notably, the acid concentration stays below the 2000 ppm requirement of the ASTM D6751–08 standard (i.e., acid number < 0.50 mg-KOH per g-biodiesel).

Remarkably, both design alternatives (base case and heat-integrated flow sheet) presented here are suitable for a large range of fatty acids and alcohol feedstock. These processes based on RD have no additional separation steps and produce no waste salt streams as water is the only by-product (Kiss and Bildea, 2012).

10.8 CASE STUDY: FATTY ESTERS SYNTHESIS BY DUAL RD

Fatty esters are key products of the chemical process industry, being incorporated in a wide variety of high-added value products, from cosmetics to plasticizers and bio-detergents. A key problem in the synthesis of fatty esters by RD is effective water removal in view of protecting the solid catalyst and avoiding costly recovery of the alcohol excess. This case study presents a novel approach based on dual esterification of lauric acid with light and heavy alcohols, namely, methanol and 2-ethylhexanol. These two complementary reactants have an equivalent reactive function but synergistic thermodynamic features. The setup behaves rather as a reactive absorption combined with reactive azeotropic distillation with heavy alcohol as co-reactant and water-separation agent. Super acid solid catalyst based on sulfated zirconia, whose activity is comparable for the two alcohols, can be used at temperatures of 130–200 °C and moderated pressure. The inventory of alcohols is controlled by fixing the reflux of heavy alcohol and the light alcohol column inflow. This strategy allows the achievement of both stoichiometric reactant feed rate and large flexibility in ester production. The distillation column for recovering light alcohol from water is no longer necessary. The result is a compact, efficient, and easy-to-control multiproduct reactive setup, as clearly illustrated by Figure 10.14 (Dimian *et al.*, 2009). Other important design details of the RD column are provided in Table 10.3 (Dimian *et al.*, 2009).

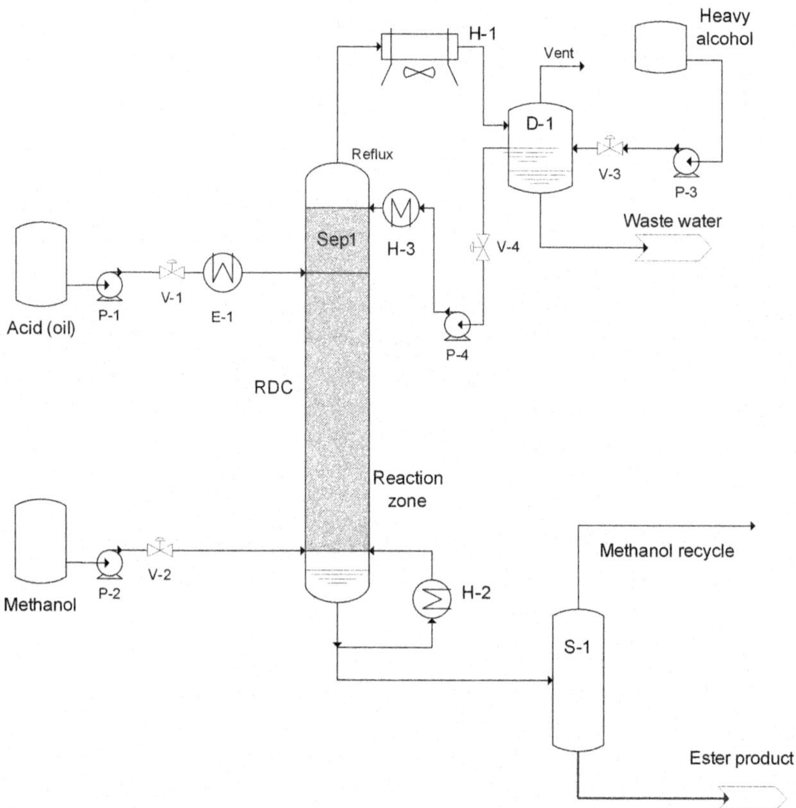

Figure 10.14 Flow sheet configuration and control strategy of the dual reactive distillation setup

In this control structure, the reactants are fed into the process in a ratio that satisfies the overall mass balance imposed by the reaction stoichiometry and the phase equilibrium at the top and the bottom of the RD column. In contrast, control structures fixing the feed rates of all reactants (acid and light- and heavy-alcohol) will not work in the presence of small control implementation errors, with the failure being manifested by the accumulation or depletion of one reactant (Kiss, Bildea and Dimian, 2007; Kiss and Bildea, 2012).

Figure 10.15a compares the temperature profiles for the base case and for a 10% increase of the lauric-acid flow rate with and without temperature control (Dimian *et al.*, 2009). Accurate control of the lauric acid concentration in the bottom stream is achieved by using a concentration controller that prescribes the set point of the temperature controller in a cascade structure. For this control configuration, the change of

Table 10.3 Design parameters for simulating the dual reactive distillation column

Parameter (units)	Value
Number of theoretical stages	25 (reactive from 5 to 24)
Lauric acid feed (on stage 5, at 3.5 bar, 150 °C) (kmol h^{-1})	100
Methanol feed (directly in reboiler, at 4 bar, 100 °C) (kmol h^{-1})	130
2-Ethylhexanol (fed in decanter, at 3.5 bar, 130 °C) (kmol h^{-1})	13.48
Catalyst bulk density (kg m^{-3})	1050
Volume holdup per stage (m^3)	0.050
Mass catalyst per stage (kg)	55
Reflux flow rate (kmol h^{-1})	2500
Column diameter (m)	1.2
HETP (m)	0.5
Fatty acid conversion (%)	> 99.99
Reboiler duty (kW)	1750
Condenser duty (kW)	−1492
Production rate (kg-ester h^{-1})	22333
Productivity of RD column (kg-ester kg-cat^{-1} h^{-1})	20.3
Bottom product composition (mass fraction) (kg kg^{-1})	650 ppb acid 11 ppm water 0.058 methanol 0.788 methyl ester 0.174 2EH ester
Specific energy requirements (kW·h per ton ester)	166.8

lauric acid feed flow rate leads to a change of the methyl-ester production rate. In contrast, when both flow rate ratios (i.e., lauric acid feed/methanol entering the column and lauric acid feed/heavy alcohol reflux) are constant, the change of lauric acid feed flow rate leads to changes of both the methyl ester and ethyl hexyl ester production rates.

Figure 10.15b (Dimian *et al.*, 2009) presents the performance of the control system for the following scenario: the simulation starts from the steady state (feed rate of lauric acid: 100 kmol h^{-1}) which is maintained for 0.5 h. Then, the feed rate of lauric acid is increased from 100 to 110 kmol h^{-1}, and after 1 h it is decreased to 90 kmol h^{-1}. Finally, the initial flow rate of 100 kmol h^{-1} is restored. This change of acid feed flow rate leads to a change of the light ester production rate having the same magnitude of change, while the production rate of the heavy ester is constant. The dynamics are fast, only 20 min was necessary to achieve the new production rate. The amount of water obtained at the top of the column reflects the amount of ester formed. During the entire transient period, the concentration of water on the reactive trays remains below the 2 wt% limit.

Figure 10.15c presents the temperature and the concentration of lauric acid in the bottom stream, for the same scenario. Both variables remain

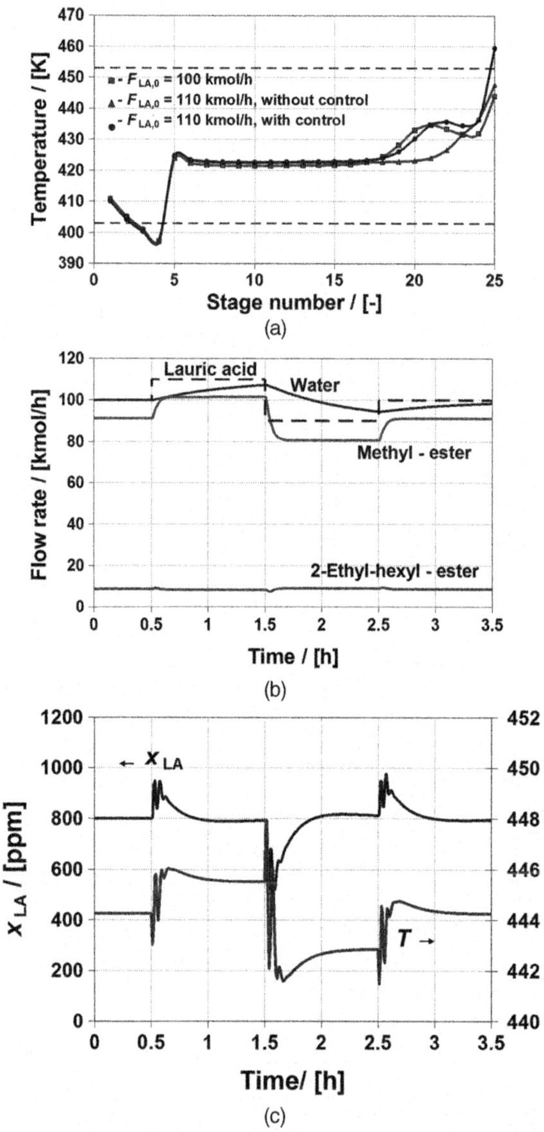

Figure 10.15 Temperature profiles along the reactive distillation column (a); dynamic simulation results showing the increase and decrease of the production rate (b); temperature and the concentration of lauric acid in the bottom stream during production rate changes (c)

very close to the nominal values. Notably, the tuning of the controllers is not critical with respect to the performance of the control system. In this case study, the parameters of the controllers were set as follows. The range of the controlled variable was set to the nominal value ±10 °C for the temperature control loops, and the nominal value ±50% of it for the level control loops. For all loops, the range for the manipulated variable was set to twice the nominal value. The gain of the feedback controllers was set to 1%/%. An integration time of 20 min was selected for the temperature controllers.

In conclusion, the control structure presented here achieves stable operation and is able to modify the throughput while keeping the characteristics of the products at their design values (Dimian *et al.*, 2009).

At optimal operation the highest yield and purity can be achieved by using stoichiometric feeds in the desired ratio of fatty esters. At this point the amount of methanol lost in the top is practically negligible. The heavy ester plays the role of a solvent and prevents escaping methanol in the top product. On the top stages the heavy alcohol enhances water concentration in the vapor phase, from which it is separated by condensation and decanting, while heavy ester is produced in an amount proportional to the reflux flow rate. The optimal operation is based on controlling the inventory of reactants by using the principle of fixed recycle flows of co-reactants, in this case the reflux of the organic phase and the methanol inflow to the RDC. This strategy allows large changes in the production rate. The control strategy is generic and can be employed for esterifications involving formation of azeotropes, such as for ethanol and (iso-) propanol. The overall result in integrating design and process control is a compact, efficient, and easy-to-control multiproduct reactive setup.

10.9 CONCLUDING REMARKS

The main business drivers for RD applications are economic (over 20% reduction of variable cost, capital expenditure, and energy use), environmental (lower CO_2 emissions, no or reduced salt waste), and social (improved safety and health due to reduced footprint and lower reactive content and run away sensitivity). The industrial-scale implementation of RD by co-operation of partners in research, scale-up, design, and operation can be considered as a model for rapid implementation of other process intensification techniques in the chemical process industry (Harmsen, 2007). Taking into account the remarkable progress in hardware development, modeling for design and simulation, control strategies, real time optimization, and also considering the pace at which new applications are being explored, RD is certainly one of the most important tools for process intensification using green chemistry and engineering (Sundmacher and Kienle, 2003; Harmsen, 2007, 2010).

Despite the more complex design, control, and equipment, RD is a process intensification technology that fulfills the principles of green engineering—prevention instead of treatment, design for separation, maximize efficiency, output-pulled versus input-pushed, meet need and minimize excess, integrate local material and energy flows, design for a commercial afterlife, renewable versus depleting—by unique features such as the reduced number of processing units, enhanced overall rates, overcoming unfavorable equilibrium, avoidance of difficult separations, improved selectivity, reduced energy requirements, and low or no solvent use (Malone, Huss, and Doherty, 2003; Sundmacher and Kienle, 2003).

Reactive distillation is nowadays unit a proven technology and an established operation in chemical process technology, being also the front-runner in the field of process intensification (Harmsen, 2007, 2010). At present, there are various models—available now in the open literature—for the screening, analysis, design, and optimization of RD columns. Moreover, residue curve maps (RCMs) are an invaluable tool for the initial screening and flow sheet development, and equilibrium (EQ) models have their place for preliminary designs, while the non-equilibrium (NEQ) models are used for the final RD design, development of control strategies, and commercial RD plant design and simulation (Taylor and Krishna, 2000; Schoenmakers and Bessling, 2003; Harmsen, 2007, 2010; Kiss and Bildea, 2012).

A major focus of future research and development should be the methodology by which to choose the most suitable equipment, as well as the scale-up of equipment since this is where great deficiencies still lie. Moreover, considering the green opportunities created by the use of reactive separations, the future plans aim to spread the use of integrated reactive distillation technologies in the production of bulk or specialty chemicals such as ethers, esters, polyesters, and biofuels—as, for example, in biodiesel manufacturing by green processing of waste materials with high free fatty acids content.

REFERENCES

Agachi, P.S., Nagy, Z.K., Cristea, M.V., and Imre-Lucaci, Á. (2006) *Model Based Control: Case Studies in Process Engineering*, Wiley-VCH Verlag GmbH, Weinheim.

Agreda, V.H., Partin, L.R., and Heise, W.H. (1990) High-purity methyl acetate via reactive distillation. *Chemical Engineering and Processing*, *86*, 40–46.

Aspen Technology (2009a) Aspen Dynamics – Reference & User Guide.

Aspen Technology (2009b) Aspen Plus: User Guide - Volume 1 & 2.

Aspen Technology (2009c) Aspen Physical Property System - Physical Property Models.

Baur, R., Higler, A.P., Taylor, R., and Krishna, R. (2000) Comparison of equilibrium stage and nonequilibrium stage models for reactive distillation. *Chemical Engineering Journal*, *76*, 33–47.

Baur, R. and Krishna, R. (2004) Distillation column with reactive pump arounds: An alternative to reactive distillation. *Chemical Engineering and Processing*, *43*, 435–445.

Bemporad, A., Morari, M., and Ricker, N.L. (2009) Model Predictive Control Toolbox 3: User's Guide, MathWorks, Inc.

Bildea, C.S. and Kiss, A.A. (2011) Dynamics and control of a biodiesel process by reactive absorption. *Chemical Engineering Research and Design*, *89*, 187–196.

Buzad, G. and Doherty, M.F. (1994) Design of three-component kinetically controlled reactive distillation columns using fixed-points methods. *Chemical Engineering Science*, *49*, 1947–1963.

Chen, Y.H. and Yu, C.C. (2003) Design and control of heat-integrated reactors. *Industrial & Engineering Chemistry Research*, *42*, 2791–2808.

Damartzis, T. and Seferlis, P. (2010) Optimal design of staged three-phase reactive distillation columns using non-equilibrium and orthogonal collocation models. *Industrial & Engineering Chemistry Research*, *49*, 3275–3285.

Daza, O.S., Perez-Cisneros, E.S., Bek-Pedersen, E., and Gani, R. (2003) Graphical and stage-to-stage methods for reactive distillation column design. *AIChE Journal*, *49*, 2822–2841.

Dimian, A.C. and Bildea, C.S. (2008) *Chemical Process Design - Computer-Aided Case Studies*, Wiley-VCH Verlag GmbH, Weinheim.

Dimian, A.C., Bildea, C.S., Omota, F., and Kiss, A.A. (2009) Innovative process for fatty acid esters by dual reactive distillation. *Computers & Chemical Engineering*, *33*, 743–750.

Doherty, M.F. and Malone, M.F. (2001) *Conceptual Design of Distillation Systems*, McGraw-Hill, New York.

Egorov, Y., Menter, F., Klöker, M., and Kenig, E.Y. (2005) On the combination of CFD and rate-based modelling in the simulation of reactive separation processes. *Chemical Engineering and Processing*, *44*, 631–644.

Grosser, J.H., Doherty, M.F., and Malone, M.F. (1987) Modeling of reactive distillation systems. *Industrial & Engineering Chemistry Research*, *26*, 983–989.

Güttinger, T.E. and Morari, M. (1997) Predicting multiple steady states in distillation: Singularity analysis and reactive systems. *Computers & Chemical Engineering*, *21*, S995–S1000.

Harmsen, G.J. (2007) Reactive distillation: The front-runner of industrial process intensification: A full review of commercial applications, research, scale-up, design and operation. *Chemical Engineering and Processing*, *46*, 774–780.

Harmsen, G.J. (2010) Process intensification in the petrochemicals industry: Drivers and hurdles for commercial implementation. *Chemical Engineering and Processing*, *49*, 70–73.

Hernandez, S., Sandoval-Vergara, R., Barroso-Munoz, F.O. *et al.* (2009) Reactive dividing wall distillation columns: Simulation and implementation in a pilot plant. *Chemical Engineering & Processing*, *48*, 250–258.

Higler, A., Krishna, R. and Taylor, R. (1999) Nonequilibrium cell model for multicomponent (reactive) separation processes. *AIChE Journal*, *45*, 2357–2370.

Hiwale, R.S., Bhate, N.V., Mahajan, Y.S., and Mahajani, S.M. (2004) Industrial applications of reactive distillation: Recent trends. *International Journal of Chemical Reactor Engineering*, 2, R1.

Lee, J.-H. and Dudukovic, M.P. (1998) A comparison of the equilibrium and nonequilibrium models for a multicomponent reactive distillation column. *Computers & Chemical Engineering*, *23*, 159–172.

Kaymak, D.B. and Luyben, W.L. (2004) Effect of the chemical equilibrium constant on the design of reactive distillation columns. *Industrial & Engineering Chemistry Research*, *43*, 3666–3671.

Kaymak, D.B. and Luyben, W.L. (2007) Optimum design of a column - Side reactor process. *Industrial & Engineering Chemistry Research*, *46*, 5175–5185.

Kenig, E., Jakobsson, K., Banik, P., Aittamaa, J., Górak, A., Koskinen, M. and Wettmann, P. (1999) An integrated tool for synthesis and design of reactive distillation *Chemical Engineering Science, 54*, 1347–1352.

Kenig, E.Y., Gorak, A., Pyhalahti, A. *et al.* (2004) Advanced rate-based simulation tool for reactive distillation. *AIChE Journal*, *50*, 322–342.

Kiss, A.A., Dimian, A.C., and Rothenberg, G. (2006a) Solid acid catalysts for biodiesel production - towards sustainable energy. *Advanced Synthesis & Catalysis*, *348*, 75–81.

Kiss, A.A., Rothenberg, G., Dimian, A.C., and Omota, F. (2006b) The heterogeneous advantage: biodiesel by catalytic reactive distillation. *Topics in Catalysis*, *40*, 141–150.

Kiss, A.A., Bildea, C.S., and Dimian, A.C. (2007) Design and control of recycle systems by non-linear analysis. *Computers & Chemical Engineering*, *31*, 601–611.

Kiss, A.A., Dimian, A.C., and Rothenberg, G. (2008) Biodiesel by reactive distillation powered by metal oxides. *Energy & Fuels*, *22*, 598–604.

Kiss, A.A. (2009) Novel process for biodiesel by reactive absorption. *Separation & Purification Technology*, *69*, 280–287.

Kiss, A.A., Pragt, H., and van Strien, C. (2009) Reactive dividing-wall columns - how to get more with less resources? *Chemical Engineering Communications*, *196*, 1366–1374.

Kiss, A.A. (2010) Separative reactors for integrated production of bioethanol and biodiesel. *Computers and Chemical Engineering*, *34*, 812–820.

Kiss, A.A. (2011) Heat-integrated reactive distillation process for synthesis of fatty esters. *Fuel Processing Technology*, *92*, 1288–1296.

Kiss, A.A., Singh, P., and van Strien, C.J.G. (2011) A systematic approach towards applicability of reactive distillation. *Computer Aided Chemical Engineering*, *29*, 191–195.

Kiss, A.A. (2012) Applying reactive distillation. *NPT Procestechnologie*, *19* (1), 22–24.

Kiss, A.A. and Shah, M. (2012) Mastering reactive distillation. *NPT Procestechnologie*, *19* (2), 26–28.

Kiss, A.A. and Bildea, C.S. (2012) A review on biodiesel production by integrated reactive separation technologies. *Journal of Chemical Technology and Biotechnology*, *87*, 861–879.

Klöker, M., Kenig, E.Y., Schmitt, M. *et al.* (2003) Influence of operating conditions and column configuration on the performance of reactive distillation columns with liquid-liquid separators. *Canadian Journal of Chemical Engineering*, *81*, 725–732.

Klöker, M., Kenig, E.Y., Hoffmann, A. *et al.* (2005) Rate-based modelling and simulation of reactive separations in gas/vapour–liquid systems. *Chemical Engineering and Processing*, *44*, 617–629.

Kreul, L.U., Górak, A., and Barton, P.I. (1999) Modeling of homogeneous reactive separation processes in packed columns. *Chemical Engineering Science*, *54*, 19–34.

Krishna, R. (2002) Reactive separations: more ways to skin a cat. *Chemical Engineering Science*, *57*, 1491–1504.

Kumar, A. and Daoutidis, P. (1999) Modeling, analysis and control of ethylene glycol reactive distillation column. *AIChE Journal*, *45*, 51–68.

Lin, Y.-D., Chen, J.-H., Cheng, J.-K., Huang, H.-P. and Yu, C.-C. (2008) Process alternatives for methyl acetate conversion using reactive distillation. 1. Hydrolysis. *Chemical Engineering Science*, *63*, 1668–1682.

Luyben, W.L. and Yu, C.C. (2008) *Reactive Distillation Design and Control*, John Wiley & Sons, Inc., Hoboken.

Maciejowski, J.M. (2002) *Predictive Control with Constraints*, Pearson Education, Harlow.

Malone, M.F. and Doherty, M.F. (2000) Reactive distillation. *Industrial & Engineering Chemistry Research*, *39*, 3953–3957.

Malone, M.F., Huss, R.S., and Doherty, M.F. (2003) Green chemical engineering aspects of reactive distillation. *Environmental Science & Technology*, *37*, 5325–5329.

Morari, M. and Lee, J.H. (1999) Model predictive control: past, present and future. *Computers & Chemical Engineering*, *23*, 667–682.

Mueller, I., Pech, C., Bhatia, D., and Kenig, E.Y. (2007) Rate-based analysis of reactive distillation sequences with different degrees of integration. *Chemical Engineering Science*, *62*, 7327–7335.

Mueller, I. and Kenig, E.Y. (2007) Reactive distillation in a dividing wall column – rate-based modeling and simulation. *Industrial & Engineering Chemistry Research*, *46*, 3709–3719.

Nagy, Z.K., Klein, R., Kiss, A.A., and Findeisen, R. (2007) Advanced control of a reactive distillation column. *Computer Aided Chemical Engineering*, *24*, 805–810.

Noeres, C., Kenig, E.Y., and Gorak, A. (2003) Modelling of reactive separation processes: reactive absorption and reactive distillation. *Chemical Engineering & Processing*, *42*, 157–178.

Okasinski, M.J. and Doherty, M.F. (1998) Design method for kinetically controlled, staged reactive distillation columns. *Industrial & Engineering Chemistry Research*, *37*, 2821–2834.

Roizard, C. and Wild, G. (2002) Mass transfer with chemical reaction: the slow reaction regime revisited. *Chemical Engineering Science*, *57*, 3479–3484.

Schembecker, G. and Tlatlik, S. (2003) Process synthesis for reactive separations. *Chemical Engineering and Processing*, *42*, 179–189.

Schmidt-Traub, H. and Gorak, A. (2006) *Integrated Reaction and Separation Operations*, Springer, New York.

Schoenmakers, H.G. and Bessling, B. (2003) Reactive and catalytic distillation from an industrial perspective. *Chemical Engineering and Processing*, *42*, 145–155.

Serafimov, L.A., Pisarenko, Yu.A., and Kulov, N.N. (1999) Coupling chemical reaction with distillation: Thermodynamic analysis and practical applications. *Chemical Engineering Science*, *54*, 1383–1388.

Shah, M., Kiss, A.A., Zondervan, E., and de Haan, A.B. (2012) A systematic framework for the feasibility and technical evaluation of reactive distillation processes. *Chemical Engineering and Processing*, *60*, 55–64.

Sharma, N. and Singh, K. (2010) Control of reactive distillation column - A review. *International Journal of Chemical Reactor Engineering*, *8*, R5.

Siirola, J.J. (1996) Industrial applications of chemical process synthesis. *Advances in Chemical Engineering*, *23*, 1–62.

Sneesby, M., Tade, M., and Smith, T. (1999) Two-point control of a reactive distillation column for composition and conversion. *Journal of Process Control*, *9*, 19–31.

Stankiewicz, A. (2003) Reactive separations for process intensification: an industrial perspective. *Chemical Engineering and Processing*, *42*, 137–144.

Stichlmair, J.G. and Fair, J.R. (1998) *Distillation, Principles and Practice*, Wiley-VCH Verlag GmbH, Weinheim.

Subawalla, H. and Fair, J.R. (1999) Design guidelines for solid-catalyzed reactive distillation systems. *Industrial & Engineering Chemistry Research*, *38*, 3696–3709.

Sundmacher, K. and Kienle, A. (eds) (2003) *Reactive Distillation: Status and Future Directions*, Wiley-VCH Verlag, GmbH, Weinheim.

Sundmacher, K., Kienle, A., and Seidel-Morgenstern, A. (eds) (2005) *Integrated Chemical Processes: Synthesis, Operation, Analysis, and Control*, Wiley-VCH Verlag GmbH, Weinheim.

Taylor, R. and Krishna, R. (1993) *Multicomponent Mass Transfer*, John Wiley & Sons, Inc., New York.

Taylor, R. and Krishna, R. (2000) Modelling reactive distillation. *Chemical Engineering Science*, *55*, 5183–5229.

Thery, R., Meyer, X.M., and Joulia, X. (2005) Analysis of the feasibility, synthesis and conception of processes of reactive distillation: State of the art and critical analysis. *Canadian Journal of Chemical Engineering*, *83*, 242–266.

Tuchlenski, A., Beckmann, A., Reusch, D. *et al.* (2001) Reactive distillation - industrial applications, process design & scale-up. *Chemical Engineering Science*, *56*, 387–394.

Venimadhavan, G., Buzad, G., Doherty, M.F. and Malone, M.F. (1994) Effect of kinetics on residue curve maps for reactive distillation. *AIChE Journal*, *40*, 1814–1824.

Venimadhavan, G., Malone, M.F., and Doherty, M.F. (1999) Bifurcation study of kinetic effects in reactive distillation. *AIChE Journal*, *45*, 546–556.

Index

Advanced process control
  D-K iteration procedure, 168
  LQG/LQR, 119, 165, 171
  LSDP, 168, 180
  GMC, 167
  H∞ norm μ-synthesis, 127, 168
  MPC, 124, 128, 153–156, 173, 373, 376
Antoine equation, 8
Approximate multi-component methods, 33
August equation, 8
Azeotropic distillation, 16, 198, 215–219
Azeotropic systems, 12, 18
  right triangle diagrams, 18
  ternary diagrams, 16

Batch distillation, 299, 302
Binary distillation, 44, 254
Brayton cycle, 241
Bubble-point, 9, 14
  calculations, 14
  curve, 9–10

Carnot efficiency, 231
Case studies
  aromatics separation, 97
  biodiesel production, 378
  bioethanol dehydration, n-pentane, 216
  bioethanol dehydration, glycol, 214, 219
  BTX separation, 132
  dimethyl ether (DME) synthesis, 205
  ethanol-water stripping, 338
  fatty esters synthesis, 383
  pentane-hexane-heptane separation, 129
  methanol-water separation, 264, 341
  PP (C3) splitting, 264
Catalytic distillation, 207, 313, 358
Chao-Seader method, 3
Chemical potential, 15
Cost estimation, 55
Columns sequence model, 84, 85
Compressors, 61
Control structures, 45, 117

Advanced Distillation Technologies: Design, Control and Applications, First Edition.
Anton Alexandru Kiss.

Controller tuning parameters, 175
Constraint control, 53
Cost indexes, 59
Cost of
  chemicals, 63
  energy, 63
  equipment, 59
  utilities, 62
Cyclic distillation, 246, 311
  control, 335
  design, 327, 331
  hydrodynamic model, 321
  mathematical model, 319
  modeling approach, 329
  sensitivity analysis, 323
  working principle, 317

Degrees of freedom, 26, 76, 112, 357
Design principles, 38, 278
Design procedure, 78, 283, 331
Dew-point (curve), 9, 14
Diabatic distillation, 274
Direct sequence, 68, 97, 205, 215
Distillation column
  analysis, 24
  control, 44
  packing, 40, 56, 61, 73, 91, 288, 291, 359
  reactive distillation, 195, 206, 354, 359
  shell sizing/cost, 56, 59, 61
  shortcut design, 33, 38, 106
  trays, 27, 33, 56, 60, 73, 91, 94, 246, 284, 292, 314, 317, 327, 357
Distillation boundary, 12, 19
Distillation curve map, 18
Dividing-wall column
  approximate design, 78
  azeotropic distillation, 198
  configurations, 70
  construction, 94
  control, 117, 128, 163, 171
  decomposition, 76
  design, 75
  dynamic model, 156
  dynamic simulations, 139, 176
  energy efficient control, 135
  equipment, 87
  extractive distillation, 199
  internals, 91
  modeling, 83
  multi-component separations, 188
  optimal design, 82
  optimal operation, 114
  optimization methods, 86
  revamping, 203
  sizing, 91
  ternary separations, 188
Dual-end control, 49
Dynamic simulation, 129, 139, 176, 336, 374, 382

Economic evaluation, 55
Efficiency of trays, 24, 313
Equilibrium curve, 12, 274, 282, 319, 323
Equipment cost, 59
Equipment sizing, 56
Extractive distillation, 199, 212

Feedback control, 117, 139, 378, 383
Feedforward control, 119, 126, 166
Feed pre-heating, 42
Fenske equation, 33
Fenske-Underwood-Gilliland, 33, 78
Fugacity, 15

Generic Model Control (GMC), 167
Grayson method, 3
Gilliland correlation, 34

Heat exchangers, 58, 61, 289, 291
Heat-integrated distillation column, 244, 271
  applications, 258, 300
  concentric column, 287, 288
  construction, 282
  control, 298
  design, 282

dynamics, 299
energy savings, 280
heat panels, 289, 293
heat transfer means, 293
ideal HIDiC, 281
inter-coupled columns, 284
modeling and simulation, 295
partial HIDiC, 280
partition wall, 286
plate-fin heat-exchanger, 290
shell & tube heat-exchanger, 289
thermodynamic analysis, 277
working principle, 273
Heat integration, 43, 244, 271, 379
Heat pumps
absorption heat pump, 234
compression-resorption, 235
HIDiC, 244, 271
technology selection scheme, 250, 255
thermo-acoustic heat pump, 236
vapor compression, 233
vapor recompression, 233, 234
working principle, 231
Heat pumps assisted distillation, 229
Heat transfer coefficients, 285, 287, 295
Heat transfer (working) fluids, 235, 240
Heavy component, 13, 22, 122
Heuristic optimization, 40
HiGee distillation, 248
HQP optimization, 373
Hybrid wet compression (HWC), 236

Ideal solution, 7
Indirect sequence, 68
Integral absolute error (IAE), 145, 178
Inter-coupled distillation columns, 284
Intermediate condensers, 42
Intermediate reboilers, 42

Joule cycle, 241

K-values, 4, 80
Kaibel column, 71, 86, 98, 103, 246
Kirkbride equation, 34
Kremser-Souders-Brown equation, 314

Lee-Kesler method, 4
Light component, 13, 22, 27
Linear Quadratic Gaussian control, 165
Liquid activity models, 4
Liquid/reflux splitter, 89
Local point efficiency, 319
Log mean temperature difference, 58

Malone cycle, 242
Marshall & Swift cost index, 59, 209
Mass separating agent (MSA), 213, 219, 262
Mass transfer, 91, 291, 319, 326, 345, 362
Maximum reflux ratio, 282
McCabe-Thiele method, 27, 32
Membrane distillation (MD), 248
MESH/MESHR equations, 25, 207
MIMO controllers, 55, 128, 180
Minimum reflux ratio, 31, 33, 40, 78, 282
Minimum number of trays, 31, 33, 40, 78, 282, 334
Mixing rules, 4, 17
Model predictive control, 55, 119, 124, 128, 153, 156, 173, 179, 374
Multi-component separations, 188, 249
Multiple effect distillation (MED), 43, 247
Multivariable control, 49, 54, 167
Murphree efficiency, 317, 325, 339

Newton-Raphson, 16
Non-random two liquid (NRTL), 3, 5, 12, 21

Objective function, 38, 101, 208, 357
Operating costs, 42, 203, 209, 347, 378
Operating lines, 28, 31, 274, 320, 324, 335
Operating pressure, 39, 77, 262
OptCon, 373
Osenbrück cycle, 236

Peng-Robinson EOS, 3
Perfect displacement, 322, 324
Perfect mixing, 322
Petlyuk column, 68, 75, 129, 157, 248
Phase diagrams, 16, 19
Phase split, 16, 207, 216
PID controller, 55, 132, 163, 175, 377
Pressure-swing adsorption, 242
Pressure-swing distillation, 198, 302
Pressure drop, 39, 62, 89, 93, 106, 287, 292, 332, 357
Property methods, 3, 21
Pump around model, 84

Raoult's law, 4, 7
RCM consistency, 20
Reaction kinetics, 207, 358, 365, 379
Reaction rate, 207, 356, 365, 368, 372, 379
Reactive DWC, 195, 208, 358
Reactive distillation, 353
  applications, 360
  control, 362, 371
  design, 357
  dividing-wall column, 195, 208, 358
  feasibility evaluation, 365
  modeling, 362
  technical evaluation, 370
  working principle, 354
Rectifying stages, 25
Redlich-Kwong EOS, 3
Relative Gain Array (RGA), 137, 148, 164, 172, 374
Relative volatility, 13, 33, 42, 119, 199, 263, 279, 365
RGA number, 137, 164
Removal of lights & heavies, 25, 97, 252
Residue curve maps, 18, 162, 368
Rigorous optimization, 41, 97, 208, 214
Rotating packed bed (RPB), 248
Rotating zigzag bed (RZB), 248

Saddle points, 20
Secondary reflux and vaporization, 277, 280
Sensitivity analysis, 97, 101, 137, 214, 323
Separate phase movement (SPM), 246, 317, 346
Sequential quadratic programming, 97, 208, 214
Simultaneous models, 86
Single-end control, 47, 52
Singular Value Decomposition (SVD), 48
SISO controllers (see also PID), 124, 128
Soave-Redlich-Kwong EOS, 3
Solid-sorption cycle, 242
Stable nodes, 20
Stage efficiency, 24, 317, 323, 326
Stirling cycle, 238, 240, 243
Stoichiometry, 365, 384
Stripping stages, 25

Temperature driving force, 44, 53, 58, 230, 248, 254, 277, 296
Temperature lift, 230, 238, 242, 254, 277
Temperature-swing adsorption, 242
Ternary diagrams, 16, 134, 162
Thermally coupled column, 67, 101, 118, 217, 246

Thermal vapor recompression, 234
Thermo-acoustic engine, 235
Thermodynamic efficiency, 43, 229, 271, 278
Thermodynamic models, 3
T-xy diagram, 9, 12

UNIFAC, 3, 21, 378
UNIQUAC, 3, 21, 214
Underwood equations, 33, 40, 78, 117
Unstable nodes, 20
Utilities, 58, 62, 105, 251, 263, 276

Van Laar model, 3
Vapor (re-)compression, 232
Vapor-Liquid Equilibrium, 2, 8, 329, 356, 365
Vapor pressure, 6, 8, 15, 199, 248, 366
Venturi effect, 234
VLE non-ideality, 11, 16
Vmin diagram, 79, 100, 115
Vuilleumier cycle, 241, 243

Waste disposal, 62, 341, 378
Waste heat, 239
Wilson method, 3, 21

xy diagram, 10

Zeotropic mixtures, 19, 80, 235

CPSIA information can be obtained at www.ICGtesting.com
Printed in the USA
BVOW09*2007311215

R6670300001BA/R66703PG431389BVX1BA/1/P

9 781119 993612